Optical Fibre Communication Systems

Optical Fibre Communication Systems

Edited by
C. P. SANDBANK
Standard Telecommunication Laboratories,
Harlow Essex
(now at BBC Research Department,
Kingswood Warren, Surrey)

A Wiley–Interscience Publication

JOHN WILEY & SONS

Chichester · New York · Brisbane · Toronto

***British Library Cataloguing in Publication Data*:**

Optical fibre communication systems.
1. Optical communications
2. Fibre optics
I. Sandbank, C P
621.38'0414 TK5103.59 79-40822

ISBN 0 471 27667 7

Filmset in Northern Ireland at The Universities Press (Belfast) Ltd.
Printed by Pitman Press, Bath

Contents

Preface

There are several ways of treating a topical subject of this type. One way is to find an author who can personally cover the whole subject. It is hard to find someone who has both ready access to the wide range of activities which optical communications covers and the time to write a book. Another way is to invite experts on each of the major topics from various organizations (and countries) to contribute chapters on their own subject. This provides a broad coverage but may lack cohesion. I have chosen an intermediate approach by collecting the views of a group of specialists, many of whom have been working together as a team covering most aspects of fibre optic communication since the concept was first proposed.

The treatment is unashamedly parochial and this has enabled each author to relate his subject to the activities of his colleagues thereby producing chapters that hang together. Where appropriate, worked numerical examples are given to assist the reader to perform calculations associated with fibre experiments. Sufficient data is given to provide the basis for the design of representative optical communications systems.

Throughout the description of the various technologies there runs a thread which tells the story of the research, development, and engineering culminating in the installation of one of the world's first wideband optical links for trunk telephony.

We would like to thank the management of STL, STC, and ITT for supporting the activity, encouraging the publication of the work, and giving permission for the reproduction of the illustrations in this book. We would similarly like to thank the BPO and the UK MOD for support. We particularly appreciate the helpful discussions which we have had with staff from our Company and the other organizations with whom we have worked.

Finally, I personally would like to thank all the authors for their splendid cooperation and to thank Mrs Rita Craft for her help in the preparation of the manuscripts.

C. P. Sandbank
Standard Telecommunication Laboratories Ltd
Harlow, Essex, England

CHAPTER I

The Basic Principles of Optical Fibre Communication

INTRODUCTION

In 1964 some experiments were carried out at Standard Telecommunication Laboratories to study the propagation of light in optical waveguides with a view to the transmission of wideband signals for telecommunications. For historic interest, a photograph of some of the light patterns indicating guiding in the crude fibres available at the time is shown in Figure 1. An analysis of propagation in dielectric guides was carried out and the results published.[1] Although the theory predicted that optical fibres could provide a viable transmission medium, the fibres available at the time were not sufficiently transparent to allow transmission over any significant distances. In the fifteen years following the original research, sufficient progress has been made to enable practical optical communications systems to be produced and installed in applications such as trunk telephony, data links, TV distribution, and military communication.

Some of the reasons why the concept of fibre optic communication has captured the imagination of electronic engineers[2–5] are illustrated by Figure 2. This shows three cables having approximately equivalent information-carrying capacity emerging from standard $3\frac{1}{2}$-in. British Post Office ducts. It will be seen at once that the cable containing the optical fibre, by its simplicity and economy in raw materials, promises cost savings over the much larger conventional cables containing substantial amounts of copper. The flexibility resulting from the small size eases many installation problems, particularly in cities, where space in overcrowded ducts is limited. Since light, the carrier of the information in the optical waveguide, is an electromagnetic radiation, having a frequency around 10×3^{14} Hz (i.e. about 10^5 times the UHF frequency used for TV transmission), the information-carrying capacity of this transmission medium is potentially very high. Since the wavelength of light is very short compared to the wavelength of electromagnetic radiation used in, say, microwave systems, the optical waveguide can be very small and has, in fact, usually about the same diameter as a human hair. Not only can these fine glass threads carry very wide band information such as several TV channels, but because of the low transmission loss signals can be sent up to 100 times further along optical

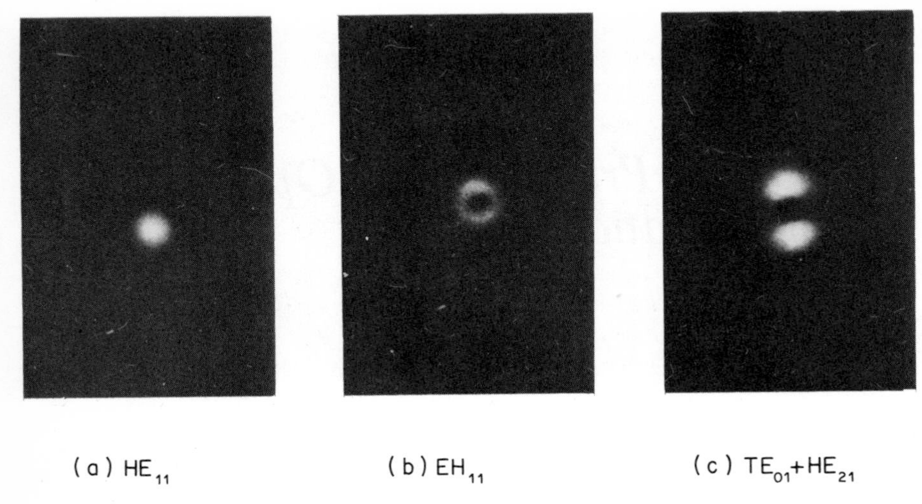

Figure 1. Fibre waveguide patterns studied at Standard Telecommunication Laboratories in 1964

Figure 2. Comparison between optical and metallic telephone cables

fibres than along copper cables without the need for intermediate amplifiers.

This chapter explains some of the differences between electrical and optical communications systems and provides the reader with background which may be helpful in relation to the more detailed descriptions in the later chapters of the book.

TRANSMISSION MEDIUM

The simplest way to consider transmission over optical waveguides is to think in terms of total reflection in a medium of refractive index n_1 at the boundary with a medium n_2, where n_1 is greater than n_2. This is the situation in a typical multimode fibre such as shown in Figure 3 which would have a circular core of diameter d and uniform refractive index n_1 surrounded by a cladding layer of refractive index n_2. Light launched into the core at angles up to θ_1 will be propagated within a core at angles up to θ_2 to the axis. Light launched at angles greater than θ_1 (see broken line in Figure 3) will not be internally reflected and will be refracted into the cladding or possibly even out of the cladding into the air at the second boundary if the launching angle is large enough, and n_1 and n_2 are small enough. The maximum launch and propagation angles are given by the numerical aperture NA:

$$\mathrm{NA} = (n_1^2 - n_2^2)^{1/2} = \sin\theta_1 = n_1 \sin\theta_2 \tag{1}$$

Since this is an electromagnetic waveguide propagation, only certain modes, which may be regarded as rays corresponding to specific quantized values of θ_2, can propagate.

The number of modes N for light of wavelength λ is given by

$$N \simeq 0.5 \left(\frac{\pi d \mathrm{NA}}{\lambda}\right)^2 \tag{2}$$

where d is the diameter of the core.

Thus, for a given combination of refractive indices, as the diameter of the core is reduced, fewer modes propagate. When eventually the diameter becomes of the same order of magnitude as the wavelength of light, then only a single mode will propagate.

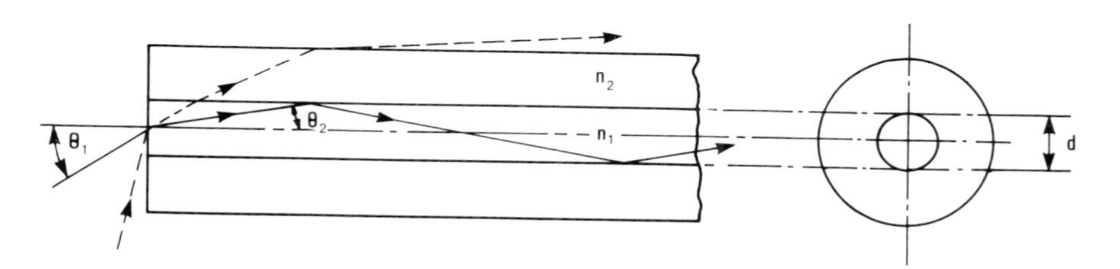

Figure 3. Ray diagram for optical waveguide

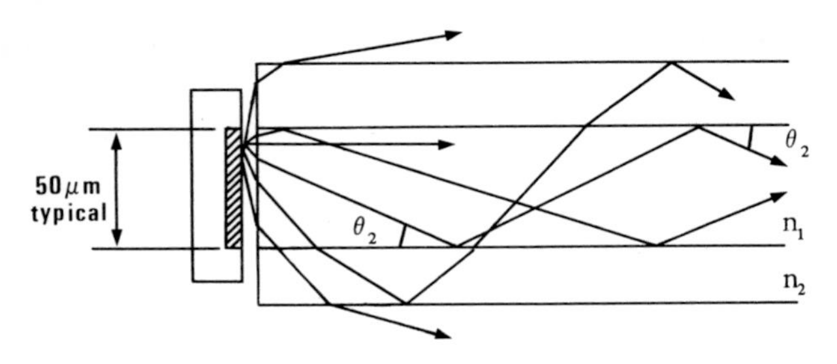

Figure 4. Simple ray diagram showing launching conditions with a LED into multimode fibre

This simple geometrical representation gives a surprisingly accurate picture of the behaviour of multimode guides. Figure 4 shows the typical ray diagram for a light-emitting diode in close proximity to a multimode fibre. This representation illustrates several points. First, to couple as much as possible of the light from the diode into the fibre the active area of the diode should have a diameter no greater than that of the fibre. This is because most of the light, coming from areas outside this circle, is most likely to enter the cladding and would not contribute usefully to the power of the signal propagated. Even if the light source is butted close up against the fibre core and matches its area exactly, only a portion of the light emitted from the diode would propagate, namely those rays which strike the fibre at angles less than θ_1. Thus, the higher the numerical aperture, the more efficient the coupling. The penalty for the improved coupling is dispersion caused by the larger path difference between the extreme modes propagated. Considering the maximum difference in the path length for a ray propagating parallel to the axis and a ray propagating at the maximum angle θ_2, it is possible to calculate the pulse spreading (i.e. the dispersion) due to the different times taken by the energy propagating along the two different paths (see equation (27) in Chapter VI). Thus, the bandwidth capability of a high NA guide is less than that of one able to support fewer modes. Although the majority of rays will adopt a skew propagation through the fibre, it can be shown that by taking the extremes of the meridional section, as shown in Figure 4, the maximum difference in the delay is identified. By definition, no dispersion is introduced in a single mode guide due to this process.

Figure 5 shows the three most commonly used types of fibre used for optical communications. The theory of propagation in such guides is discussed more fully in the next chapter. We present here a more general description to give the reader a quantitative feel for the construction of these dielectric guides. The power density is shown as a continuous line for the HE_{11} mode on the right and an estimate of the envelope of the power in many modes is shown by the broken line on the right.

The multimode guide, Figure 5(a), is essentially the same as that shown in Figure 3. It can be seen that most of the energy is transmitted in the core but in order to ensure that the cladding immediately adjacent to the core is of a high quality, thus giving overall low loss transmission, a thin optical cladding of refractive index n_2 is sometimes located adjacent to the core of refractive index n_1.

The construction shown in Figure 5(b) is designed to achieve the best compromise between a multimode and single mode fibre. In this case the core has a graded index which decreases radially from a maximum value n_0 at the centre of the core according to a law

$$n_r = n_0(1 - \alpha r^2)$$

where α is a constant determining the effective numerical aperture.

The rays representing different modes which are shown symbolically in the diagram illustrate that because modes corresponding to larger angles of θ spend a greater time in a region of the core where the refractive index is low, they travel faster than the axial modes, and hence this fibre design compensates for the mode dispersion found in the step index fibre. These fibres, therefore, give lower mode dispersion than the step index fibres without the requirement that light must be launched into a very small diameter core or the difficulties of splicing fibres with very small cores. Graded index fibres normally have a core in which the graded region extends to about half the diameter of the fibre and then remains at a constant level. The energy is thus confined to this 'core' which is again usually surrounded by a high quality optical cladding region.

In the single mode fibre illustrated in Figure 5(c), the diameter of the core is only a few microns, i.e. comparable with the wavelength of light, and it can be seen from the power distribution in this single mode that a substantial amount of the energy is propagated in the cladding region. In this case it is therefore particularly important that the material for the cladding has the same high quality optical properties as the core. In these fibres the only dispersion is that due to the differences in the propagating velocities of different colours. The disadvantage of this type of fibre stems principally from the fact that exact alignment of these small cores presents much greater problems in joining and splicing than with the other types of fibre. This aspect is discussed in Chapter V.

Figure 6 illustrates the three most commonly used methods of optical waveguide fabrication. The first consists of a composite rod which may have either a step index or graded index core. The rod will be several millimetres in diameter and several metres long. Such rods can be produced by depositing pure silica, doped to varying degrees with, for example, GeO_2, Al_2O_3, TiO_2, or P_2O_5 to produce higher refractive indices, or B_2O_3 to produce a lower refractive index. The silica and the dopant can be deposited from the vapour phase to produce the vitreous starting material. The fibre is then produced by pulling at a controlled rate from the rod. The second method is

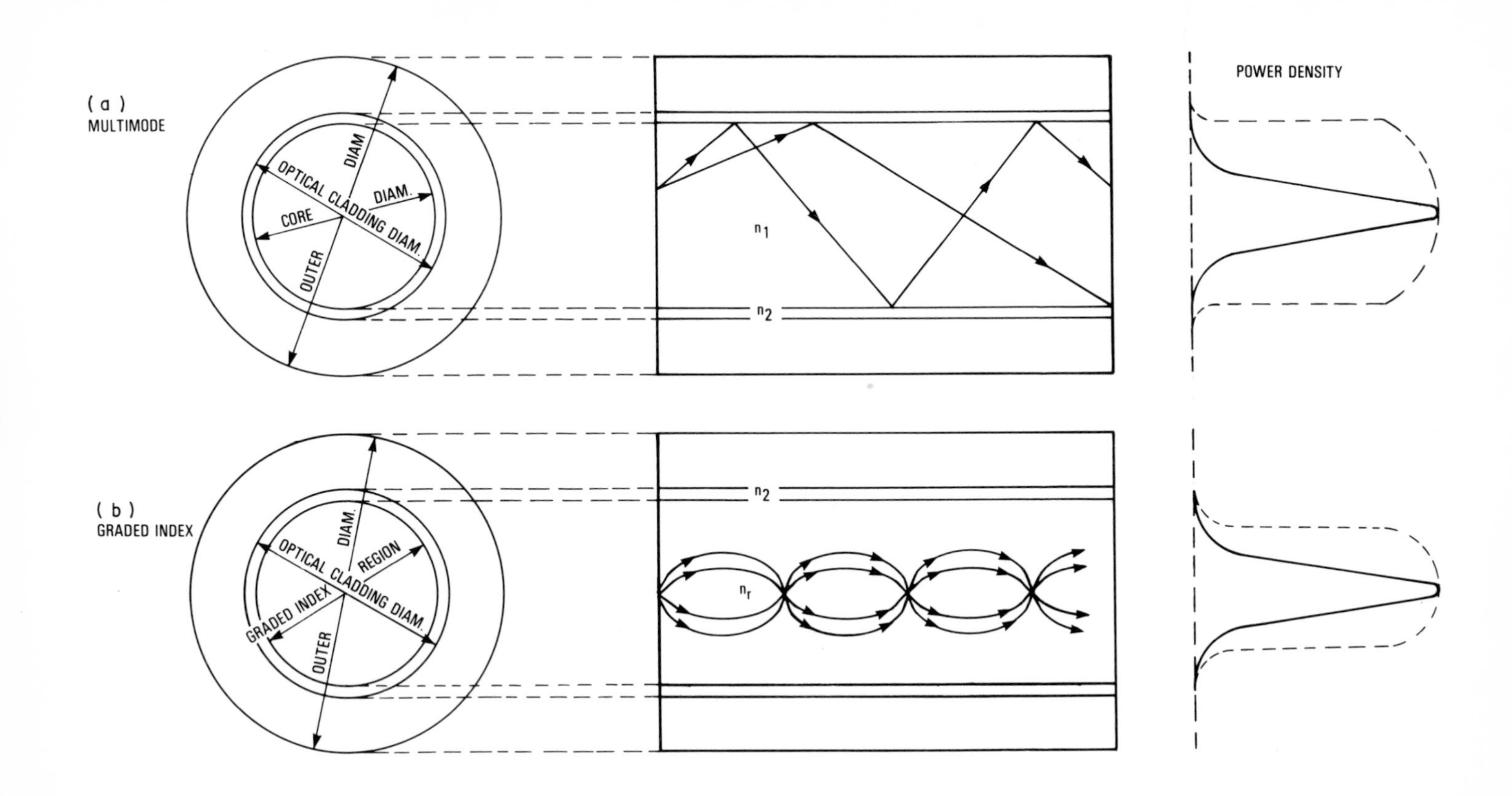

(a)
MULTIMODE
(b)
GRADED INDEX
POWER DENSITY
OUTER DIAM.
OPTICAL CLADDING DIAM.
CORE DIAM.
GRADED INDEX REGION DIAM.
n_1
n_2
n_r

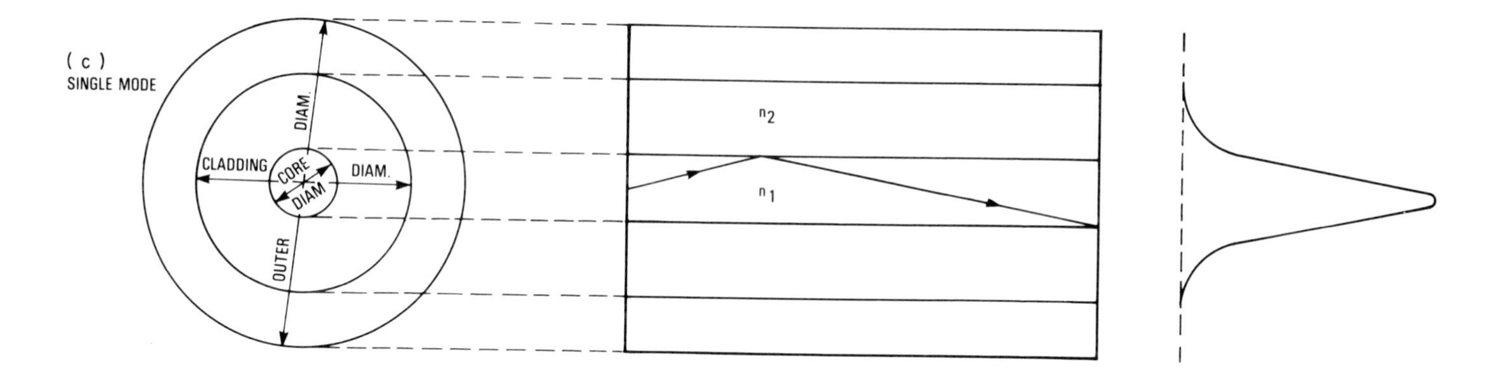

Figure 5. Three typical optical fibre cables

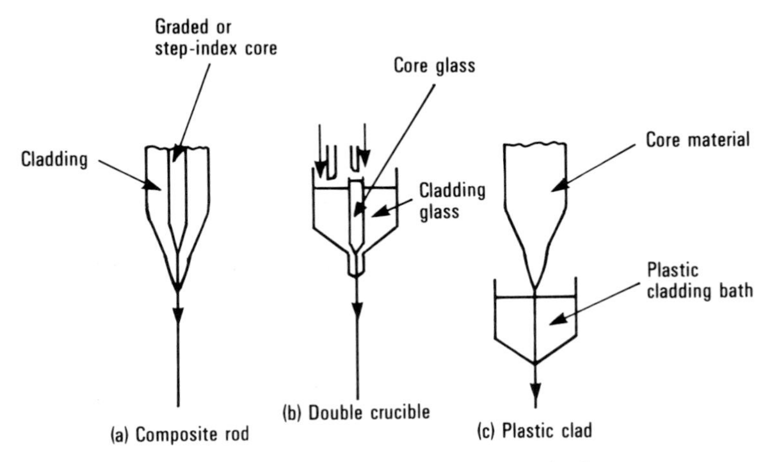

Figure 6. Some fibre fabrication methods

to pull the fibre from a double crucible containing the core and cladding materials, using ion exchange to obtain a graded index where required. The third method, for step index fibres only, is to pull the fibre through a rod consisting of homogeneous low loss starting material and then to apply a plastic cladding such as a thin layer of silicon resin which has a lower refractive index than the material of the rod, thus confining the light. Since very little of the energy is propagated in the cladding it is possible to use plastic cladding which does not have a very low loss propagation and still achieve sufficiently low attenuation fibres for some applications. Fibres with attenuations of less than 10 dB/km can readily be produced by all three methods, but the first method has tended to produce the lowest attenuations. These three techniques of fibre fabrication are discussed in detail in Chapter III, where performance figures are quoted.

Some typical values of the fibre parameters discussed so far are given in Table I. It can be seen that the three step index fibres differ principally in the magnitude of the acceptance angle, θ_1. The fibre with the lead glass core enables the highest numerical aperture to be obtained but the penalty paid for this is a very high modal dispersion (and, incidentally, with this material also a somewhat higher attenuation). The value of modal dispersion quoted for the GeO_2/P_2O_5 doped multimode silica fibre is the theoretical worst case calculated from equation (27) in Chapter VI. In practice a value of about a one-third of this would be obtained, since the energy is not uniformly distributed amongst all the possible modes and is biased towards the lower orders.

This effect is even more marked in the case of the plastic clad silica fibre, where the high order modes are attenuated by the proximity of the high loss plastic cladding. For plastic clad fibres of the type illustrated in Table I the modal dispersion would be reduced from the theoretical value of 165 to an effective value of about 30, but the price paid for this is a substantially higher attenuation than that which one would associate with the silica core (see Chapter III).

It will be seen that the value of n_1 for the multimode graded index fibre is higher than that for the step index fibre using similar materials. In this case the value refers to the maximum refractive index of the core at its centre and the value of 0.26 for the numerical aperture and 15° for θ_1 applies also only to the central region of the core.

Although the graded index profile has not eliminated modal dispersion, the value quoted in Table I for this fibre is a typical experimental result and is, of course, very much less than would be obtained for the comparable step index fibre. It should be noted that the amount of light that can be launched into a single mode fibre such as the one illustrated in the table, is very much less than the multimode examples, not only because of the lower acceptance angle but because the area of the core is very much smaller. Here there is, of course, no modal dispersion; but because the light sources are not monochromatic there is some material dispersion which will depend on the type of light source used.

At the wavelength of 850 nm used to calculate the parameters in Table I, the material dispersion would be around 90 ps/km for every nanometer

Table I
Some optical fibre characteristics

Fibre type	Core Index n_1, 850 μm	Cladding Index n_2, 850 μm	Core diam. (μm)	Optical cladding diam. (μm)	Fibre outer diam. (μm)	Max. Theoretical numerical aperture	θ_1	Δt Modal Dispersion (ns/km)
Multimode step GeO_2/P_2O_5 doped core SiO_2 or SiO_2/B_2O_3 clad	1.469	1.452	50	55–60	125	0.22	13	57
Plastic clad silica Step index	1.452	1.405	150	230	230	0.37	22	165.0
Lead glass borosilicate cladding step index	1.62	1.55	80	100	100	0.47	28	249.0
Multimode graded index GeO_2/P_2O_5 doped core SiO_2 or SiO_2/B_2O_3 clad	1.475	1.452	50	55	125	0.26	15	<0.5
Single mode GeO_2 doped core SiO_2 cladding	1.462	1.452	3.3	45	125	0.17	10	0

spread in the wavelength spectrum of the source. Thus, for a light-emitting diode having a spread of 20 nm, the pulse spreading due to material dispersion would be 1.8 ns/km. This would largely invalidate the advantage of the absence of modal dispersion in the single mode fibre and in fact would also dominate over the 0.5 ns/km modal dispersion obtained with the graded index fibre. For such high bandwidth fibres a source with a much narrower spectrum of radiation such as the GaAs laser must be used. For example, a laser with a 3-nm wavelength spread around 850 nm would give a pulse spread of 0.27 nm/km due to material dispersion. Even this is still a significant addition to the modal dispersion in the graded index fibre. With the fibre compositions illustrated in Table I, material dispersion reaches a negligible minimum value at longer wavelengths and it will be seen from the discussion in later chapters that operation at a wavelength of 1.3 μm permits propagation with very low attenuation and negligible dispersion.

SEMICONDUCTOR LIGHT SOURCES

In most optical communications systems semiconductor light sources are used to convert the electrical signals to light launched into the fibres.

When the p–n junction in a direct gap semiconductor diode is forward biased, spontaneous photon emission takes place due to recombination of the electron–hole pairs. The photon energy, and hence its wavelength, are determined by the energy gap between the valence and conduction bands in the semiconductor. In GaAlAs this corresponds to about 850 nm, and can be slightly varied by changing the percentage of aluminium. It can be seen from Figure 38 in Chapter III that this wavelength is in the low attenuation range of the fibres.

Figure 7 shows four types of light-emitting source used in optical communications. The diode in Figure 7(a) has essentially the same type of construction used in devices designed for visible display, but operates at the infrared wavelength. This type of device would be used in conjunction with fibre bundles having a total diameter of up to about 1 mm, but is not suitable for use with single fibre systems such as described in this book.

The high radiance diode[6] shown in Figure 7(b) is specially designed for the efficient launching of a relatively intense beam into a single strand multimode fibre of about 50 μm diameter.

Lasers enable the largest amounts of optical power to be launched into the core of the fibres. A semiconductor laser is formed by confining the photons emitted in the p–n region by mirrors deposited on the cleaved ends of the device. The junction is heavily forward biased, thus creating a very dense population of electrons in the conduction band. When a spontaneously emitted photon of energy $h\nu$ is reflected by one of the mirrors and meets an electron about to emit another photon of energy $h\nu$, then stimulated emission of this second photon takes place. The two photons will then stimulate the emission of further photons by meeting electrons of the

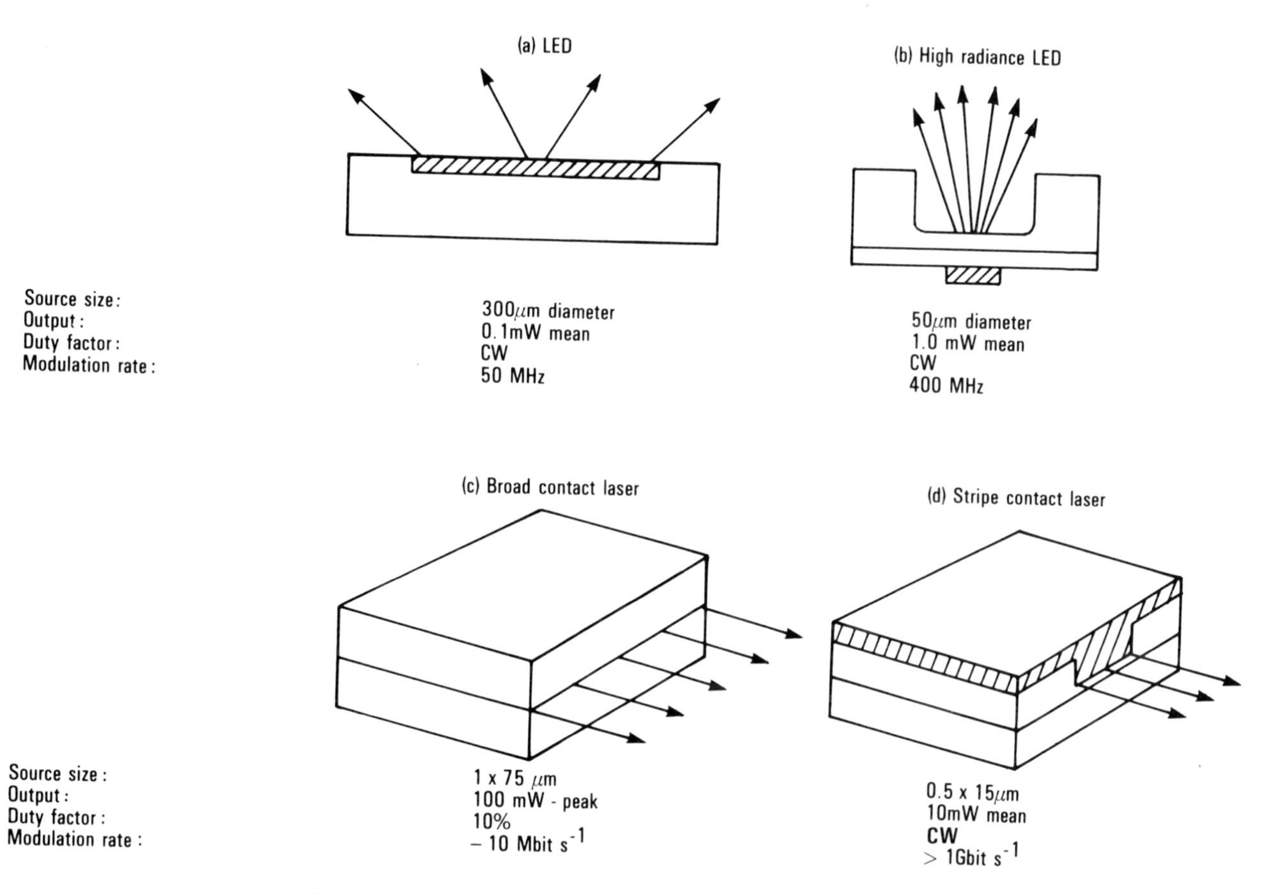

Figure 7. Laser and LED characteristics

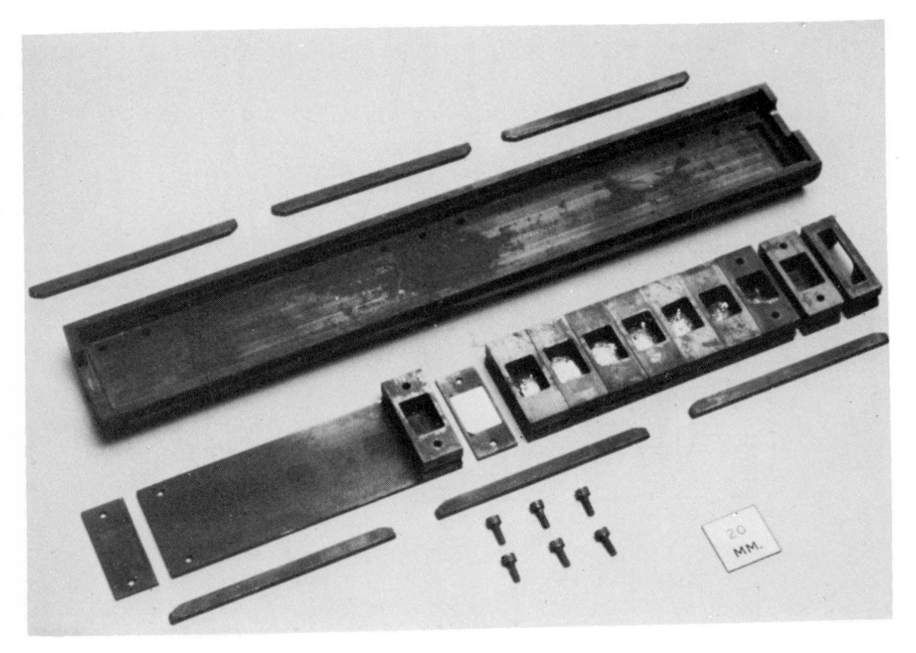

Figure 8. Sliding graphite boat for liquid epitaxy

appropriate energy level in the densely populated conduction band. Thus, the laser gain action is maintained once a threshold current has been exceeded to give a sufficiently high electron density. Because the photons will stimulate emission of the same wavelength, the gain mechanism serves to narrow the emission spectrum to a width of less than 1 nm compared with more than 20 nm for LEDs.

To obtain continuous wave or high duty cycle devices, very efficient heat sinks must be used. Furthermore, special methods of device construction have to be developed. Multilayer heterojunction technology[7] has provided the design flexibility enabling this to be realized. By using the sliding boat multilayer epitaxy technique shown in Figure 8, the refractive index of the crystal on either side of the central GaAs region indicated in Figure 9 can be modified by alloying aluminium. The graphite boat containing a series of suitably doped molten alloys is slid over the GaAs substrate to create the series of layers shown in Figure 9 by the process of liquid epitaxial deposition. The optical guide created by the interface between the p-type GaAs and the p- and n-type GaAlAs establishes a photon–electron confinement that enables sufficiently low threshold current densities for efficient continuous wave operation to be obtained, as discussed in detail in Chapter VII. The use of heterostructures such as GaInAsP/InP enables the band structure to be further modified to shift the wavelength to the region 1.1–1.3 μm, where fibre attenuation and material dispersion is lower than at

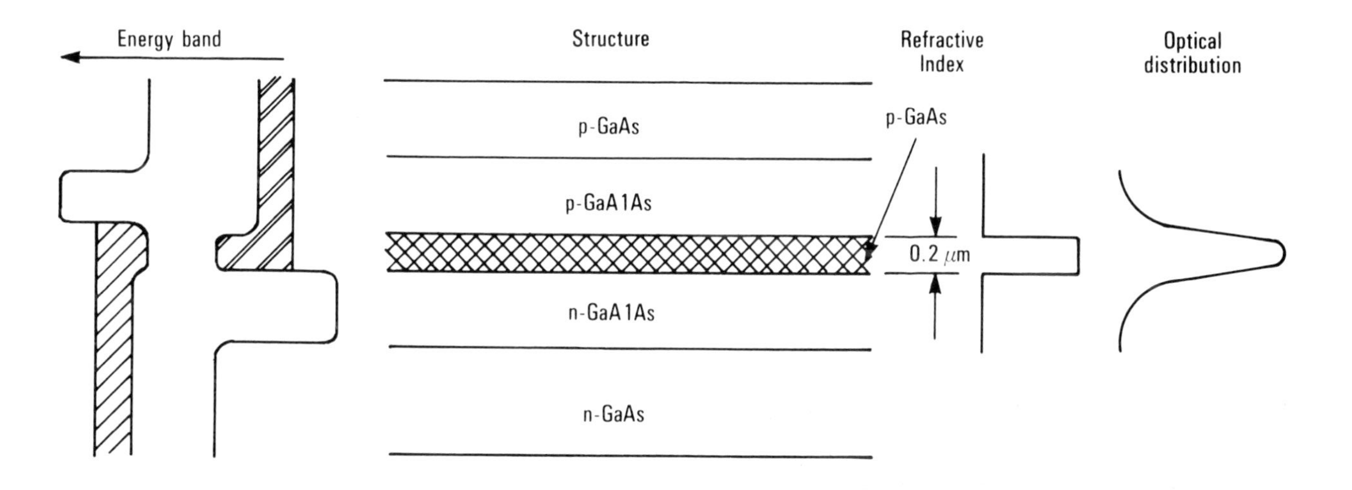

Figure 9. Double heterostructure laser

850 nm. These devices also have a good lattice match across the heterostructure which provides the low mechanical strain necessary for the long-term stability of semiconductor lasers.

SYSTEM CONSIDERATIONS

The basic elements of the fibre transmission system are shown in Figure 10. All systems would follow essentially the form shown in the diagram, the difference would lie principally in the nature of the elements in the blocks indicated by the broken lines. In the case of the simplest system, the broken lines represent merely through connections. The transmitter would have only a light-emitting diode, there would be no repeater, and the receiver would only have a photodetector. The next stage of complexity might be a system using direct linear modulation of the sources with an amplifier included in the transmitter, the receiver, and the repeater (if fitted). Where the system requires the higher transmitter power of the laser, it is no longer possible to use simple amplitude modulation. This is because the very large difference between the light emitted with the laser current above and below the threshold makes this a very nonlinear device and the techniques described in Chapter IX, in the section on amplitude modulation, must be used.

For trunk communications, where a link might include several repeated sections, a more complex encoding such as 'pulse code modulation' would be used. In this case, the repeater would contain circuits for the amplification, reshaping, and retiming of signals. The transmitter and receiver would contain the appropriate circuits for, respectively, encoding and decoding the PCM signals.

Just as the analysis and design of optical waveguides has its roots in microwave transmission theory, so the design of optical transmission systems is derived from classical line transmission. Having said this, it is important to stress that in designing systems due consideration must be given to the fact that the energy is transmitted by relatively high energy photons rather than much lower energy electrons. This is one of the factors which biases the designer of optical systems towards digital or pulse-analogue techniques rather than direct linear modulation of the light sources.

To appreciate some of the differences between an optical communication system and a conventional one using metallic conduction, it is useful to

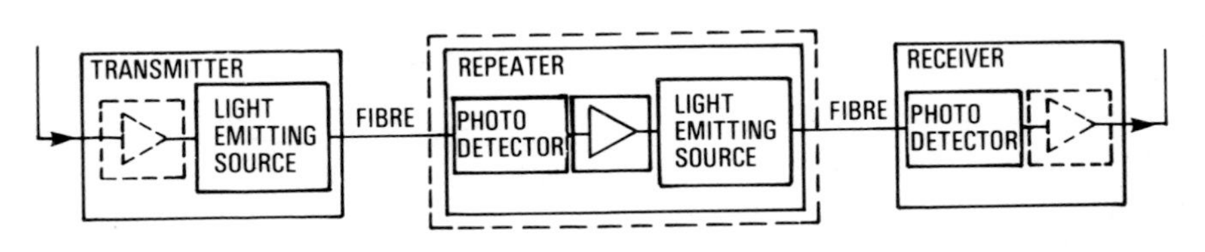

Figure 10. Basic fibre transmission system

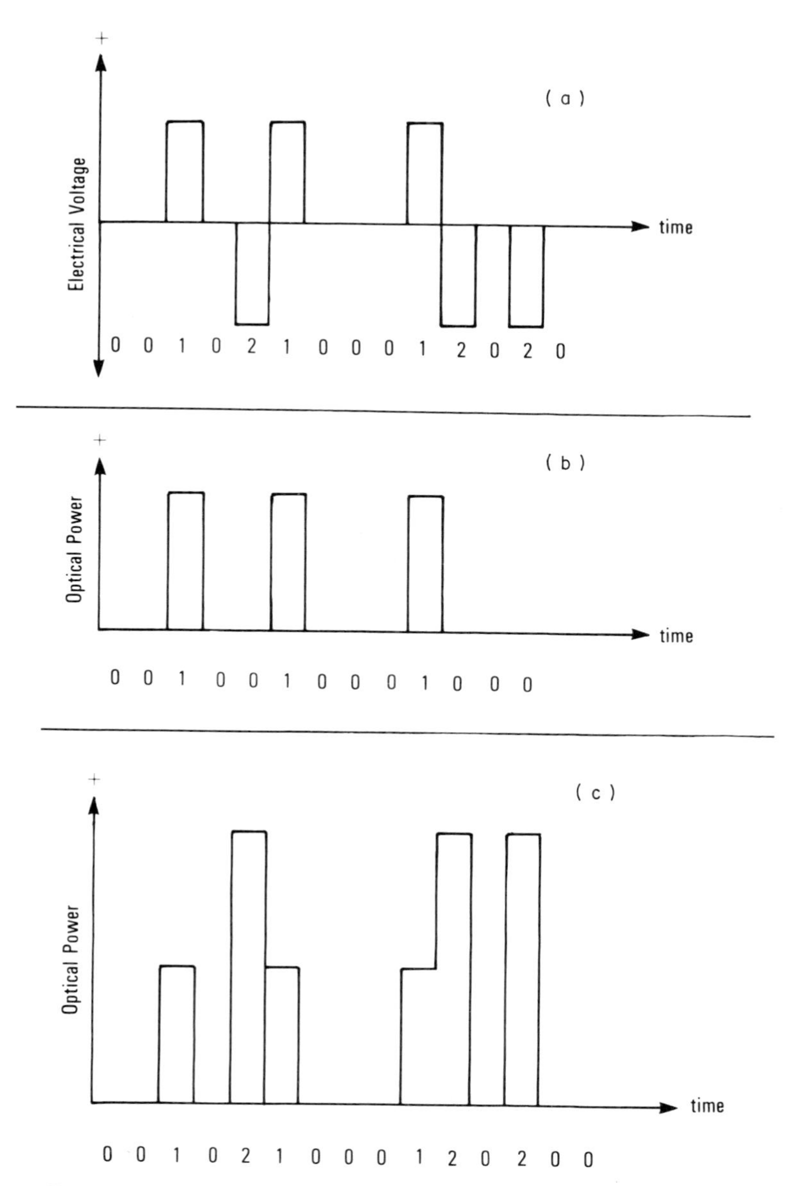

Figure 11. Comparison of electrical and optical digital signals

consider the transmission of a PCM bit stream, Figure 11(a) is an idealized representation of the voltage which might be detected at the input to a conventional PCM repeater connected to a metallic cable. In such a system it is quite usual to use a three level digital code where the bits are denoted as '0', '1', and '2'. To simplify the discussion the effects of dispersion will be ignored.

A '1' is detected if a positive voltage is received which is sufficiently above a threshold set to distinguish the '1' from the '0' level, which is effectively the noise at the receiver due to the thermally agitated electrons in the metallic system. This threshold is set such that the probability of a '1' not being distinguished from a '0' is the very low 'bit error rate' required, which in a trunk telephony system is usually 10^{-9}. To achieve the third level in such a system a negative voltage of sufficient magnitude to be distinguished from the noise with a high probability is detected.

In an optical communication system it is not possible to reproduce the situation described in Figure 11(a) in the same way, for the simple reason that, unlike electrons, photons do not go backwards. The detector in an optical system detects the power generated by the received photons and since this cannot be negative a binary optical bit stream would be as illustrated in Figure 11(b). To achieve a three level code similar to that used in Figure 11(a) in an optical system it would be necessary to modulate the amplitude of the pulses using, for example, the method illustrated in Figure 11(c). The power detector at the input of the optical receiver would have an intermediate threshold level set to distinguish the '1' from the '2'.

Another important difference between optical and electrical systems stems from the different statistical situation which applies when considering photons in comparison with electrons. From the discussion in Chapter IX, it appears that in the limit to achieve a bit error rate of 10^{-9} in a digital optical system it is possible to have as few as 10 photons/bit. This is because the energy associated with the photons is very much greater than that of the thermal electrons, which in an ideal optical system can be ignored even at room temperature. In such a system the noise is due to the statistical fluctuation in the emission of the photons, i.e. the 'quantum noise'. In the absence of thermal noise one can postulate that provided at least one photon is received by the detector a '1' will be registered and if no photons are received a '0' will be registered. It will be seen from Chapter IX that to achieve a practical system bit error rate, i.e. that the probability of there being less than one photon is 10^{-9}, the average number of photons per bit can be as low as 10.

These quantum noise considerations are illustrated in Figure 12, where one bit of a 140 Mbit/s stream is shown occupying an area on the power/time graph corresponding to the energy of 10 photons ($E = P \times t$). In an equivalent 140 Mbit/s coaxial system there would be about 800 thermal noise electrons associated with each bit and to achieve the required bit error rate of 10^{-9}, 8000 signal electrons would have to be detected at the receiver. Thus, in the limiting case of a conventional metallic cable system the granularity due to individual electrons can be ignored, whereas in the limiting optical case the granularity is significant. Although the ideal situation, assuming absence of thermal noise, has been considered, it will be seen from later chapters that practical optical systems are already getting quite close to such conditions.

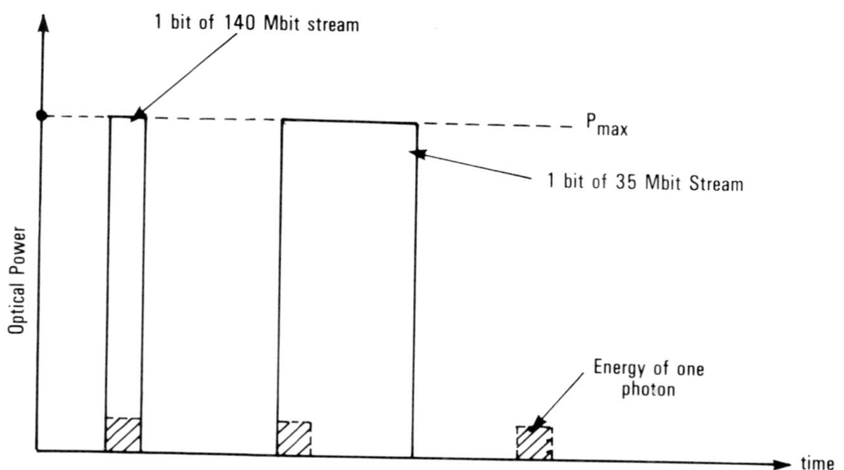

Figure 12. Energy of pulses in 140 Mbit/s and 35 Mbit/s streams compared with the energy of one photon

In metallic cables the attenuation of the signal along the line increases roughly with the square of the carrier frequency for a given cable design. That is why coaxial cables, such as the one shown at the top left-hand corner of Figure 2, designed to have low attenuation at high rates, are fairly large and costly. Since the 'carrier frequency' of the optical signal is 10^{14} Hz, which is very much higher than the information bit rates used to modulate the optical 'carrier', it might seem that the distance between repeaters required to amplify the signal in an optical communications link would be independent of bit rate. However, from Figure 12 it can be deduced that this is not the case. If we assume a given laser launching power of P_{max} into a particular fibre, then from Figure 12 it is clear that the number of photons per bit in the case of a 35 Mbit/s system is four times as great as that in the 140 Mbit/s system. In the ideal case a '1' will still be registered if one photon arrives in the 35 Mbit/s case. Thus, for a fibre attenuation of a given number of dB/km (this attenuation is, of course, independent of the modulation rate of the light) the detector can be further away from the transmitting laser in the 35 Mbit/s system than in the 140 Mbit/s system.

Some of the practical implications of these system considerations are illustrated in Table II, which compares some typical metallic cable transmission systems with the equivalent optical communication systems for a number of standard PCM bit rates which are used in the European telephone system. The 2 Mbit/s system might use one of the voice frequency twisted pair wires such as those in the cable at the top right-hand corner of Figure 2. This would require repeater spacings of about 2 km. An optical system operating at this bit rate could have repeaters spaced at distances of up to 20 km. The relative economics of these systems are discussed in Chapter XI. Suffice it to say at this stage that although the fibre cost may be

Table II
Typical set of digital systems and repeater spacings

		Copper cable transmission systems		Fibre cable transmission systems	
Mbit/s	Voice channels	Cable type	Repeater spacing (km)	Fibre type	Repeater spacing (km)
2.048	30	voice frequency	1.8	step index	15–20
8.448	120	0.9 mm screened pair	3.6	graded index	12–15
34	480	2.9 mm coax.	2	graded index	8–12
140	1920	4.4 mm coax.	2	graded index	7–10
560	7680	4.4 mm coax. 9.5 mm coax.	1 2	single mode	4–6

greater than the cost of a simple copper wire, the attraction of a 2 Mbit/s optical system lies in the possibility of eliminating the need for buried repeaters with their associated installation costs, such as manholes and power feeding. At the higher bit rates the conventional systems have to use progressively larger and more costly coaxial cables to achieve acceptable repeater spacings. Thus, at the highest bit rate considered in the table the attraction of an optical fibre cable compared with the 9.5-mm coaxial cable shown at the top left-hand corner of Figure 2 used in the transmission of 560 Mbit/s PCM, lies in the potentially lower cost of the optical cables and the reduction in the number of repeaters required. The repeater spacings shown in this table are based on the fibre and cable results discussed in Chapters III and IV. The design criteria are examined in detail in Chapter IX and these will explain why a range of repeater spacings is given in the final column of Table II. The actual value depends on the choice of route and compromises between cost and system performance.

If the technology continues to improve along the lines suggested by recent experiments[245] future optical systems may well be practical with repeater spacing ten times greater than those shown in the last column of Table II.

OPTICAL SIGNAL PROCESSING

In all the systems described in the subsequent chapters of the book the optical fibre is used as the transmission medium, but any processing of the signal, such as switching, amplification, and regeneration is carried out in the electrical regime. It is possible that in the future system designers may prefer to process some of the signals in the optical regime rather than take the signal, process it electrically and then convert it back to optical before onward transmission. One could imagine the repeater section in Figure 10 being replaced by an all-optical device, illustrated symbolically in Figure 13. The transmission medium would include a section of guide having suitably doped glass with some periodic perturbations which form the partially reflecting mirrors of a laser cavity. The laser would be pumped externally with a powerful optical source so that photons entering from the left would find themselves in a region of gain and emerge on the right as an amplified signal. In practice, no satisfactory technique which would provide a noise advantage over the arrangement in Figure 10 has been devised. Furthermore, no practical methods have been suggested for carrying out the essential functions of regeneration and retiming which would have to be carried out in an optical repeater for digital systems in addition to amplification.

Despite these limitations, work is being carried out on integrated optic technology which is analogous to the integrated circuits that are being produced to replace conventional wired electrical circuits. It is likely that these will find application in systems where it may be advantageous to switch a signal in the optical regime rather than convert to electrical, e.g. to avoid interference in an environment of high electrical noise.

The fabrication of integrated optic circuits, like their electrical counterpart, is carried out by photolithographic processes which define optical guides in much the same way that the metallic conducting tracks are defined on electrical circuits. Thus, in the integrated optic circuits the photolithographically defined waveguides replace the fibres in the same way that the metallic tracks of an electronic integrated circuit replace wires.

Figure 14 is a photograph of an integrated optic power divider with a sketch drawn to the same scale as the photograph. The optical waveguide indicated in the sketch is produced by defining the track with a photolithographic mask and then modifying by diffusion the refractive index of the substrate under the defined area. Light launched from an optical fibre on the left centres the guide and is then split into two diffuse guides to emerge at the right-hand side of the substrate. Although the guides themselves are not

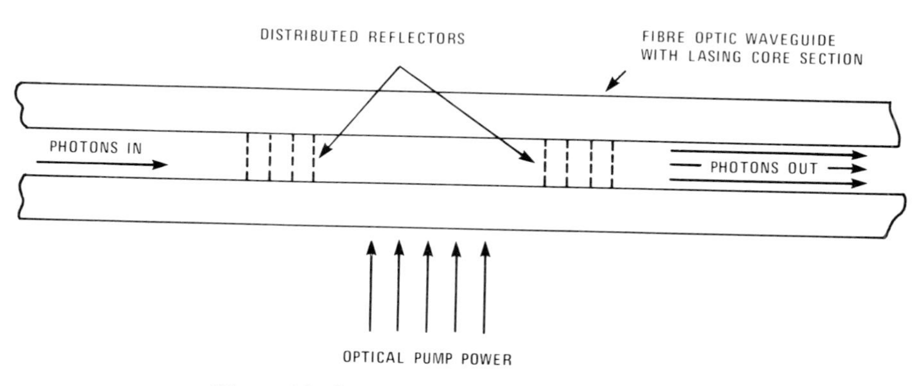

Figure 13. Symbolic integrated optical repeater

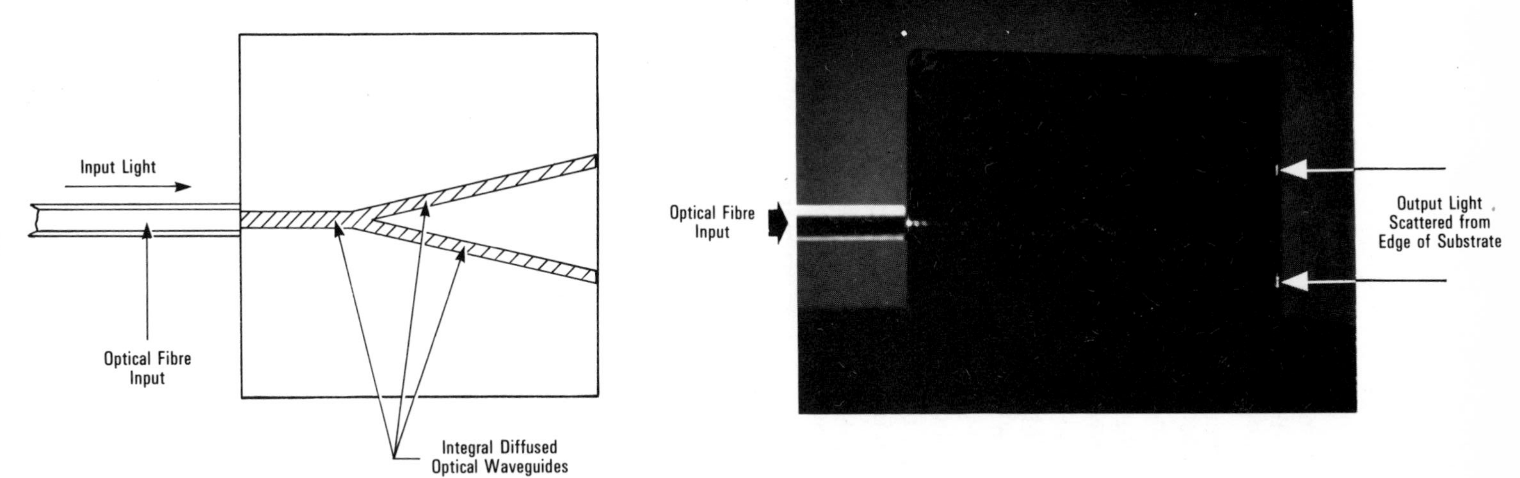

Figure 14. Integrated optic power divider

visible in the photograph, some scattered light can be seen at the left-hand side near the optical fibre and the light emerging from the ends of the guide can be clearly seen, as indicated by the arrows on the right. Because of the problems of transition between the optical fibre and the planar substrate, single devices of this type are not attractive compared to the types of power divider described in Chapter IX, Figure 209, which do not use planar technology. The planar technology would come into its own when large numbers of tracks have to be interconnected in a way which can be achieved on one substrate.

Figure 15 illustrates an active version of the device described in Figure 14 in which the signal can be switched from one leg to the other by the application of an electrical signal.[8] This device relies on the modification of the refractive index of an electro-optic material such as lithium niobate by the application of an electric field. In the same way as with the passive device in Figure 14, the light is confined to the planar tracks by locally modifying the refractive index of the lithium niobate by diffusion and then further changing this to achieve switching by applying a voltage to the adjacent metallic electrodes. The photographs show the mode lines produced for a device fabricated at Standard Telecommunication Laboratories when the light is divided into either 'A' or 'B'. The availability of such integrated optic technology will give the designers of future optical communication systems the freedom to decide whether to process some of the signal optically in circumstances where an overall system advantage can be obtained by avoiding the electrical regime.

Perhaps the largest area of application where more than just the transmission medium will be in the optical regime, is in the field of telemetry, or instrumentation. It is very common in such applications for the basic parameters that need to be measured to be essentially optical in nature, but because of the difficulty of transmitting these faithfully, conventional systems use a transducer which locally converts the parameters into electrical signals. With the availability of cables containing low loss optical fibres it now becomes sensible either to transmit the optical data to a region where it can be analysed more readily or to use a transducer which modifies the photon stream rather than the electron stream if the former is more convenient or accurate.

Consider, for example, the remote measurement of a parameter such as pressure in a vessel containing an explosive material. There would be some hazards in using a transducer, which could convert pressure into an electrical signal, because of the possibility of sparking. In such a case it might be advantageous to let the pressure transducer operate a local optical attenuator and place this device in an optical transmission link to enable remote measurement to be made.

A more sophisticated example is the measurement of the composition of a fluid which requires optical analysis but where access to the fluid is restricted. The oxygen content of the blood determines the relative absorption

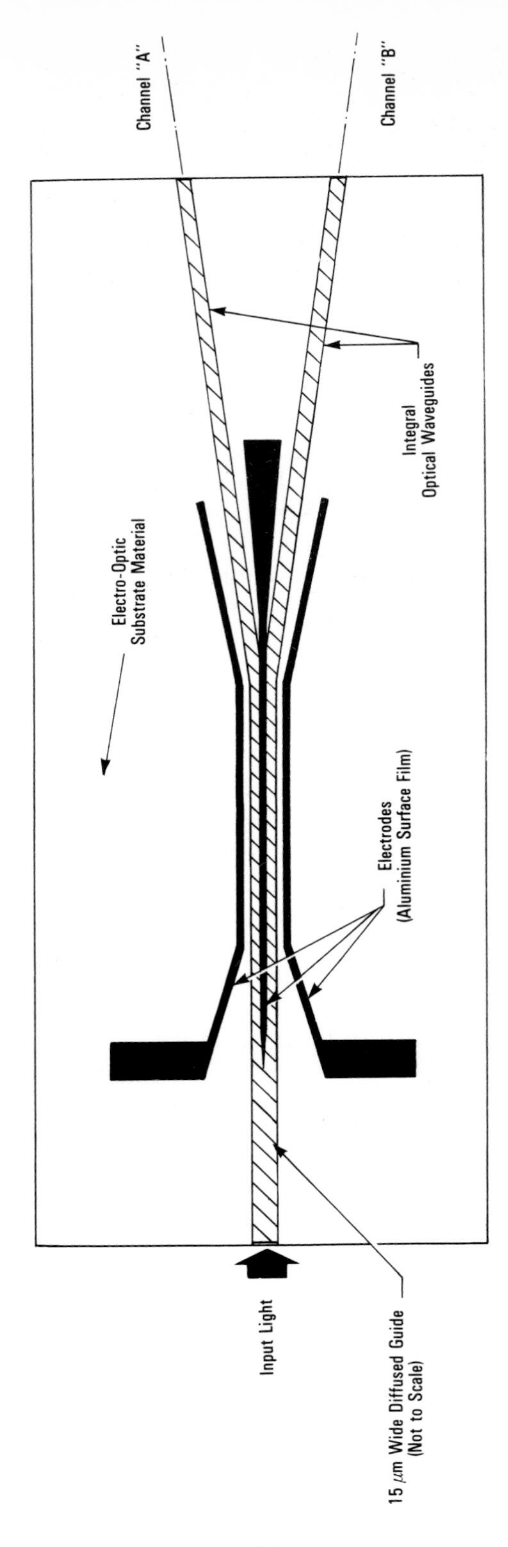
Channel "A"
Channel "B"
Integral
Optical Waveguides
Electro-Optic
Substrate Material
Electrodes
(Aluminium Surface Film)
Input Light
15 μm Wide Diffused Guide
(Not to Scale)

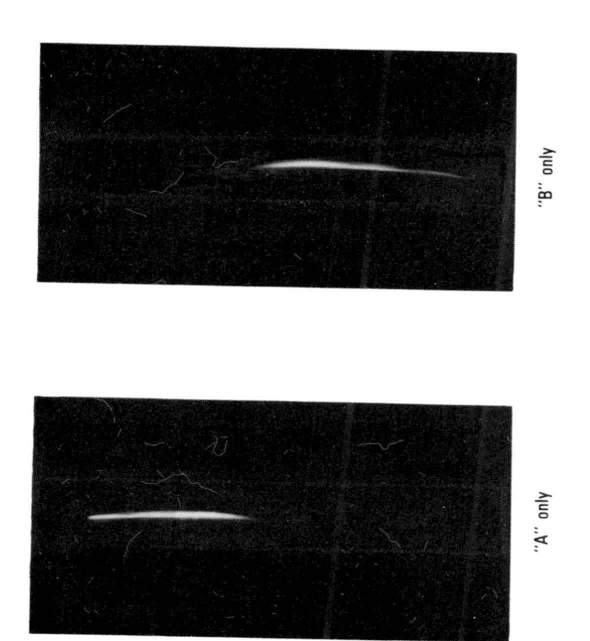

Figure 15. Integrated optic switch

of light at different wavelengths. Thus, by using optical fibres to transmit light to and from a particular region of the body (fibres can conveniently be led along arteries), it is possible to obtain a direct measurement of the oxygen content of the blood, for example in the heart itself, by remote spectral analysis. Another example is the measurement of pollution in water by oil. This entails remote measurement of optical scatter by oil particles in the bilges of ships.[9]

As optical communication technology becomes more widely accepted and available, it is likely that system designers will use transducers which modify the photon stream as readily as they now use devices which modify electron streams.

CHAPTER II

Propagation in Optical Fibre Waveguides

INTRODUCTION

The guiding of light by a dielectric medium is not a new idea. In 1870 Tyndall demonstrated to the Royal Society that light could be guided within a jet of water,[10] and less than ten years later Alexander Graham Bell studied the possibility of transmitting speech on a beam of light.[11] By 1910 Hondros and Debye[12] had reported a theoretical study of dielectric waveguides, and experimental results were reported by Schriever ten years later.[13] However, a brief investigation of a dielectric rod with a refractive index typical of glass surrounded by air, shows that either it will be highly multimoded or impracticably thin. Typically, with a refractive index n of 1.5, such a guide would have to have a diameter of 0.5 μm or less to transmit only a single mode of light from a gallium arsenide (GaAs) laser. Moreover, since not all the energy would be carried within the core, discontinuities in the surrounding air would cause unacceptable losses and make support of such a guide extremely difficult.

Both these practical limitations were overcome with the invention of the cladded dielectric waveguide, announced in 1954 by van Heel, and by Hopkins and Kapany in two letters to Nature, each describing a different application.[14,15] The use of a cladding of slightly lower refractive index than that of the core material not only enlarges the permissible diameter of the waveguide but, because the cladding can easily be made sufficiently thick to ensure that the field has decayed to a negligible value at the surface, makes support of the waveguide a practicable proposition.

Further experimental and theoretical studies of this class of waveguide followed. Attenuations were usually high and attention was concentrated on aberration-free transmission of an optical image, point by point, using a coherent bundle of optical fibres. At this stage interest in the new transmission medium was aroused at STL, a British Research Centre of ITT. When Kao and Hockham proposed that this type of waveguide could form the basis of a new communications medium, typical attenuations were over 1 dB/m, and attention became concentrated on reducing the waveguide attenuation.[1] Today, with attenuations below 1 dB/km, it is rewarding to consider other aspects of propagation in fibres.

TYPES OF FIBRE WAVEGUIDE

Some knowledge of the manufacturing processes involved in fibre waveguide production is advantageous before fibre types can be discussed sensibly. A broad outline was given in the introduction; a subsequent section will describe the processes in more detail. However, all the fibre types described below can be produced by one or other of the fabrication techniques already reviewed in Chapter I.

Figure 16(a) shows an unclad fibre. This is the simplest form of guide but is impracticable for system use. Figure 16(b) shows a typical step index multimode fibre with a core of radius a and refractive index n_1, surrounded by a cladding of lower refractive index n_2, where

$$\Delta = \frac{n_1 - n_2}{n_1} \tag{3}$$

The normalized frequency V is a useful parameter for characterizing the modes guided by such a fibre:

$$\begin{aligned} V &= ka(n_1^2 - n_2^2)^{1/2} = ka \cdot NA \\ &= n_1 ka(2\Delta)^{1/2} \end{aligned} \tag{4}$$

where $k = 2\lambda\pi^{-1}$, λ is the free space wavelength of the guided light, and NA is the numerical aperture of the fibre.

Also, the number of such modes can be calculated from

$$N_{\text{modes}} = \tfrac{1}{2}V^2 = (n_1 ka)^2 \Delta$$

For a typical multimode fibre such as that in Figure 16(b), this gives 800 guided modes. Ray analysis indicates that for a geometrically perfect fibre with $\Delta = 1$ per cent, the time interval between the fastest and slowest modes is 60 ns after propagation through 1 km of fibre. Despite the limitations of this simple analysis, which are discussed below, it points to a severe curtailment of the information carrying capacity of such a fibre. This dispersion can be reduced dramatically by using a geometry similar to that in Figure 16(c),[16] where

$$n_1 = n_1[1 - \Delta r^{\alpha} a^{-\alpha}], \qquad 0 \leqslant r \leqslant a$$

and

$$n_r = n_1[1 - \Delta], \qquad r > a$$

where α is the exponent which rises to infinity for a conventional step index fibre and n_r is the refractive index for radius r.

If α is made equal to 2, such a fibre shows periodic focusing of light launched from a point source. Its most important property for the communications engineer is its reduced modal dispersion. If only axial modes are considered, the group velocity of all modes is almost identical. However, even when skew modes are also considered, a value of α close to 2 dramatically reduces modal dispersion. For such a multimode fibre with the same Δ as a step index fibre, only half the number of modes are propagated.

If the normalized frequency of a fibre optic waveguide is reduced, either by reducing the diameter of the core or by reducing the difference in refractive index between core and cladding, then the number of modes propagated is steadily reduced. When V reaches 2.405 the last of the higher order modes is cut off and only the dominant HE_{11} mode propagates.

A convenient geometry for such a fibre is shown in Figure 17(a). Although with a 1 per cent difference in refractive index this fibre shows excellent guiding of the dominant mode, the 3 μm diameter of the core makes launching, splicing, and coupling difficult. Figure 17(b) shows a fibre with a much larger core and a correspondingly reduced refractive index difference. Such a fibre would exhibit considerable radiation losses on bends and would therefore be difficult to cable and install. Other geometries providing better compromises between core diameter and bending loss are shown in Figures 17(c) and 17(d). For instance, graded index fibres can also be reduced in normalized frequency until only the dominant mode is propagated. The W-guide (the 'W' being derived from the refractive index profile) of Figure 17(d) is an interesting example of the application of optical waveguide design principles to achieve particular transmission characteristics. In this case an excellent compromise between good guiding of the fundamental mode and discrimination against higher order modes can be obtained.

In order to take full advantage of its transmission properties, fibre should only be used with a source that emits in a single transverse mode, say a single mode semiconductor laser. However, as such it represents the ultimate in transmission bandwidth for a fibre optic communications system.

THEORETICAL ANALYSIS

Ray analysis of multimode fibres of the types shown in Figure 16 was considered in Chapter I. Although it is sufficiently accurate for cases in which a large number of modes is propagated, and is capable of extension to graded-index fibres, the limitations and inaccuracies of the ray model are apparent at lower values of the normalized frequency. For instance, this model gives a maximum dispersion between modes of

$$\Delta_t = n_1 L C^{-1}(\sec \theta_{\text{crit}} - 1)$$

where C is the free space velocity of light, l is the waveguide length, and θ_{crit} is the critical angle, governed by n_1 and n_2.

The velocity of the lowest order mode approaches that of a plane wave propagating in the core, while at cutoff a mode is approximated by a plane wave propagating in the cladding. The above formula predicts the same difference in group velocity as the mode approaches cutoff, but of opposite sign, and hence introduces a singularity at cutoff when the mode switches

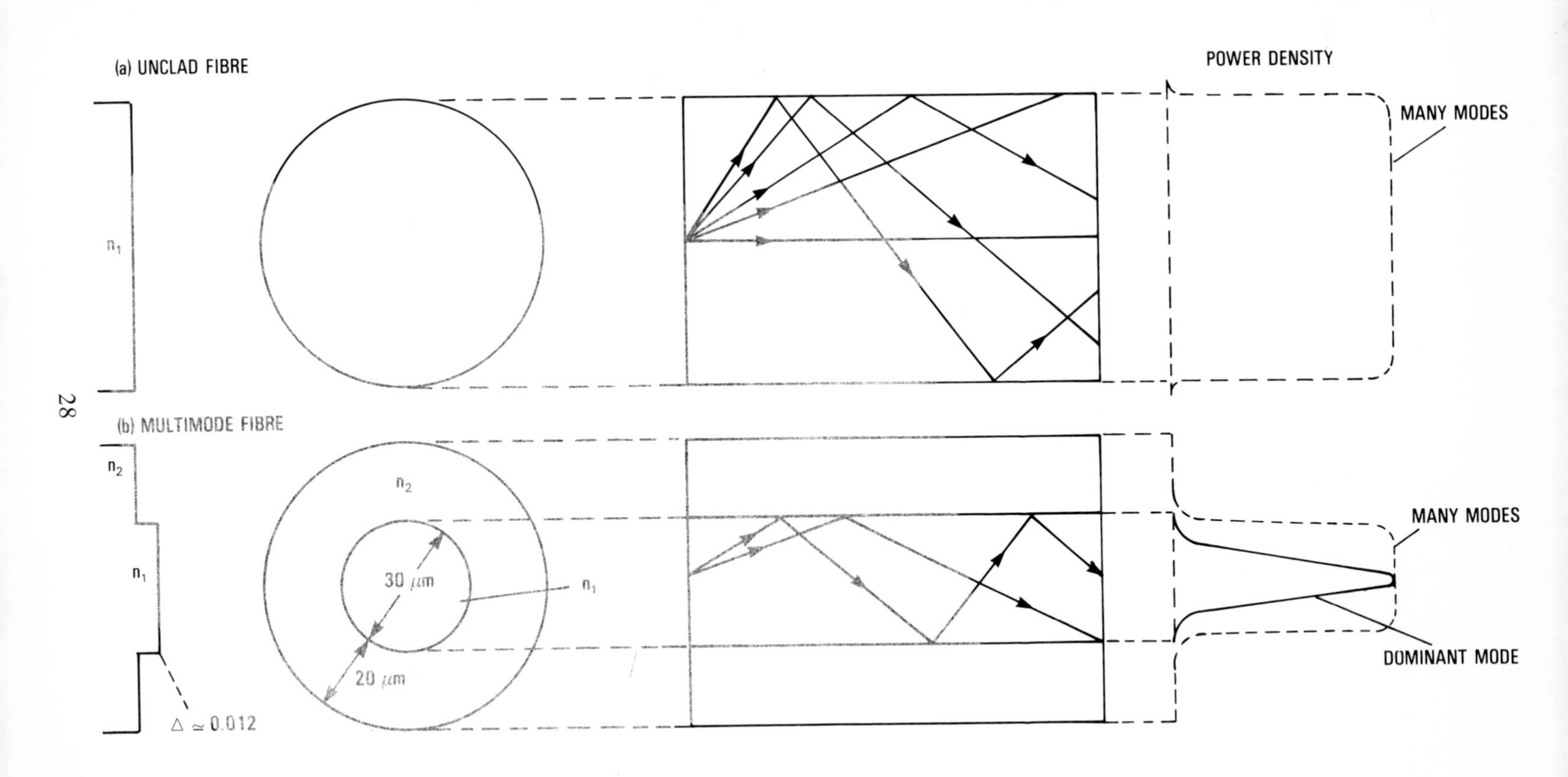

(a) UNCLAD FIBRE
n1
POWER DENSITY
MANY MODES
(b) MULTIMODE FIBRE
n2
n1
Δ ≃ 0.012
n2
30 μm
n1
20 μm
MANY MODES
DOMINANT MODE

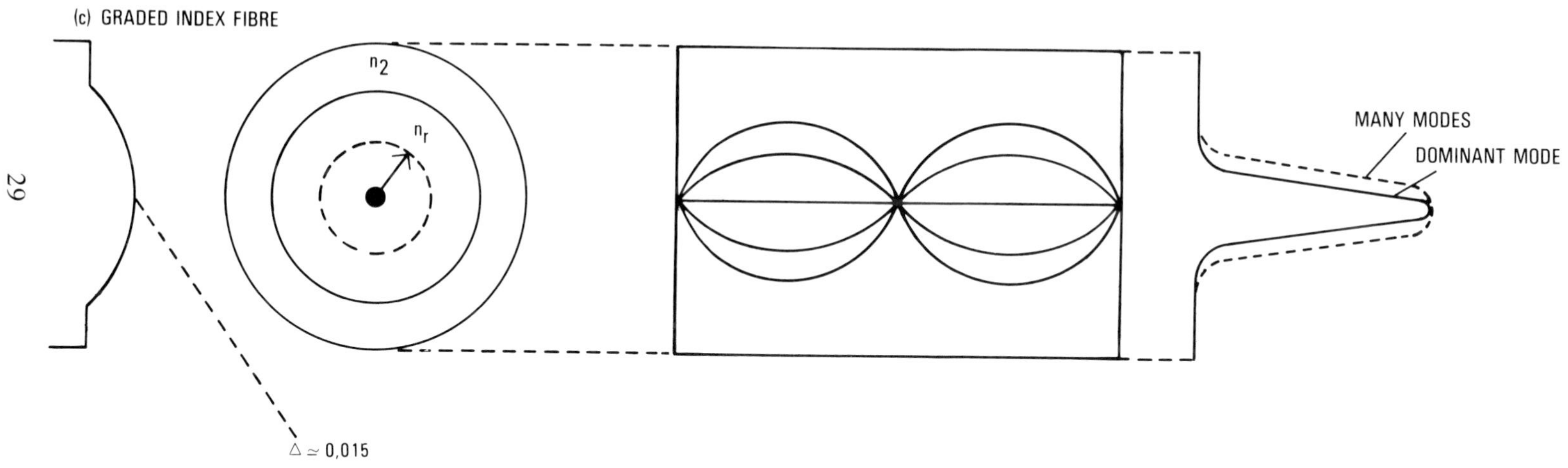

Figure 16. Features of the main types of fibre

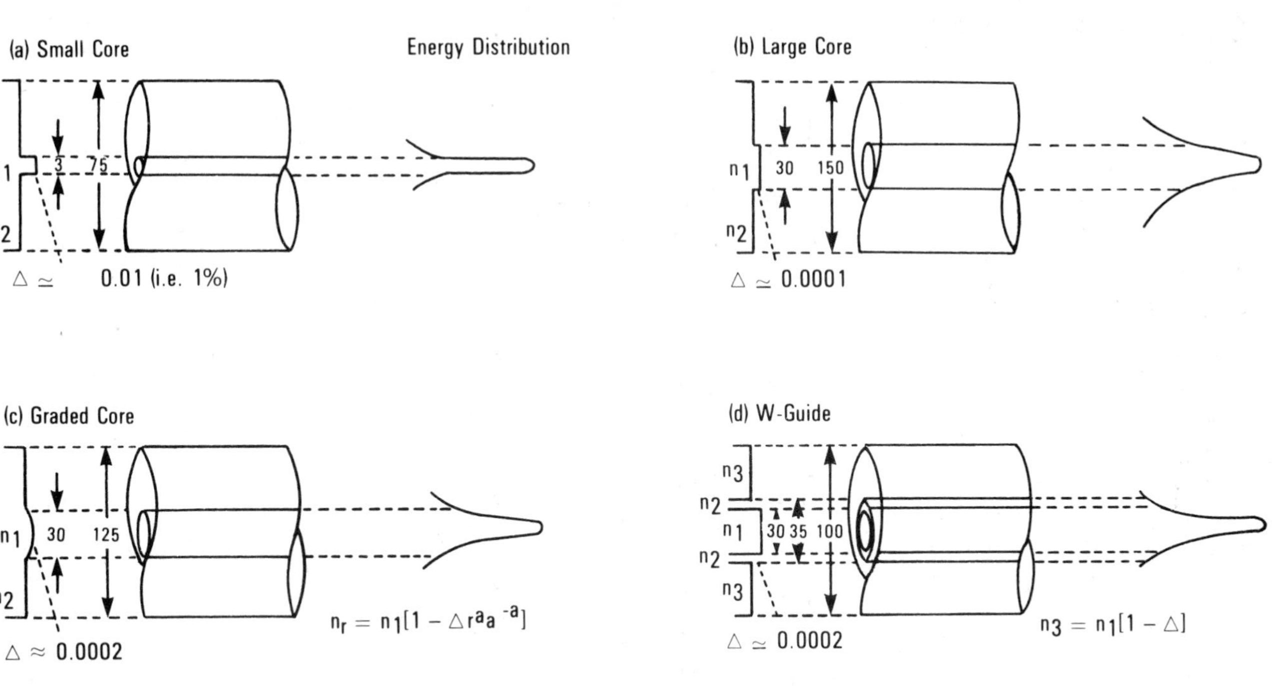

Figure 17. Dimensions and energy distributions of single mode fibres

from the slowest to the fastest propagation constant. This and other anomalies can be resolved by deriving the characteristic equation of the multilayered dielectric waveguide. This is readily formulated and solutions to the transcendental equation lead to the propagation coefficients of the modes. The two region or step index fibre is readily investigated in this manner. However, since a 4×4 determinant is required for a three layer structure with loss in the outer medium this can be tedious and does not give much insight into the processes involved.

Snyder[17] and also Gloge[18] have introduced approximations on the basis that the differences in refractive index between the regions are small and that the material dispersion is similar in all the regions. These simplifications lead to more manageable expressions that have given considerable insight into fibre waveguide behaviour, and can explain the above singularity showing that axial modes rapidly progress from the slowest to the fastest propagation constants as cutoff is approached. Many nonaxial modes cut off before this condition is reached, a result not predicted by the ray model. The analysis can be extended from step index fibres to graded index fibres with a variety of profiles,[19] and also to other fibre types. However, there exist a variety of dielectric profiles for which closed form solutions cannot be achieved. Three methods of circumventing this difficulty can be adopted in principle.

The first method is to express the radial refractive index profile as a staircase approximation. In this manner use is made of the fact that the arbitrary profile is approximated by multiple concentric dielectric layers the thickness and refractive index differences between adjacent layers of which approach zero as the number of layers becomes larger. In this method the roots of the characteristic equation are difficult to track numerically for the range of parameters used in an optical communication system.

The second method adopts a ray-optical solution based on real phase functions which is essentially the WKB approximation. In this case the assumption is made that the refractive index variations are small and, furthermore, that the variations occur over distances of many wavelengths. This leads to an approximate characteristic equation of the following form for the (ν,μ)th mode

$$\int_{\nu_1}^{\nu_2} \left[k^2 n^2(\nu) - \beta^2 - \frac{\nu^2}{r^2}\right]^{1/2} \mathrm{d}r = (2\mu - 1)\frac{\pi}{2}$$

where ν_1 and ν_2 are the turning radii. This approach is extremely powerful in its generality, but fails to account for the fields at or beyond the turning radii. A theory which alleviates this difficulty, whilst maintaining the advantage of generality, has been proposed by Felsen and Choudhary.[20] They pointed out that the use of complex phase functions instead of real ones could lead to simple solutions for the field beyond the turning points.

In the third method the direct numerical solution of the differential wave equation is used. This method is particularly attractive since it offers the real advantage of high accuracy in a relatively straightforward manner, and does not possess the inherent difficulties of the first method or the approximations implicit in the second method, although it should be borne in mind that the use of a digital computer is essential to obtain numerical results. Highly accurate results can then be obtained for arbitrary profiles, thereby permitting an optimum distribution to be derived for minimum pulse dispersion. This comprises material dispersion, due to the frequency dependence of the material refractive index, and waveguide dispersion caused by the fact that β is a function of ka, where a is the effective radius of the fibre.

The group delay τ is given by

$$\tau = L\frac{\partial\beta}{\partial\omega}$$

where L is the fibre length and $\partial\beta/\partial\omega$ is the total rate of change of β as a function of the angular frequency ω. If it can be assumed[12] that, to first order, $\partial n/\partial\omega$ is constant throughout the material, then the optimum profile is given by

$$\eta(r) = \eta_1\left\{1 - \Delta\left(\frac{r}{\alpha}\right)^{\alpha}\right\}$$

where α is approximately equal to 2. If the profile extends to infinity then the field solutions for the parabolic index curve can be expressed analytically by Laguerre–Gaussian polynomials. However, in practice the cladding permittivity beyond $r = a$ is typically uniform, making it impossible to obtain closed form solutions. Clearly, the optimum index α will be close to 2 and the general analysis of using the third method, albeit numerical, does permit accurate results to be obtained for all modes, even for those in the vicinity of cutoff.

As an example, the W-guide is considered. This appears to offer advantages for single mode operation, giving good guiding properties while being less susceptible to radiation losses due to bending, particularly the microbends introduced during cabling. The configuration is illustrated in Figure 17(d). The presence of the inner ring of refractive index n_2, which is less than that of the outer infinite medium n_3, can give rise to modes with phase velocities less than that of medium 3. Consequently, all modes of this structure possess nonzero cutoff frequencies. This can be explained physically with reference to Figure 18. The solution to the wave equation for the mode HE_{mp} in region 1 is of the form

$$\begin{Bmatrix} E_z \\ H_z \end{Bmatrix} = \begin{Bmatrix} A_p \\ B_p \end{Bmatrix} J_p(u_1 r a^{-1}) \exp{(jp\theta - j\beta z a^{-1})} \tag{5}$$

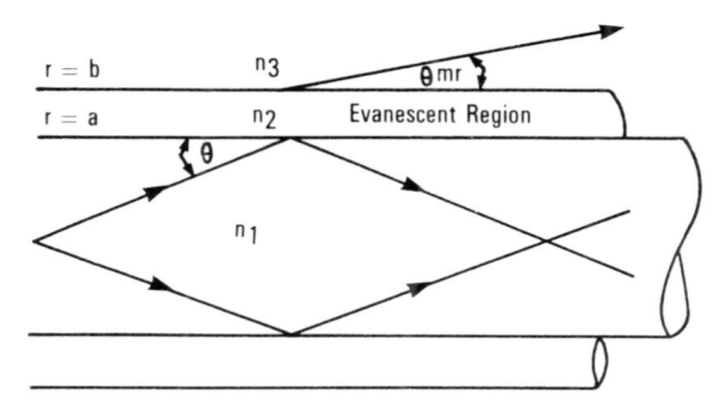

Figure 18. Geometry of the three-layer model of a single mode fibre

where

$$u_1^2 = k^2 n_1^2 - \beta^2. \tag{6}$$

$E_z, H_z = z$ components of electric and magnetic fields, respectively,

A_p, B_p, etc = constants,

β = mode propagation constant,

and where k, β are normalized to the radius of the central core region $r = a$.

For the case $u_1 r a^{-1} \gg 1$, which is valid provided r is not too small, the field given by equation (5), is of the form of plane waves, illustrated by the rays in Figure 18, which propagate at angles $\pm\theta$ to the interface between regions n_1 and n_2. Provided $\theta < \theta_{\text{crit}}$, where θ_{crit} is the critical angle defined by

$$\cos\theta_{\text{crit}} = n_2 n_1^{-1}, \tag{7}$$

then the rays are totally internally reflected. It has been assumed that the discontinuity caused by the presence of region 3 has no effect, which is obviously incorrect. However, the field within region 2, which must exist to maintain continuity of the field at the boundary $r = a$, is evanescent and is of the following form

$$\begin{Bmatrix} E_z \\ H_z \end{Bmatrix} = \left[\begin{Bmatrix} E_p \\ G_p \end{Bmatrix} K_p(u_2' r a^{-1}) + \begin{Bmatrix} F_p \\ D_p \end{Bmatrix} I_p(u_2' r a^{-1}) \right] \exp(+jp\theta - j\beta z) \tag{8}$$

where

$$u_2'^2 = \beta^2 - k^2 n_2^2. \tag{9}$$

Digressing for a moment and assuming that region 2 extends to infinity, where the fields must approach zero, leads to the case $F_p = D_p = 0$, and for $u_2' r a^{-1} \gg 1$ the modified Bessel function behaves asymptotically exponentially. The second term above, containing F_p and D_p, will be nonzero owing to the presence of the secondary boundary at $r = b$, and can be regarded as a reflected evanescent field. Imposing the boundary conditions for the two layer model (i.e. $b \rightarrow \infty$) we find that a finite number of angles $\theta_m < \theta_{crit}$ exist, each corresponding to a particular guided mode of the structure. The number of modes depends on the core diameter, wavelength, and n_1 and n_2; however, as the dominant HE_{11} mode has a zero frequency cutoff, there is always at least one guided mode. Returning to the three layer model, where n_2 is less than both n_1 and n_3, this is no longer true.

The field in region 3 is of the form

$$\begin{Bmatrix} E_z \\ H_z \end{Bmatrix} = \begin{Bmatrix} G_p \\ H_p \end{Bmatrix} K_p(cu_3 rb^{-1}) \exp(+jp\theta - j\beta z) \tag{10}$$

where

$$c = ba^{-1} \quad \text{and} \quad u_3^2 = \beta^2 - k^2 n_3^2 \tag{11}$$

In this region it has been assumed that $\beta^2 > k^2 n_3^2$ and the field exhibits an exponential decay for $cu_3' rb^{-1} \gg 1$. In this case the propagation coefficient β is purely real and the mode is a pure surface wave. The presence of a low permittivity region n_2 permits the propagation coefficient β to lie in the region $kn_2 < \beta < kn_3$, and it can be seen that for a sufficiently low frequency, u_3^2 can be negative for all modes. Since the field must satisfy the radiation condition at infinity it is necessary to take the positive imaginary root for u_3.

$$u_3 = j(k^2 n_3^2 - \beta^2)^{1/2} = +ju_3' \tag{12}$$

and the field in region 3 has a radial dependence described by

$$\begin{Bmatrix} E_z \\ H_z \end{Bmatrix} = \begin{Bmatrix} G_p \\ H_p \end{Bmatrix} H_p^{(2)}(cu_3' rb^{-1}) \exp(+jp\theta - j\beta z) \tag{13}$$

In this case the field is no longer radially evanescent, but is approximately of the form of a plane wave propagating at an angle θ_{mr} to the waveguide axis given by

$$\cos \theta_{mr} = \beta k^{-1} n_3^{-1} \tag{14}$$

as illustrated in Figure 18.

In this situation β is now complex since the mode is leaky and must be of the form

$$\beta = \beta_{mr} - j\alpha_{mr} \tag{15}$$

where α_{mr} is the attenuation coefficient of the mode due to radiation loss.

If the width of region 2, $(b-a)$, is large enough so that the field decays to a low value at the boundary $r = b$, then the phase constant β_{mr} is close to that for the corresponding two region model and α_{mr} approaches zero as the gap becomes theoretically infinite. Nevertheless, provided α_{mr} is nonzero the mode will be attenuated and at sufficiently large distances from the source only pure surface waves will exist.

The preceding discussion has given a conceptual view of the W-guide. The characteristic equation for the three layer region was derived by Kao and

Hockham[21] in 1968 for the case of a lossy outer medium and, more recently, using the approximations introduced by Snyder[17] and Gloge,[18] by Kawakami and Nishida.[22] Using this characteristic equation a more rigorous analysis of the W-guide has been carried out. This illustrates the use of the analysis for the characterization of the optical waveguide and similar methods can of course be applied to other structures.

MODE CUTOFF

It is convenient at this stage to derive the normalized cutoff frequencies of the modes, which are defined here as $u_3^2 = 0$ and correspond to the transition region separating the leaky modes from the pure surface wave modes.

Therefore, imposing the condition $\beta = kn_3$, and noting that $u_1 = V_{13}$ and $V_{32} = k(n_3^2 - n_2^2)^{1/2}$, the characteristic equation derived by Kao and Hockham[21] can be simplified. In a typical fibre $k = 2\pi a\lambda_0^{-1}$, where $2a\lambda_0^{-1}$ may be greater than 10, and $(n_3^2 - n_2^2)$ is relatively large ($u_2' \gg 1$), so use can be made of the asymptotic expansions of the modified Bessel functions of the first and second kind, leading to the following equation for the values of V_{13} at cutoff

$$\frac{1}{V_{13}} \frac{Jp'(V_{13})}{Jp(V_{13})} = \frac{1}{V_{32} \tanh (V_{32}c - 1)} \tag{16}$$

To obtain the normalized cutoff frequencies, V_{13}, for the two region model, put $c = 1$; from equation (16)

$$Jp(V_{13}) = 0$$

$$V_{13} = 0, 3.83 \ldots, \quad \text{for} \quad p = 1$$

and

$$V_{13} = 2.405, \qquad \text{for } p = 0$$

As expected, the dominant mode corresponding to $p = 1$ has a zero cutoff frequency, while the first higher order modes corresponding to modes TM_{01}, TE_{01}, and HE_{21} have a cutoff value of $V_{13} = 2.405$.

However, when $c > 1$ the right-hand side of equation (16) is no longer infinite so that at cutoff $V_{13} > 0$ for all modes. Figure 20 shows the relationship at cutoff between V_{13} and c for two typical values, $V_{23} = 4$ and 8, for both the dominant HE_{11} mode and the first set of higher order modes.

Now consider a fibre waveguide with parameters as defined in Figure 19. In practice the value of c would be chosen primarily to achieve a sufficiently high attenuation coefficient for the leaky modes. Although this has not been considered in this chapter, it is assumed that $c = 1.1$, giving $b - a = 0.9$ μm. This gives a value of $u_2' = 4$ and Figure 19 shows that V_{13} must lie in the range

$$1.05 < V_{13} < 2.86$$

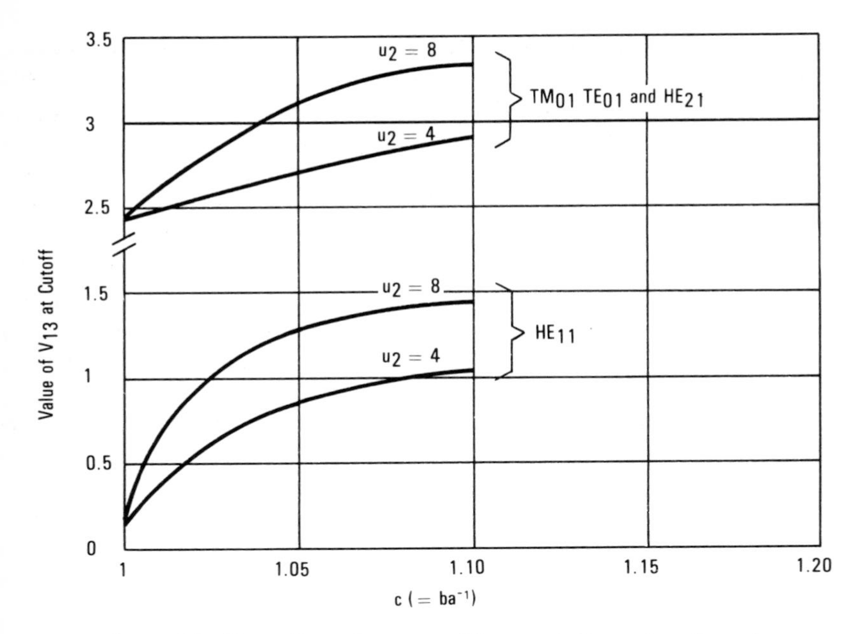

Figure 19. Normalized frequency V_{13} at cutoff as a function of c

in order that all modes, with the exception of the dominant HE_{11} mode, are leaky. For the present fibre

$$V_{13} = 2\pi a \lambda_0^{-1}(n_1^2 - n_3^2)^{1/2} = 1.79$$

which is clearly within the required range.

A perturbation method of solution to calculate the leakage coefficient has been discussed by Kawakami and Nishida.[23]

Waveguide Dispersion

The characteristic equation (16) has been solved for βk^{-1} as a function of k using the above parameters. However, since $n_3 < \beta k^{-1} < n_1$ the dispersion is very small, which is particularly advantageous in a long distance communications link. To display the information on a suitable scale it is convenient to plot β, defined as

$$\bar{\beta} = (\beta k^{-1} - n_3)(n_1 - n_3)^{-1} \tag{17}$$

as a function of k. In this case $0 < \beta < 1$ and is directly proportional to βk^{-1}. Figure 20 illustrates β as a function of V_{13} for the lowest order modes.

ATTENUATION MECHANISMS

Signal level dependent losses (nonlinear losses) have been considered by Kao and Hockham.[1] Even for small core single mode fibre waveguides, these can be ignored when light is launched from modern semiconductor injection lasers. More recently, the limitations imposed by nonlinear effects on the maximum usable power in fibre optic systems have been reviewed by Smith.[24] Both stimulated Raman scattering[25] and stimulated Brillouin scattering have been experimentally and theoretically studied.[26] The indications are that when laser line widths are comparable with the Brillouin line width (<100 MHz) power of a few milliwatts will lead to nonlinear effects in small core single mode fibres. Since very large information bandwidths will be exploited when narrow spectral line widths are available, and since large core single mode fibres are preferred for ease of launching, connecting, and splicing, we can safely ignore such effects for the present and concentrate on the linear losses.

Several mechanisms lead to transmission loss in fibre waveguides. These are:

(a) material absorption,
(b) material scattering,
(c) mode coupling to the radiation field,
(d) radiation due to bends, and
(e) leaky modes.

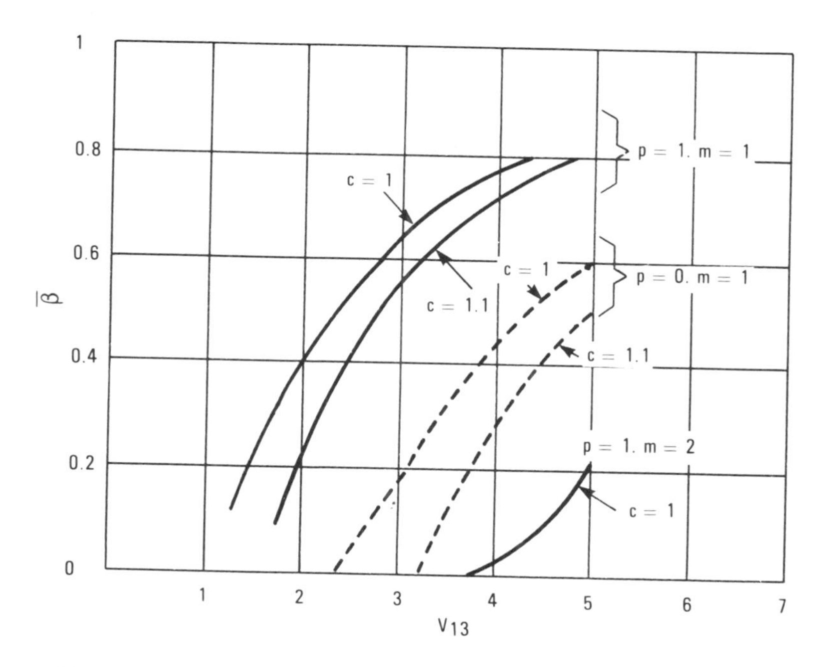

Figure 20. Modified propagation coefficient $\bar{\beta}$ as a function of the normalized frequency V_{13}

Material Absorption

Possibly the most important loss mechanism, i.e. material absorption, is discussed in detail in relation to the manufacturing processes in Chapter III. Briefly, many glass compositions would show negligible material absorption in the wavelength regions of interest if no foreign elements were present. However, traces of transition metal or hydroxyl ions have a profound effect. Both the glass type and the state of oxidation of these impurities influence the absorption. However, many fibres have been measured in which scattering losses predominate, showing that the very high purities required can now be achieved.

Material Scattering

Several linear scattering mechanisms can be distinguished in most fibre waveguides. The most fundamental, Rayleigh scattering, is always present due to the inhomogeneities, small in comparison to the guided wavelength, which are produced in the guide during glass melting and fibre drawing. In general one might expect this scatter to be higher in a fibre than in a carefully made glass, but indications are that this is not necessarily so. In fact, some modern fibres show losses below those earlier predicted for Rayleigh scattering alone.[27] Rayleigh scattering can normally be identified by its proportionality to λ^{-4} and by its angular dependence proportional to $(1+\cos^2\theta)$.

When the inhomogeneities are comparable in size to the guided wavelength, Mie scattering can be observed. This is predominantly forward scatter and is not easily separated from that due to tunnelling out of the highest order modes propagated in the fibre under the conditions in which scattering is usually measured. Gross imperfections, however, show wavelength independent scatter with quite random angular dependence. Below the threshold Raman and Brillouin scatter have little effect on transmission.

Some absorption mechanisms, for instance the fibre drawing peak reported by Kaiser[28] or absorption exhibited by the fibre (Figure 21), are accompanied by a resonance scatter which is easily distinguished by its wavelength correspondence with the associated absorption mechanism.

Mode Coupling Scatter

Variations in core diameter or core/cladding refractive index difference along the length of a fibre waveguide can influence the transfer of power from one mode to another and hence to the radiation field. Marcuse has studied these effects in slab waveguides[29] and in round fibres.[30] For any two modes with propagation constants β_x and β_y there will be a coupling parameter $\theta_{xy}=(\beta_x-\beta_y)$ which will be particularly effective in inducing

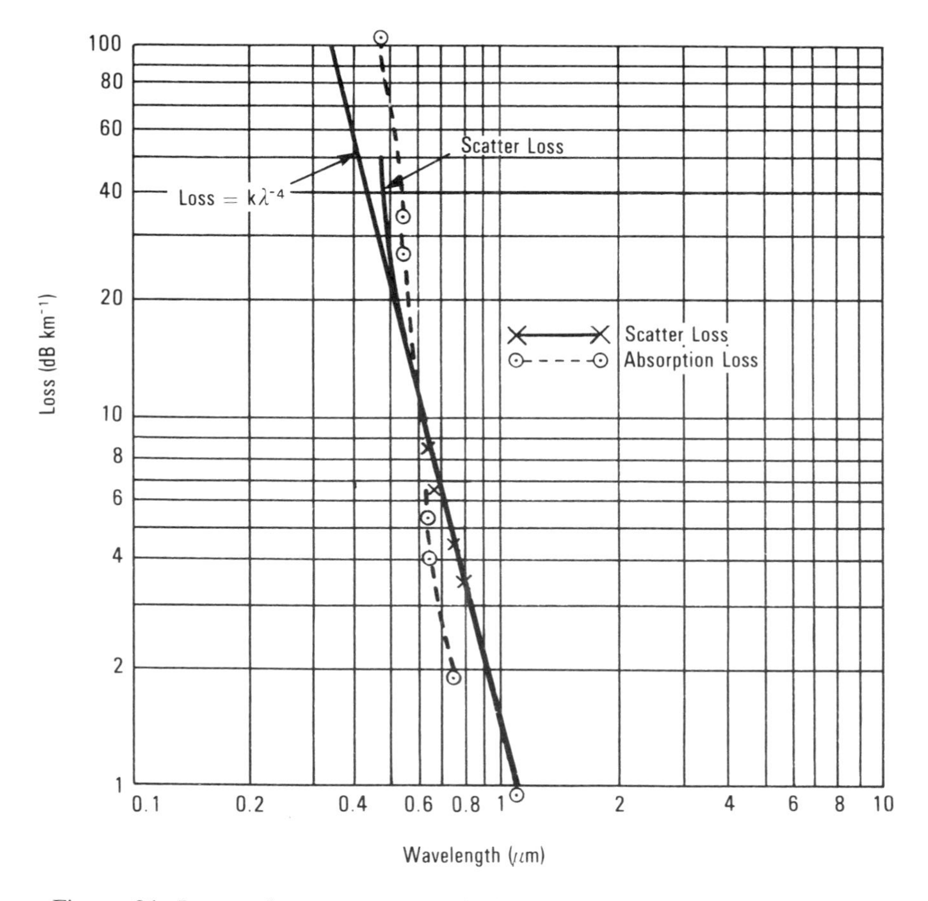

Figure 21. Losses due to scatter and absorption in a fibre waveguide as a function of wavelength

coupling. For a step index fibre there is a steady increase in the difference between nearest neighbour propagation constants from the lowest order modes to the radiation field. A spectrum of mechanical imperfections $C_{(z)} = \sum A_n \sin \theta nz$ can therefore couple modes very effectively. Fortunately, the processes employed in manufacturing fibres and the mechanical properties of glass ensure that the most critical correlation lengths do not occur, and mode coupling usually leads to an equilibrium distribution of power between modes in lengths of hundreds of metres or even kilometres rather than centimetres.

For the special case of parabolic index fibres, Ikeda has shown[31] that the differences between nearest neighbour propagation constants are identical. Therefore, if this particular frequency can be removed from the mechanical spectrum, a guide without mode coupling can be obtained. Practical results confirming this have been noted for near parabolic index fibres,[32] and by Midwinter and co-workers for fibres which are near step index.[33]

Radiation due to Bends

All dielectric guides, other than those that are absolutely straight, will radiate. This was considered theoretically by Marcatili[34] and quickly confirmed when low loss fibres were first constructed. Figure 22 shows why this loss should occur. It shows the effect of a bend on the dominant mode in a low V-value fibre for two different core diameters, or the equivalent, two radii of curvature. Since the bend is gradual its influence on the transverse field is negligible. Now, the field in the cladding extends to infinity, so at some displacement x_r the radius of curvature R implies energy propagating at greater than the velocity of light, and so guiding ceases. The radiation attenuation coefficient has the form

$$a_r = c_1 \exp(-c_2 R)$$

where c_1 and c_2 are constants which are not dependent on R. The exponential dependence of a_r on R, which arises from the steep reduction of field with penetration into the cladding, means that at a critical radius (dependent on the waveguide design) a reduction of R by a factor of two can change the loss from a quite negligible to a totally prohibitive value.

Leaky Modes

Snyder has drawn attention to a class of modes which are not completely guided, but will slowly leak.[35] These modes are in two classes, both of which can be represented by highly skewed rays. In one class owing to the skew of the propagating rays, the critical angle is exceeded but only in either the

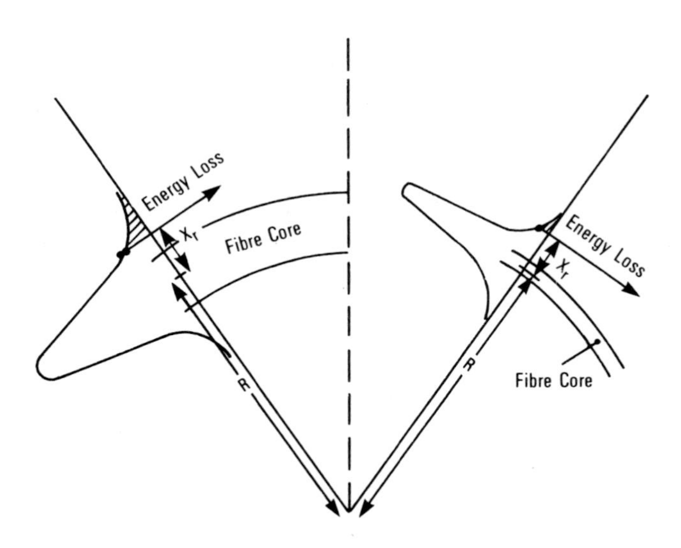

Figure 22. Energy loss caused by bends in a fibre

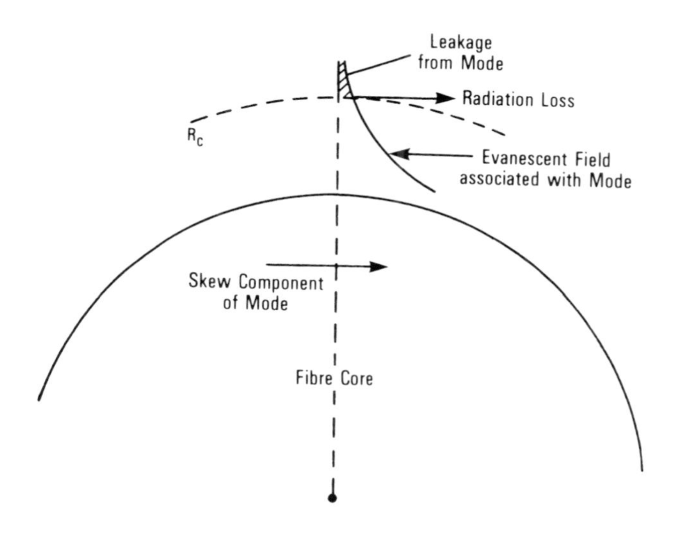

Figure 23. Leakage from a skew mode

meridianal or circumferential plane. This results in partially trapped radiation and a slowly leaking mode. In the other class, although the critical angle is not exceeded in any plane, radiation appears to tunnel through a barrier region in the cladding and propagate from beyond this barrier. The mechanism is exactly analogous to the radiation due to curvature and can be understood from Figure 23. Here we consider the cross section of the fibre waveguide. Skew rays have a circular component with a field extending to infinity. For the leaky modes this field extension implies a radius beyond which energy is propagating faster than the velocity of light and radiation occurs from this critical radius which defines the barrier region. For fibres with a small normalized frequency the leaky modes radiate rapidly and can only be observed much closer to the source than the region in which measurements are normally made. For large normalized frequencies, up to 50 per cent of the propagating modes can be leaky and may radiate over a distance of 1 km or more. For practical values of the normalized frequency, say up to $V = 100$, the leaky modes only amount to a few per cent of those propagating.

CHAPTER III

Fabrication of Optical Fibre Waveguides

INTRODUCTION

Having considered the implications of optical systems and discussed the fundamental propagation principles for some possible structures, we now discuss the problems encountered in producing material with the desired properties. Some of the practical methods which have been used to achieve the necessary structures and performances are described in this chapter.

In their 1966 paper, Kao and Hockham[1] considered that the main obstacle to the production of optical waveguides for communication lay in the quality of the glasses available at that time, and predicted that improvements in the techniques of glass preparation would make it possible to achieve optical losses below 20 dB/km. As a result of this early work a programme for the development of glass fibres was started in conjunction with the British Post Office. At first, STL concentrated on developing techniques for the production of fibres and their evaluation. This work centred on the double crucible pulling method using commercially available glasses, but later changed to specially developed glasses from the BPO Research Department and other laboratories. As interest grew, a new technique for melting pure glass was developed at STL and this soon became the main source of material.

During measurements on the bulk properties of a number of materials, it was discovered that commercially available fused silica could have losses below 5 dB/km.[36] This result stimulated a new area of investigation into fibre preparation methods at STL, in the US and in Japan; it was a fibre based on fused silica which led to the first fibre loss of less than 20 dB/km, reported by Kapron *et al.* in 1970.[37] Workers at Corning Glass,[38], Bell Laboratories,[39] STL,[40] and Southampton University[41] had all reported fibre losses below 4 dB/km by 1974, using techniques based on fused silica.

A broad distinction between fibres based on fused silica and those based on low softening point glasses lies both in the intrinsic properties of the materials and in the techniques adopted for processing them into low loss fibres. Here, therefore, the term 'silica fibre' is used to refer to fibres consisting predominantly of fused silica glass, while the low softening point glasses such as the sodium borosilicates, sodium calcium silicates, and lead silicates are called 'glass fibres'.

As new glass compositions are tried and new processes evolved, however, there are some occasions where such a generalization is less directly applicable. In what follows the main techniques for the manufacture of glass and silica fibres are described together with a section on plastic clad silica fibres in which the optical cladding is a reasonably transparent polymer rather than a glass.

MATERIAL REQUIREMENTS FOR OPTICAL FIBRES

The three main waveguide structures were described in Chapter I. They are the single mode, multimode step index, and multimode graded index structures and are chosen with reference to the source to be used and the bandwidth and length requirements of the system. All three waveguide structures require a core material with high optical transmission at the emitting wavelength of the proposed source. This core must be surrounded by a cladding material with a refractive index in the range 0.5–2 per cent below that of the core material, though for a few applications a higher index difference may be required. The choice of index difference depends on the normalized frequency requirement for the system application. Optical properties of the cladding material must be similar to those of the core material in single mode fibres since a significant proportion of the transmitted energy propagates in the cladding region. However, transmission requirements are less stringent for the cladding in multimode fibres with high normalized frequencies.

Graded index multimode fibres create an additional problem since the core region must be produced with a refractive index that varies in near parabolic manner to minimize dispersion due to differences in the delays of the propagating modes. The optimum profile has been predicted[19] to have the form

$$n_r = n_{\max} \left|1 - 2\Delta(r/a)^{\alpha}\right|^{1/2}$$

where

n_r = refractive index at a distance r from the axis of the fibre,

$n_{\max}$ = index on the axis,

a = radius of the core,

Δ = given by $n_{\max} - n_{cl}$ and

n_{cl} = cladding index.

The value of α is normally close to 2, but depends on the composition of the fibre core and on the design wavelength. However, by judicious choice of the core composition the wavelength dependence may be minimized.

The operating wavelength region for most of the early optical systems has been the emission wavelength range of the $Ga_xAl_{1-x}As$ family of laser- and light-emitting diode sources. This lies between 820 and 910 nm. At these wavelengths a second source of pulse dispersion is material dispersion, owing to the finite linewidth of a source at a point where the refractive index of the core varies rapidly with wavelength. The effect of material dispersion varies with composition and wavelength. For silica based fibres[42,43] this effect goes to zero at wavelengths in the region of 1.25–1.32 μm, and has led to development of new sources, detectors, and fibre structures aimed at exploiting the much higher system bandwidths possible at these wavelengths. In choosing a fibre composition, therefore, both wavelength regions must be considered for minimizing attenuation and for the optimization of the profile exponent α.

In addition to the basic optical characteristics and the ease of manufacture of very long lengths at reasonable cost, a number of other factors need to be considered for particular requirements. These include radiation hardness, durability, strength, stability over a wide temperature range, as well as compatibility with other aspects of optical system design such as cabling, splicing, and connector assembly. Thus, the liquid-filled fibres which gave low losses in 1972,[44,45] have not been pursued by most of the laboratories who investigated them because of their incompatibility with other processing stages.

SOURCES OF LOSS IN GLASSES AND FIBRES

An earlier section described some of the attenuation mechanisms which arise in fibres and, in particular, radiative losses arising from the fibre waveguide structure such as mode coupling, radiation on bends, and losses through leaky modes. These losses are largely due to either the fibre design or, in the case of mode coupling, mechanically induced. The types of loss considered in more detail here are the material dependent losses which arise from the material or from the manner in which it is processed. These losses may be separated into absorption and scatter loss components.

Absorption Loss

A pure glass exhibits two main intrinsic absorption mechanisms for optical wavelength energy. Absorption bands occur at infrared wavelengths due to the interaction of photons with molecular vibrations within the glass. The fundamentals of these bands normally occur at wavelengths above 6 μm. A second absorption mechanism is the fundamental absorption edge of the glass which arises from the stimulation of electron transitions by higher energy excitation. This type of absorption occurs at an ultraviolet wavelengths for the typical glass structures chosen for fibre manufacture. The tails of both the infrared vibrational absorption peaks and of the UV absorption edge can both extend into the wavelength ranges of interest, however, and care must be taken in the choice of both core and cladding compositions to minimize these effects.

Most glasses exhibit further absorption due to the presence of small quantities of impurities. These absorption effects are induced by transition metal ions from the first series (Ti, V, Cr, Mn, Fe, Co, Ni, and Cu) and are due to electronic transitions between the energy levels associated with the incompletely filled inner subshell. Absorption due to these effects is normally a broad peak in the visible spectrum and is the source of colouration in many glasses and crystals. Familiar examples are the green colour of thick slabs of most commercial glasses, due to ferrous ions, and ruby and sapphire gem stones which have small quantities of chromium and titanium, respectively, as impurities in aluminium oxide.

The loss at any wavelength due to impurity ions depends upon the concentration of the impurity and the oxidation state of the element. Both the composition of the host glass and its method of preparation significantly influence the coordination state of an impurity ion and therefore its effect on absorption. Table III shows examples of the measured contributions of the most significant impurities in three of the most commonly used glass systems. Table III indicates that to achieve low optical absorption in the 800–900 nm wavelength range, considerable attention must be paid to minimizing the impurity content of the glass reaching the fibre stage; it is this constraint which has led to the need for processing similar to that used for semiconductors to achieve low loss optical fibres.

In addition to transition metal ion absorption, two other effects may influence the absorption of a glass and are significant in the fibre form. The first of these effects is absorption due to hydroxyl ions, often referred to as the 'water absorption loss'; this contributes a vibrational absorption associated with the hydroxyl ion. In fused silica, for example, the fundamental vibration occurs at 2.73 μm with overtones at 1.37 and 0.95 μm, and a series of smaller bands corresponding to the combination frequencies from the various overtones to the hydroxyl ion vibration and the fundamental

Table III
Reported impurity ion contributions to absorption loss in different glasses

Inpurity Element	Absorption induced at 850 nm dB/km for each part per million		
	Na_2O–CaO–SiO_2[46] (sodium calcium silicate)	Na_2O–B_2O_3–TlO_2–SiO_2[47] (sodium borosilicate)	SiO_2[48] (fused silica)
Fe	125	5	130
Cu	600	500	22
Cr	<10	25	1300
Co	<10	10	24
Ni	260	200	27
Mn	40	11	60
V		40	2500

vibrations of the silicon–oxygen bonds in silica.[38] The second effect is absorption arising from a defect state in the material. Such effects, often known as 'colour centres', are normally generated by high energy irradiation of the glass,[49] but may also be induced by the fibre manufacturing process.[28] This type of absoprtion is normally accompanied by a corresponding scattering loss peak.

Scatter Loss

Light energy passing through glass may be scattered and lost through two main mechanisms: Rayleigh scatter, an intrinsic property with an ultimate limit, and scatter due to defects in the glass or introduced by the processing technique which it should be possible to eliminate. Rayleigh scattering arises from the variations in refractive index which occur within the glass over distances that are small compared with the wavelength of the scattered light. Such index variations are caused by local fluctuations in composition and density within the glass and are very dependent on the basic glass and its method of preparation. There is, however, a lower limit to the Rayleigh scatter which also dominates the theoretical limit for the reduction of loss in the material. Rayleigh scatter exhibits a characteristic variation as λ^{-4} and typical measured values for glasses of interest are in the range 0.7–2 dB/km at 850 nm.

The second major scattering effect is due to defects in either the processing or the structure of the glass. Processing defects may be in the form of bubbles of gas trapped during cooling or released from the glass in a reboil effect, phase separated regions, devitrified sections, or unreacted materials. Such effects may be eliminated by careful handling and by a sensible choice of working compositions. Structural defects of the type described in the previous section on absorption may occur; these may be annealed out of bulk glass but can be a serious problem in fibres and should be minimized both during processing and by the choice of glass composition.

Other potential scattering effects are Brillouin and Raman scattering which are nonlinear effects at high power densities. Such effects are not expected to be significant at the power levels at present anticipated in fibres. Mie scattering due to inhomogeneities comparable in size to the scattered wavelengths has been identified in some glasses but may be minimized in principle. Some discussion of these mechanisms is included in Chapter II.

MANUFACTURE OF GLASS FIBRES

Glass fibre manufacture is a two-stage process in which pure glass is first produced from high purity powders and converted to a form (normally a rod) suitable for fibre making. The pure glass is then loaded into a double crucible apparatus for fibre pulling. Whichever technique of glass melting is adopted, very close attention must be paid to eliminating risks of contamination of the raw materials or the glass at any stage of the processing.

The RF induction technique for glass melting illustrated in Figures 24 and 25 was developed independently by Sheffield University[50] and STL[51] after early work with high purity powders melted in platinum crucibles had given losses well in excess of the levels predicted from a knowledge of the impurity content of the powders.[47] This loss was attributed to ferrous and cupric ion absorption: the iron and copper causing the loss was thought to

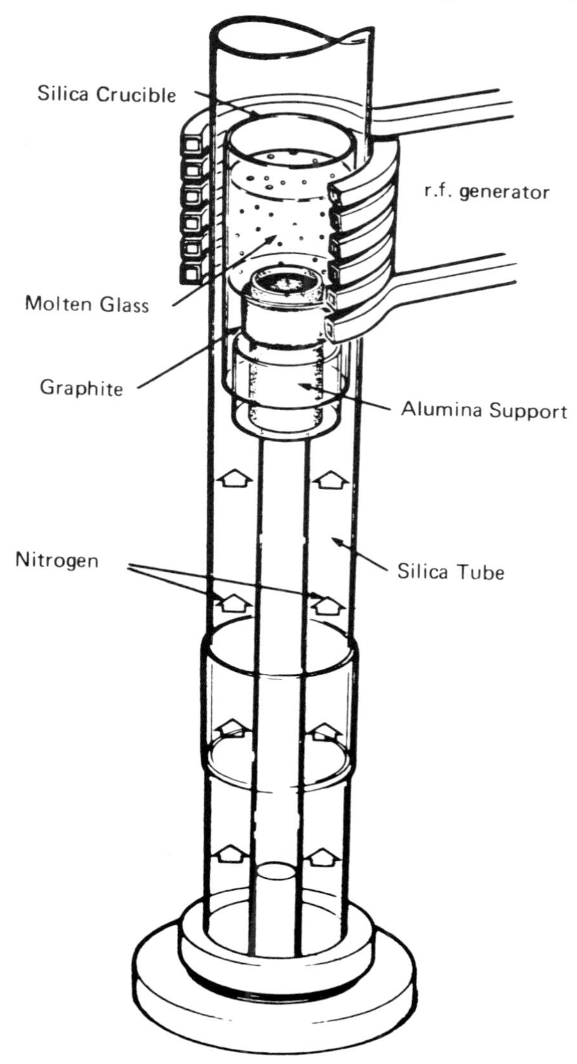

Figure 24. Ultra pure glass melting using RF heating

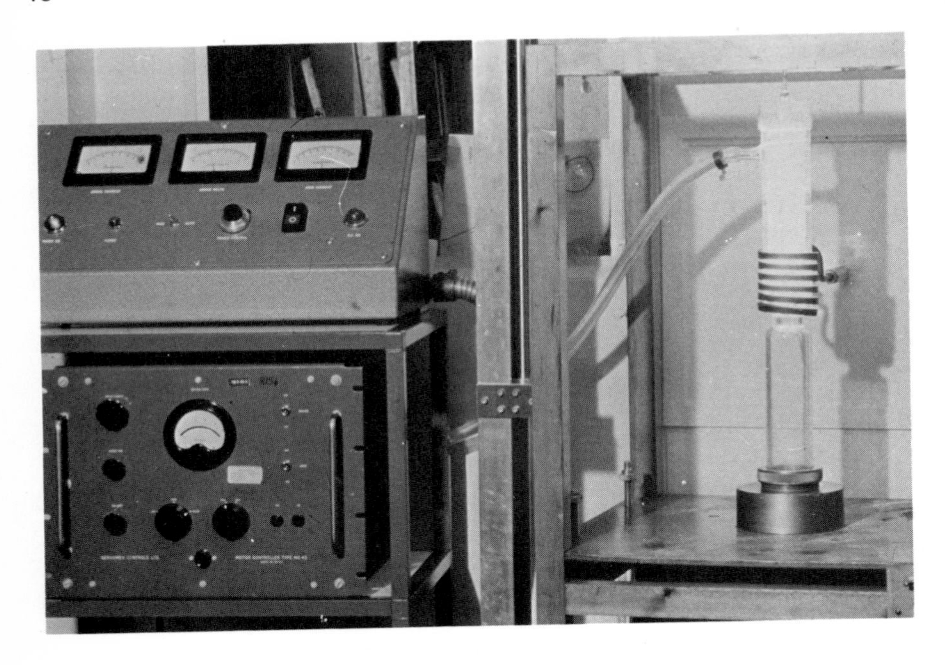

Figure 25. RF melting apparatus

have been leached out of the platinum crucible used for melting and refining the sodium calcium silicate glasses. At temperatures in excess of 1300 K most alkali glasses exhibit sufficient ionic conductivity to enable coupling between the melt and a high frequency (~5 MHz) RF field. This induction heating of the glass itself means that the surrounding crucible is no longer the hottest part of the melt, but is kept cold either by gas or water flow. A silica crucible is normally used since it has no significant conductivity at the frequencies and temperatures used. The melt is protected from impurities in the crucible by a thin layer of solid pure glass which forms between the melt and the crucible as a result of the steep temperature gradient at the interface.

In the course of a typical melt a small quantity of the premixed high purity powders is loaded into the crucible. The graphite susceptor shown in Figure 24 is coupled into the RF field to preheat the glass.

At this temperature a melt is formed which is lowered into the RF field and conditions adjusted to achieve direct coupling. The remainder of the powders is then added in stages until a full melt is achieved. After melting, the glass is homogenized and refined to produce a uniform composition with the melt. This is essential if the eventual fibre scatter loss is to be kept down to Rayleigh scatter, as described earlier. Homogenization is essentially achieved either by stirring the melt direct or by bubbling with pure gases. Refining is normally carried out by raising the melt temperature so that the viscosity falls allowing the bubbles to rise to the surface. After the melt has been suitably processed it is cooled to a temperature at which rod pulling direct from the surface may be carried out.

In converting the glasses to fibre the main technique employed has been the double crucible method of pulling. The double crucible assembly is shown in Figures 26 and 27 and consists of a pair of concentrically mounted crucibles made of high purity platinum. Each bushing has a circular nozzle of a size suitable for the required fibre geometry so that single mode or multimode fibres simply require different inner crucible nozzle diameters. As the molten cladding glass flows through the nozzle of the outer crucible it draws core glass down from the inner crucible to form the fibre structure. The glass is introduced in rod form into the crucible by a constant speed feed either directly after pulling from the melt or after very careful storage in which no handling or other contact is allowed.

After loading, the temperature of the furnace is adjusted to the level at which the glasses have a suitable viscosity for pulling into fibre and, as in the melting stage, the gas above the melt is carefully controlled. Pulling of the fibre is achieved by direct take-up on to a drum rotating at constant speed as it traverses slowly beneath the furnace. By carefully adjusting the loading and pulling conditions the apparatus is capable of producing long lengths of fibre with controlled geometry[52] and has the potential of becoming a continuous production technique.

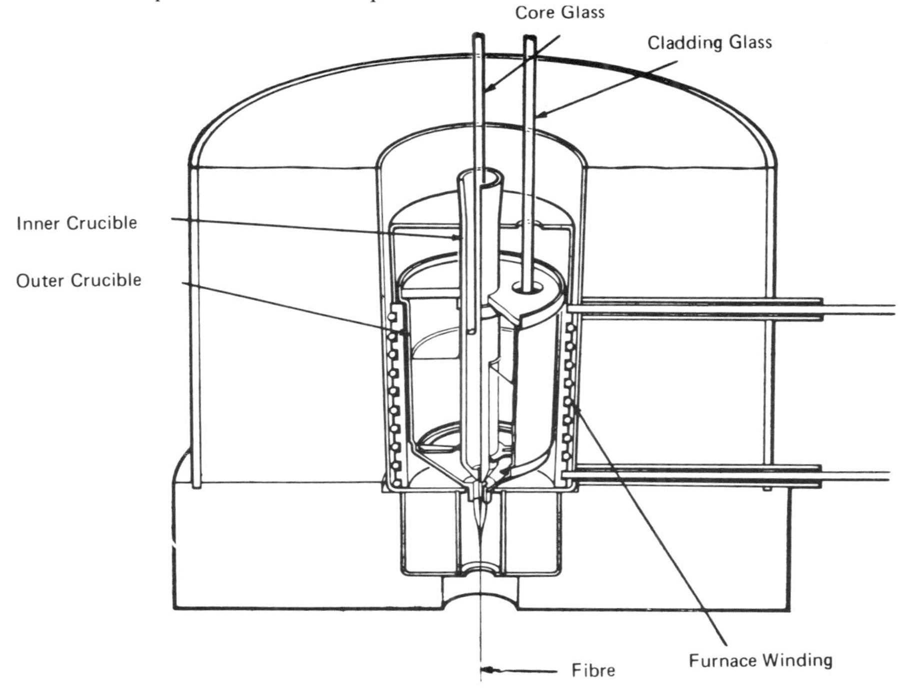

Figure 26. Diagram of double crucible pulling assembly. Core glass refractive index, n_c. Cladding glass refractive index, $n_c(1-\Delta)$; $\Delta \sim 0.01$

Figure 27. A platinum double crucible

As can be seen from this description, the possibility of impurity contamination of the material has been almost completely eliminated. The absorption loss of the glass should therefore only be limited by the quality of the starting powders while the absorption of the fibre depends upon conversion of the rods to fibre without additional contamination. Initial measurements on sodium calcium silicate glasses prepared by induction heating showed an improvement from 80–100 dB/km to 27 dB/km compared with conventional melting techniques. Fibres produced from these rods continued to show higher losses, however.

Two factors contributed to these loss increments and were demonstrated by separate experiments. The first, crucible contamination during fibre pulling, was demonstrated when a pair of silica crucibles was used instead of the platinum and gave a reduction of 30 dB/km in fibre loss. The second factor was believed to be an adverse change in the oxidation state of the main impurity, iron, during the pulling operation. This was eliminated by a clad rod experiment[53] in which a cladding layer was applied to the core glass rod as it was produced. The cladding glass was contained in a silica crucible floating on the surface of the core glass melt and the core rod was drawn through a hole in the base of this crucible and up through the cladding melt to form a rod preform which was subsequently drawn into a fibre with only a small change in loss. These results increased interest in glasses with lower processing temperatures in which the impurity oxidation states could be more readily controlled.

The sodium borosilicate glass system proved suitable, especially following the studies on control of impurity oxidation states by Newns *et al.* at the BPO,[54] and has yielded some of the best loss results obtained by the double crucible method.

The BPO have achieved minimum losses of 3.4 dB/km in step index fibres with the sodium borosilicate glass system.[55] In this case the glasses are melted in silica crucible using a resistance furnace. Considerable attention is paid to the details of controlling the oxidation state, minimizing contamination (in particular hydrogen sources which lead to the hydroxyl ion absorption peaks), and to the loading and pulling conditions. Control of the atmosphere above the melt is essential to eliminate the formation of bubbles at the core–cladding interface through an electrolysis effect. A number of other glass systems are also being investigated. Imagawa[56] has reported on melting a multiple component material successfully in platinum, following purification of their own raw materials and prefusion of the powders so as to reduce the temperature for glass melting. Germanium sodium calcium silicate[57] and germanium sodium silicate[58] systems have also been reported.

Graded index glass fibres were initially made by Nippon Sheet Glass[59] using a sodium borosilicate glass system in the double crucible. The refractive index profile is produced by adding 1.5–4 per cent Tl to the core composition and allowing diffusion to take place between the core and cladding regions as the fibre is pulled. The profile is controlled by the spacing between the crucible nozzles, the temperature of the crucibles, and the pulling speed. Koizumi[60] has reported losses of below 10 dB/km and dispersion of 0.6 ns/km for low mode launch conditions in fibres produced by this technique. Newns *et al.*[55] have used similar techniques but based on Na^+ diffusion between the two borosilicate glasses, whilst Gossink *et al.* have used K^+ and Na^+ exchange in the germanium silicate system.[58] The latter system gives faster diffusion rates and therefore either higher pulling speeds or shorter diffusion zones.

SILICA FIBRE MANUFACTURE

The main technique of silica fibre manufacture has been to utilize methods in which a glassy layer consisting predominantly of fused silica, but containing one or more dopants, is formed either on the inside of a fused silica tube, which is converted into a rod preform, or on the outside of a rod mandrel, which is subsequently removed. Fibre is then drawn either from a preform or from the tube which is collapsed as the drawing proceeds. Thus, it is a batch process, unlike the potentially continuous double crucible technique, but the batch size can be quite large, e.g. up to 40 km from a single preform.[61] More recently two other techniques have been developed for silica fibres. In one a preform is grown continuously in a longitudinal direction on the end surface of a seed rod by a vapour phase Verneuil method.[62] In the other, a porous structure consisting almost entirely of fused silica is formed by leaching out the boron-rich region of a borosilicate glass. The porous structure is then filled with dopant from solution before consolidation into a rod preform.

There was a rapid development of methods of fibre manufacture after the original announcement of a 20 dB/km loss. Two main approaches have been adopted for forming the glassy layers. Corning[48] and one group at Bell Laboratories[63] have used flame hydrolysis to produce coatings on the surface of a rod. The second approach has been to use direct oxidation chemical vapour reaction to form the required layers on the wall of a substrate tube.

Bell Laboratories reported on direct oxidation of hydride vapours in September 1973[64,65] and STL reported fibres produced by oxidation of halide vapours early in 1974.[66,40] The latter important advance offered the possibility of complete elimination of hydroxyl absorption which arises during deposition from both the flame hydrolysis and the hydride oxidation processes owing to the presence of hydrogen-containing compounds.

The choice of dopant or, perhaps more accurately, refractive index modifier for the fused silica, has varied. Since pure fused silica has one of the lowest refractive index values known in glasses (1.458 at 583 nm) most efforts have centred on techniques by which a core glass is formed inside a substrate tube. Among the dopants that have been used to produce a higher index are GeO_2, Al_2O_3, TiO_2, and P_2O_5. The use of TiO_2 has always given rise to problems of optical absorption in the glass due to the formation of reduced Ti^{III} during fibre processing which has proved almost impossible to control. Al_2O_3 is a good dopant, but leads to problems with suitable source compounds for the halide oxidation process. GeO_2 and P_2O_5 have perhaps had the widest use, with GeO_2 allowing much higher numerical apertures to be achieved than P_2O_5 (>0.35 compared with 0.2) but suffers a slightly higher scatter loss. The use of GeO_2 plus P_2O_5 in combination offers advantages of a lower glass fusion temperature and higher material growth rates.

Conversely B_2O_3[67] and F[68] have both been shown to produce a lower index when combined with silica. The borosilicate system makes use of an anomalous dip in the refractive index in the B_2O_3/SiO_2 binary system near the composition $6\,SiO_2 : 1\,B_2O_3$, which allows a maximum index difference of only 0.01. The lower index materials have been used to increase the index difference between a doped core and cladding, and as a direct cladding for pure silica.

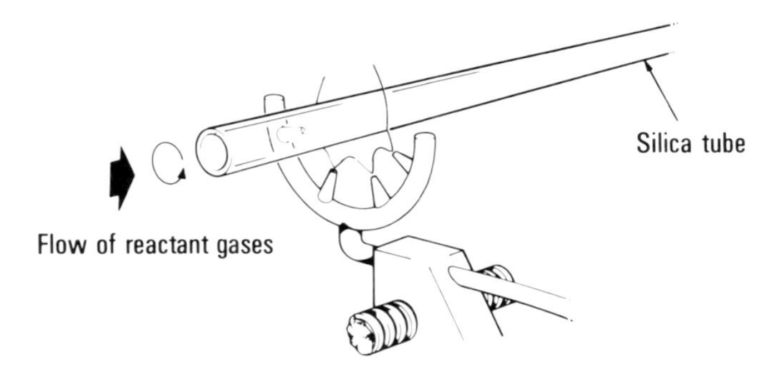

Figure 28. The deposition arrangement

Figure 28 shows the first stage in the 'hydrogen-free' direct oxidation of halides process used to produce a silica fibre. Tubing made from low optical quality silica but having very good geometrical tolerances and freedom from bubbles is mounted in a type of horizontal lathe and rotated at approximately 60 rev/min. A moving reaction zone is produced by locally heating the substrate tube to a temperature in the range 1200–1600°C with a traversing oxy-hydrogen flame. Reactant gases enter the tube at carefully controlled flow rates. At ambient temperature no reaction occurs, but within the deposition hot zone a thermally induced chemical vapour reaction occurs to form the mixed oxides which form a layer of glass on the tube surface. Typical reactions are:

$$\left.\begin{array}{r} SiCl_4 + O_2 \rightarrow SiO_2 + 2Cl_2 \\ GeCl_4 + O_2 \rightarrow GeO_2 + Cl_2 \\ 4POCl_3 + 3O_2 \rightarrow 2P_2O_5 + 6Cl_2 \end{array}\right\} \text{core material}$$

$$\left.\begin{array}{r} SiCl_4 + O_2 \rightarrow SiO_2 + 2Cl_2 \\ 4BBr_3 + 3O_2 \rightarrow 2B_2O_3 + 6Br_2 \end{array}\right\} \text{cladding material}$$

Deposition can be made to occur at lower temperatures by a surface phase heterogeneous nucleation mechanism, but is usually accomplished by a combination of heterogeneous and homogeneous nucleation mechanisms at higher temperatures. In this latter case the temperature is normally maintained at a value higher than the fusion temperature of the deposited glass, which therefore allows material to be simultaneously deposited and fused to a clear glassy layer as it is formed. As the hot zone is traversed along the tube, a uniform layer of glass is deposited and a thick layer built up by repeated passes of the hot zone. The composition of the layer may be varied on each pass to allow the formation of a composition and hence a refractive index profile across the deposited core region. A single mode step index multimode fibre may also be produced by varying the structure, composition, and dimensions of the deposited material. This 'hydrogen-free' vapour deposition process, as well as producing very high purity glass free from transition metal impurities, has the potential for producing completely OH free glass if all hydrogen-containing compounds are precluded from entering the deposited material. In practice, however, a very small residual impurity of a few parts per thousand million will remain due to trace impurities such as water vapour in source gases, chlorosilane impurities in source compounds, and OH diffusion from the substrate tubing.

In the alternative 'soot' process using flame hydrolysis, layers of doped silica soot are deposited onto the surface of a rod mandrel, and are then separately sintered to consolidate the soot material into a fused glass blank. This process is necessary to avoid the loss of volatile dopants such as GeO_2,

Figure 29. A deposition burner in operation

which occurs if the material is deposited directly as a glass.[69] During this sintering stage, particularly if carried out in a helium atmosphere, the removal of a large part of the hydroxyl incorporated in the 'soot' during the flame hydrolysis deposition is possible. Hydroxyl levels can be as high as 0.1 wt per cent in the soot, reducing to several p.p.m. during the sintering step.[70] Because the process starts off with material of very high OH content which is then dried, it is difficult to foresee it reaching the same ultimate levels as the hydrogen-free process.

Figure 29 shows a close-up of one of the burners used for the inside tube deposition process, and Figure 30 an overall view of the deposition apparatus. The silica substrate tube can be seen mounted on the special deposition lathe, with the vapour train for generation of the source vapour species shown on the left. It is essential to control very accurately the flows of oxygen and reactant species entering the reaction zone. This is normally achieved by containing the liquid source halide materials in a vaporizer held at a constant temperature of $\pm 0.05°C$ to maintain a constant source vapour pressure. The oxygen carrier gas is metered through electronic thermal mass flow controllers which deliver a constant mass flow of oxygen independent of system pressure fluctuations. By flowing this gas through the vaporizers a constant quantity of source halide vapour plus oxygen is delivered to the reaction zone. Because the vaporizers have an electronic sensor and control

Figure 30. Apparatus for the vapour deposition process

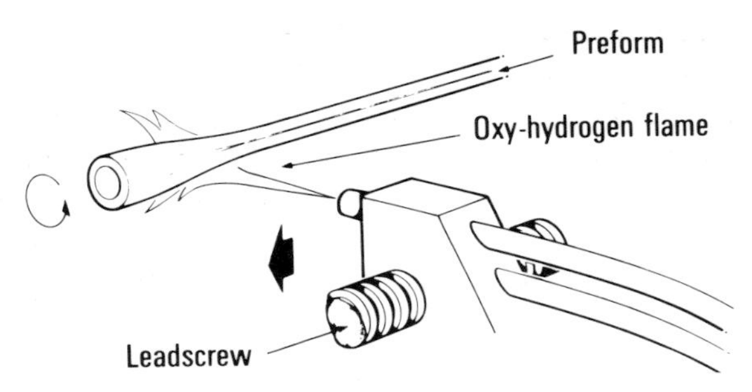

Figure 31. Diagram of tube collapse process to form preform.

valve, it is relatively easy to vary flow rates for producing graded refractive index profiles, e.g. by using control programmes in an automated electronic system.

After depositing the glassy core and cladding layers the tube is collapsed to form a rod preform, as shown in Figure 31. The tube is heated to the softening temperature by the traversing flame, in the region of 2,300 K and collapses under surface tension to form a uniform rod, as shown on the right-hand side of Figure 32. The central core and optical cladding materials are formed from the deposited layers on the inner tube wall. The control of the atmosphere inside the tube during collapse is essential to avoid two problems. The first is the tendency for coated tubes to collapse with

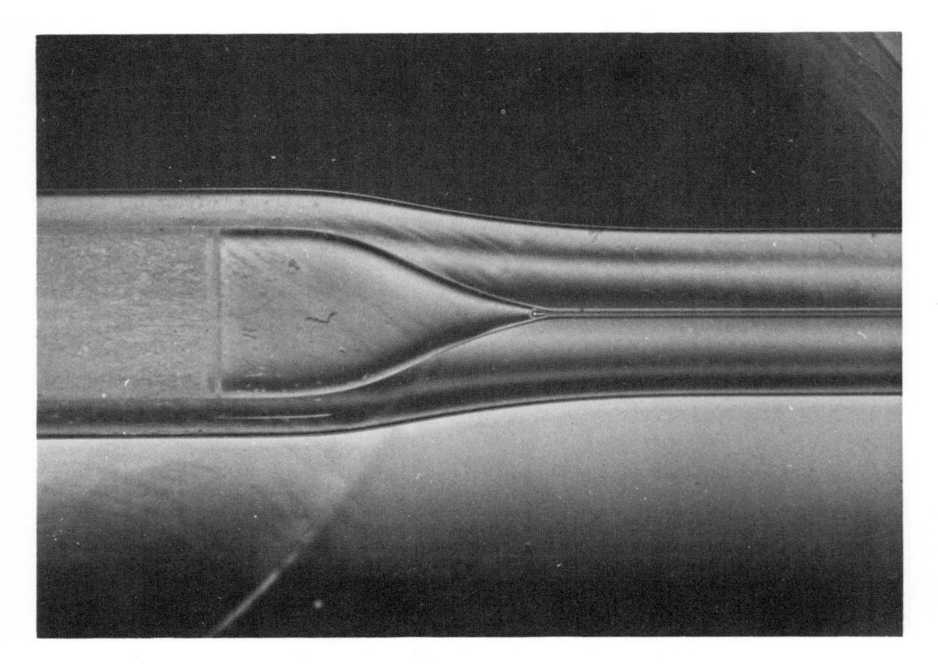

Figure 32. The Preform showing part before and after collapse

noncircular symmetry, particularly at high dopant levels where thermal expansion and viscosity mismatches exist. This can be controlled by slightly pressurizing the interior of the tube during collapse. The second problem is of dopant vaporization at the higher temperatures used in collapse, e.g. $GeO_2 \rightleftharpoons GeO + \frac{1}{2}O_2$. This results in a dip in the refractive index at the centre of the core, but can be compensated by flowing the source compound and oxygen through the tube during collapse. This has the effect of suppressing the evaporation equilibrium, e.g.

$$\begin{array}{c} GeCl_4 + O_2 \rightleftharpoons GeO_2 + 2Cl_2 \\ \upharpoonleft\downharpoonright \\ GeO_2 \rightleftharpoons GeO + \frac{1}{2}O_2 \end{array}$$

Fibre is produced by drawing from the heated tip of a preform as it is lowered into a graphite resistance furnace (Figures 33 and 34). Other heat

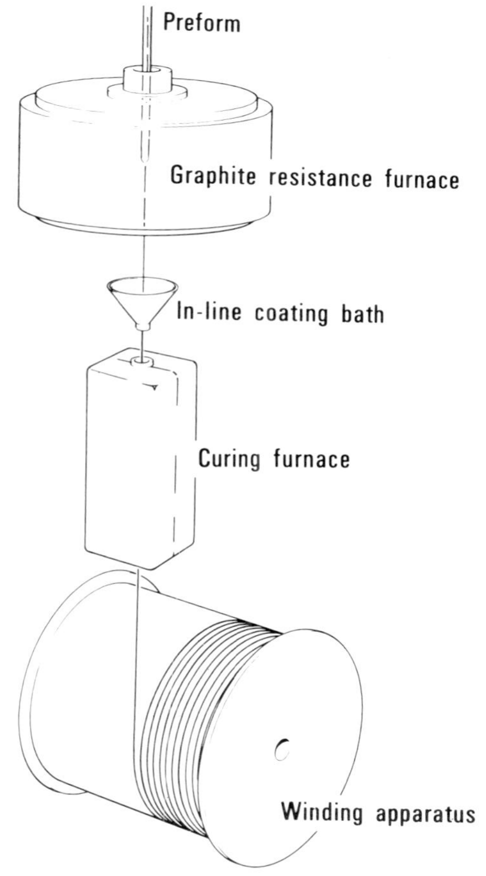

Figure 33. Diagram of silica fibre pulling process

Figure 34. The pulling tower

sources can be used, such as oxy-hydrogen flames, CO_2 lasers, or RF inductively heated susceptors, but the resistance furnace offers good temperature control, stability of fibre diameter, and good strength provided care is taken with its cleanliness of operation. The use of a diameter monitor which detects fibre diameter just after it is drawn, and a feedback system to the fibre take-up system allows the overall diameter to be controlled to better than $\pm 1\ \mu m$ over fibre lengths in excess of 3 km.

During the fibre drawing process a thin layer of polymer is usually applied as a primary coating layer to preserve the pristine strength of the fibre. A number of materials and application methods have been considered. The solution coating method has proved successful. When applying material by solution coating a thin layer of polymer in liquid form is applied to the fibre, which must then be cured to form a solid protective layer. Methods which rely on solvent evaporation tend to result in very thin layers (<5 μm). Attempts to produce thicker layers are frequently limited by blister and bubble formation. Coating materials which involve a curing process with no vapour generation offer the best results. Figure 35 shows fibre being drawn through a bath of two component heat curing silicone resin. The resin is cured by passing through the heating ovens such that a uniform layer of resin, typically 35 μm in thickness, is completely cured before contacting the winding drum. Similar coating quality in thick layers can be achieved with epoxy acrylates and UV curing. Variations to the coating process are

Figure 35. Close-up photograph showing cone for applying plastic coating (with mirror showing fibre emerging from cone)

possible; for example, a two stage tandem coating process applying a thin layer of primary silicone resin can be followed by extrusion coating of a second polymer to form a packaged fibre for further cabling use, or the application of extra solution coating materials to achieve particular surface characteristics (e.g. low friction) for the packaged fibre.

Alternative silica fibre fabrication techniques are also under study. The main objective of such work is the achievement of high bandwidth, high material growth rates, good geometrical tolerances, and high fibre strength on a routine basis with low cost. Among the techniques under investigation is the use of radio frequency initiated gas plasmas to enhance the thermally initiated vapour deposition process. Low pressure discharges at microwave frequencies[71] have allowed the deposition of material at much lower temperatures, thus eliminating tube distortion. However, this regime is not consistent with high material growth rates. The use of high temperature inductive discharges at several megahertz frequencies[72] has allowed higher production rates to be achieved, but with some risk of tube distortion. The study of plasma augmentation offers the combined possibilities of increased growth rate and wider compositional options to achieve the desired performance.

The technology of fibre production discussed so far is concerned with the need to produce a low loss core–cladding structure in vitreous materials. A type of fibre that avoids these problems is plastic clad silica, where the core is fused silica but the lower refractive index cladding is provided by a plastic coating applied in a manner similar to that used for the primary coating process shown in Figure 35. This process is suitable for producing large core step index fibres where very little of the energy is carried in the cladding. Such fibres are particularly attractive for short distance, low bandwidth systems where cost is a major consideration. Low cost, large area LEDs may be used with simpler launch conditions for large core high NA fibres, and connector alignment problems are much reduced. A further consideration is that the received power on a short length system will be higher for a high moded medium loss fibre than for low moded, low loss fibres. (For example, on a 200-m system length the received power from a 150-μm core diameter 0.39 NA fibre of 50 dB/km loss can be equivalent to that from a 50-μm, 0.2 NA fibre of 5 dB/km loss, when using a large area source.) Thus, adequately low loss for many applications can be obtained by this process, although it is not capable of providing the very low loss and high bandwidth achieved with the all-glass structures.

The most satisfactory plastic cladding has been found to be RTV silicone resin. It can be obtained in a clear, transparent form, is readily applied by dip coating, and has found widespread use as an on-line protective coating material for other fibre types. Polyvinylidene fluorides, such as ‘Kynar’, can also be applied by dip coating, but they are less transparent and, because they are applied in solution, give a much thinner coating. Other suitably low refractive index claddings are the extrudable fluorinated plastics such as

'Teflon', FEP, and PFA.[73] Because of their semicrystalline structure, however, they give rise to excessive scattering losses, unless applied as a loose-fitting tube with minimum contact area with the core. Such a fibre would be susceptible to mechanically induced losses and, lacking an intimate protective coating, would be weak.

Termination of the plastic clad silica fibre has required different approaches to connector design owing to the need to remove the cladding for centring and fixing purposes, but these problems have been overcome (see Chapter V). It has also been important to demonstrate that packaging of such a fibre structure into cables in order to meet environmental constraints is possible before incorporating such fibres into systems. Some results are quoted in the following section.

FIBRE TRANSMISSION PERFORMANCE

Figure 36 is a photograph of the cross-section of a step index fibre having a phosphosilicate core. It can be seen that adjacent to the core a B_2O_3 doped optical cladding layer with a slightly lower refractive index than the jacket formed by the tube material has been deposited to ensure a high quality interface region at the step. The overall fibre diameter is 125 μm. The loss curve of such a fibre is shown in Figure 37 which represents the average loss obtained by the cut-back method described in Chapter VI for 1 km of fibre.

The curve illustrates the following key features of fibre attenuation.

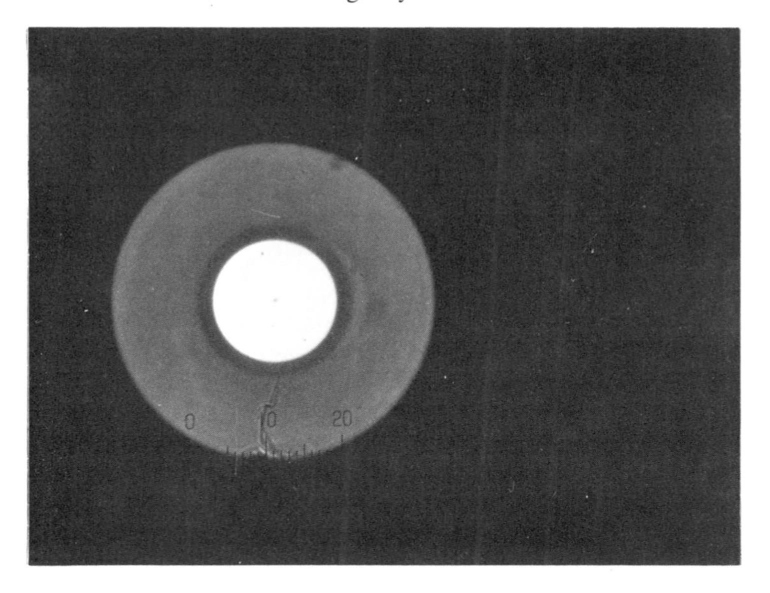

Figure 36. Photomicrograph of step index fibre cross-section showing the interface layer (dark circle) surrounding the core

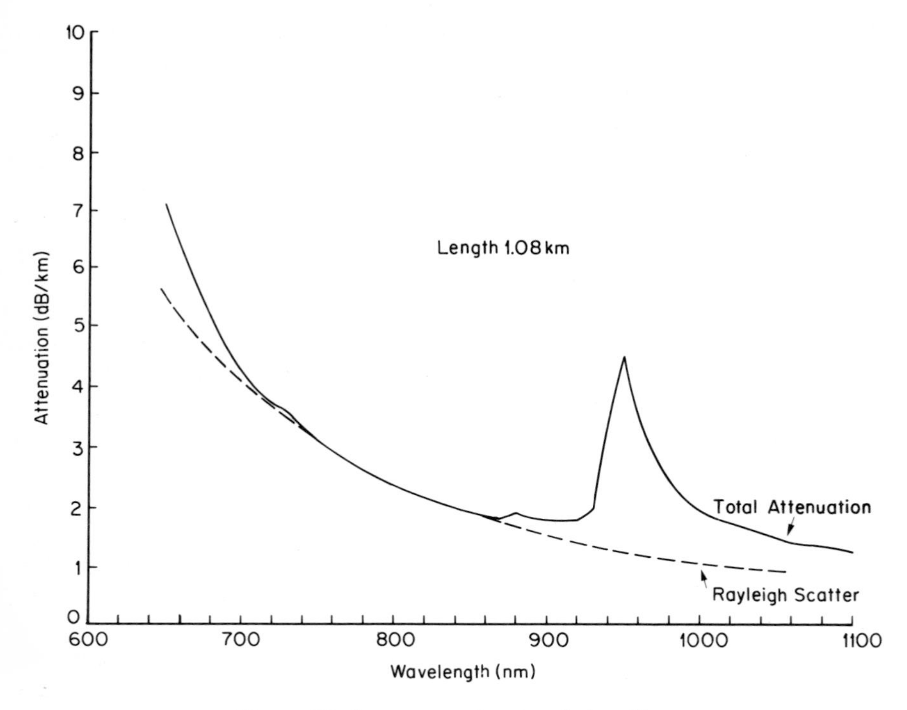

Figure 37. Loss curve for step index fibre

(1) *Fundamental material scatter* (*Rayleigh scatter*). This is inversely proportional to λ^4. To some extent one may regard the attenuation curve as the sum of the absorption at the various wavelengths added to the Rayleigh scattering. Rayleigh scattering tends to dominate the shape of the curve, as indicated by the broken line which represents the λ^{-4} relationship.

(2) *Fundamental material absorption of the glass.* At wavelengths below 500 nm the rising edge to the loss spectrum is partly caused by the fundamental UV absorption of the glass itself due to electronic transitions. At wavelengths above 1.5 μm there is additional loss caused by the IR absorption edge associated with vibrational energy transitions. For example, the wavelength of the fundamental vibrational absorptions of Ge–O, Si–O, P–O, and B–O are at 11.0, 9.0, 8.0, and 7.3 μm, respectively. The tail of these absorption peaks reaches back to approximately 1.0 μm. Neither of these effects causes a significant (>0.2 dB/km) contribution to the illustrated curve.

(3) *Impurity absorption.* The absorption due to transition metals is very low but there remains an absorption due to OH radicals. The fundamental OH vibrational absorption is 2.73 μm and this has a first harmonic at 1.39 μm, a second at 0.95 μm, and a third at 0.73 μm. Additional Si–OH combination overtones occur at 1.24 and 0.88 μm. In order for the attenuation of these vibrational absorptions to become negligible, the hydroxyl

content of the core would have to be of the order of a few parts per thousand million. The peak at 950 nm in Figure 37 corresponds to a hydroxyl content of approximately 2.5 parts per million. However, it can be seen that on both sides of this peak the attenuation of the fibre is below 2 dB/km. The excess loss at the short wavelength end of the curve is due to structural defects in the core material arising through incomplete oxidation of the core dopant. A residual impurity content of Fe^{2+} at a level of a few parts per thousand million gives rise to the excess loss at 1.1 μm over that due to OH impurity and the fundamental Rayleigh scatter.

Figure 38 is typical of the type of graded refractive index multimode fibre which was used in the cables made for the 140 Mbit/s field demonstration described in Chapter XI. This has a loss of 3.0 dB/km at the GaAlAs laser wavelength of 850 nm and at 1.16 μm the attenuation is 0.9 dB/km. The attenuation curve for this fibre shows a very low residual absorption peak due to hydroxyl content, which in this case is less than 0.5 parts per million, superimposed upon a higher Rayleigh scatter curve. Pulse dispersion for such graded index multimode fibres can be better than 0.5 ns/km, which allows repeater spacings of greater than 10 km at 140 Mbit/s. Minimum loss values in such fibres are 2.45 and 0.70 dB/km at 0.85 and 1.18 μm, respectively.

Figure 39 is a photograph of a cross section of a single mode fibre. It can be seen that in this case the deposited cladding material occupies a much greater fraction of the total fibre diameter because in this case it is required not only to ensure a satisfactory interface at the small (3.5 μm diameter)

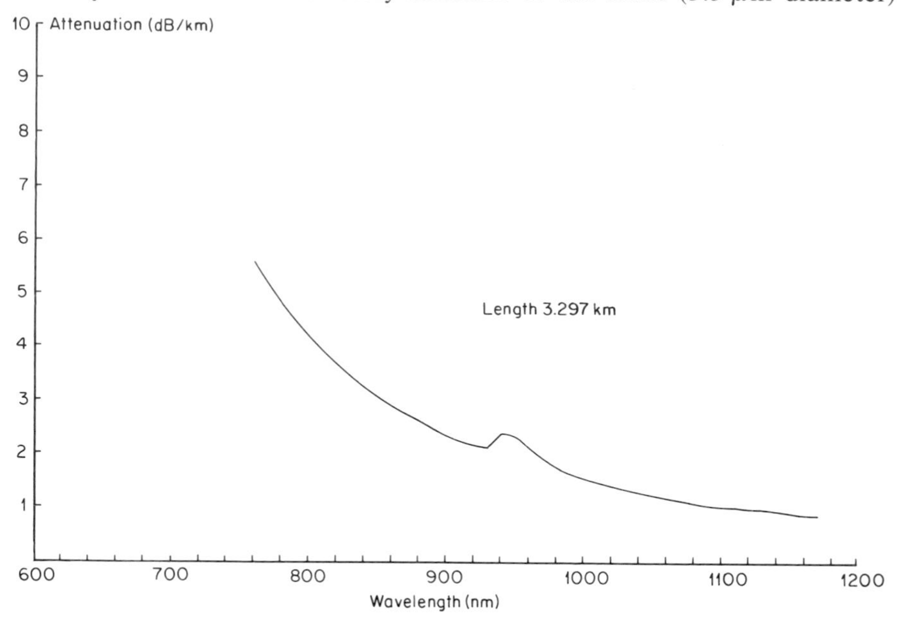

Figure 38. Loss curve of graded index fibre

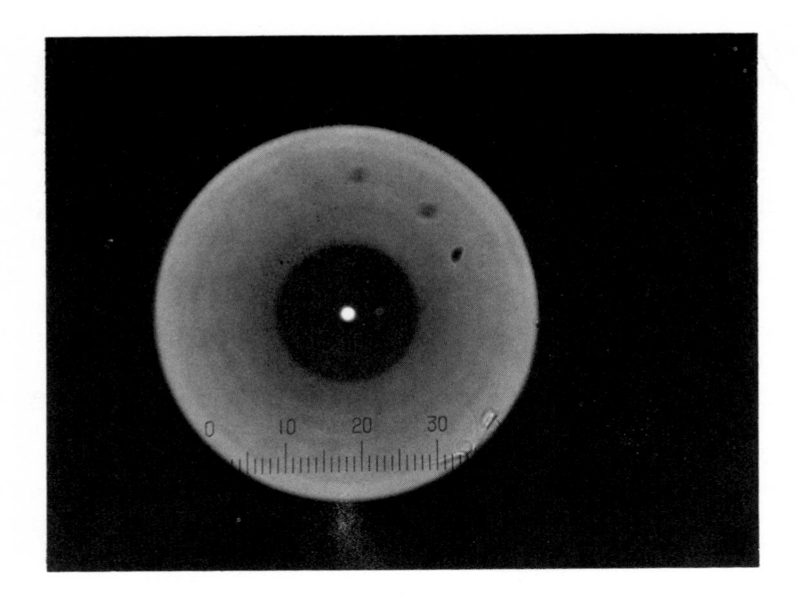

Figure 39. Photomicrograph of single mode fibre cross-section

core, but because a substantial fraction of the energy is transmitted in the cladding region around the core and therefore similar requirements for purity also apply to this region.

Figure 40 gives the attenuation curve for a single mode fibre, in this case after it has been fabricated into a cable with negligible change in attenuation. The loss curve shows the mode cutoff peak at 0.77 μm associated with the higher order modes, the fibre behaving as a single mode guide only at the longer wavelengths. This fibre was designed for operation with GaAlAs laser sources, but a suitable choice of the core size and index difference allows selection of the operating wavelength region. The bandwidth of such a fibre is dominated by material dispersion. Measurements of pulse dispersion on this fibre indicate a value of 86 ps nm^{-1} km^{-1} at the 0.84 μm wavelength. Higher bandwidths will be obtained by working at wavelengths above 1.3 μm where material dispersion effects are zero. The maximum bandwidth is achieved by operation at a wavelength where material and waveguide dispersion effects compensate one another. This wavelength is closely related to the fibre design since the former effect is dependent on core composition and the latter is related to both the V-value of the fibre and its composition.

Figure 41 is an early result for a fibre produced by plasma augmented deposition. This process enables the deposition rate to be increased over that used to produce the fibres shown in Figures 37 and 38. Initial experiments show that comparable attenuation can be obtained but at this stage it can be seen that the hydroxyl level is still rather high.

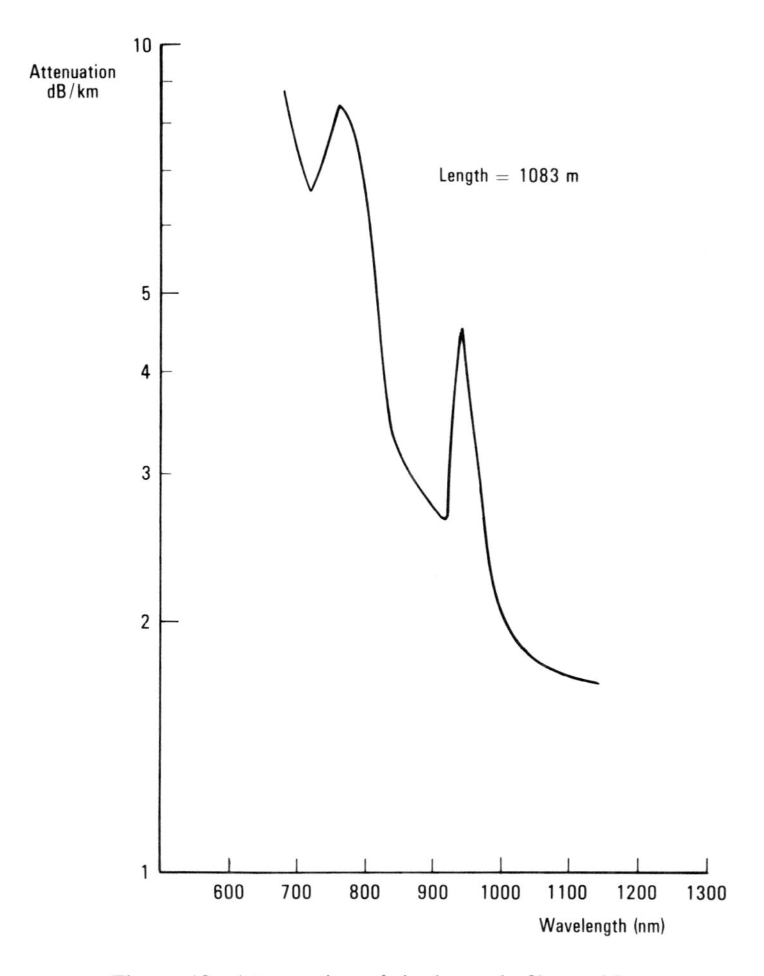

Figure 40. Attenuation of single mode fibre cable

The spectral loss curves of two silicone clad silica fibres are shown in Figure 42 and have a cross section similar to that shown in Figure 43. The peaks at 910, 1020, and 1090 nm illustrate the contribution of the cladding to the fibre loss, being due to stretching vibrations of C–H bonds in the silicone. The peaks at 950 and 880 nm are due to OH in the silica core, and the broad peak at 720 nm to overlapping OH and silicone absorption peaks.

The lowest cost plastic clad silica fibre is produced with silica prepared from selected high purity natural quartz and, as the curve for 'Herasil' 1 shows, can produce a medium loss fibre. As one would expect with a naturally occurring material, the transition metal and OH ion impurity content is variable, but the fibre loss, at 850 nm, falls mainly in the range 15–30 dB/km.

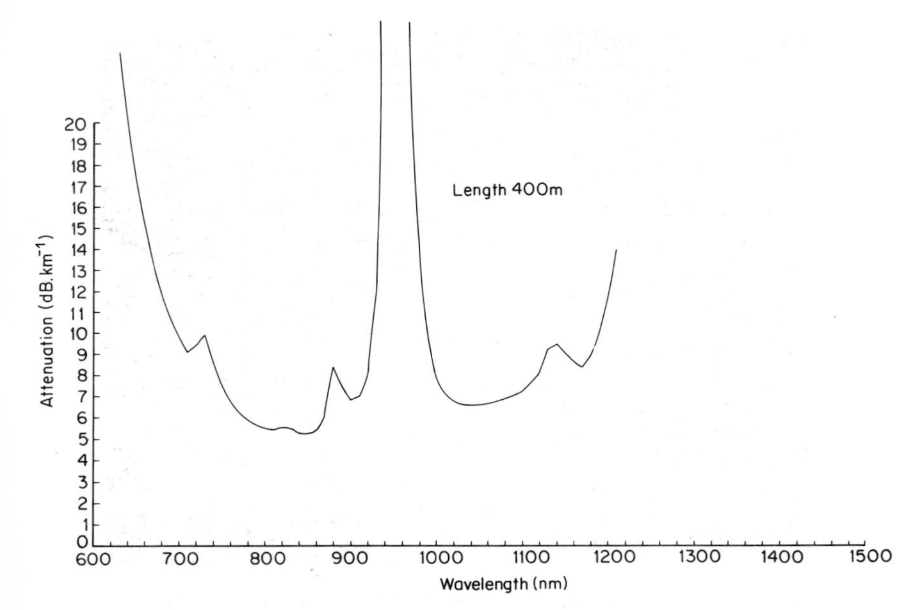

Figure 41. Attenuation curve for plasma deposited fibre

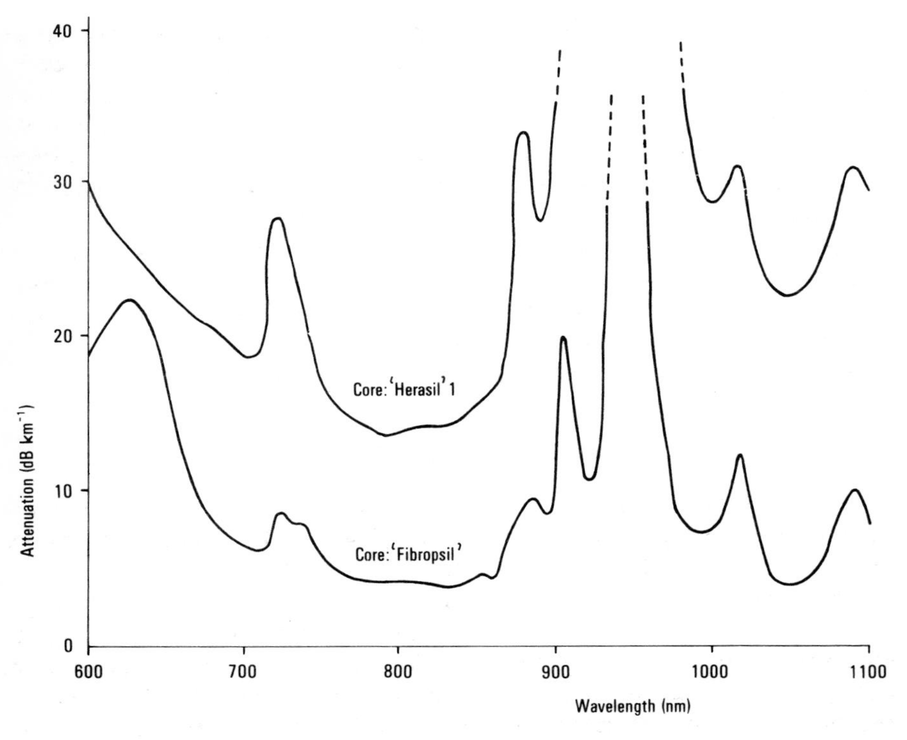

Figure 42. Attenuation curve of plastic clad silica fibres

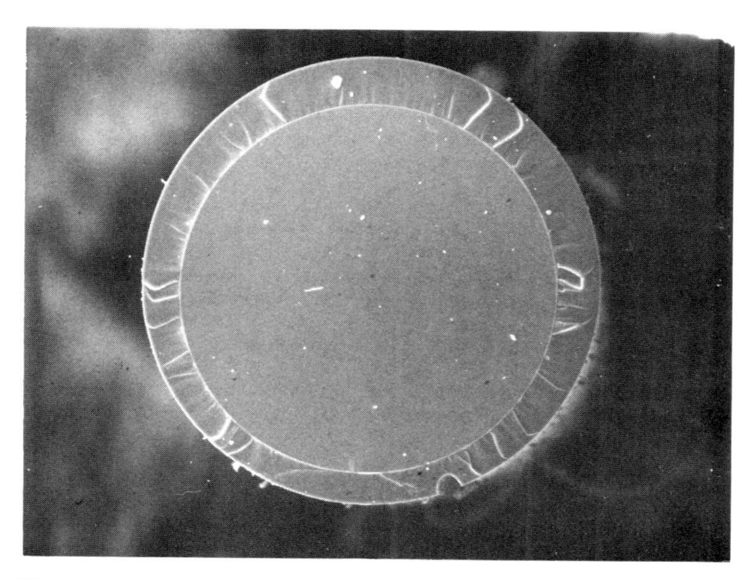

Figure 43. Photomicrograph of cross-section of plastic clad silica fibre

By using synthetic, and much more expensive, silicas, such as 'Fibropsil' or 'Suprasil' W, both of these impurities may be controlled at much lower levels. Although the silicone C–H peaks increase the loss at higher wavelengths, the curve for 'Fibropsil' demonstrates a low loss 'window' in the 750–870 nm range. Medium-priced synthetic silica produced by flame hydrolysis also give low loss at 800 nm but in a much narrower 'window' between the broad OH absorption peaks.

The C–H absorption peaks on the spectral loss curves provide the most direct evidence of the influence of silicone absorption loss on fibre properties, but it also affects the loss as a function of length, the NA, and the mode dispersion of the fibre. The attenuation constant $\chi(\gamma)$ of the γth mode of a fibre capable of accepting N modes is[74]

$$\chi(\gamma) = \chi_{\text{core}} + \frac{\gamma}{N(2N-2\gamma)^{1/2}}(\chi_{\text{clad}} - \chi_{\text{core}}) \tag{18}$$

The attenuation constant of the cladding, χ_{clad}, is 3000 dB/km at 850 nm, three orders of magnitude higher than χ_{core}, and so the higher order modes are much more rapidly attenuated than the lower ones.

Thus, while the acceptance NA, calculated from typical values of 1.458 and 1.405 for the refractive indices of silica and RTV silicone, is 0.39, the effective NA measured from the far field pattern when modal equilibrium has been established is ~0.25. Similarly, the calculated pulse dispersion on the basis of these refractive indices is ~180 ns/km but the higher order mode stripping, and mode coupling[75] effects give much lower values in

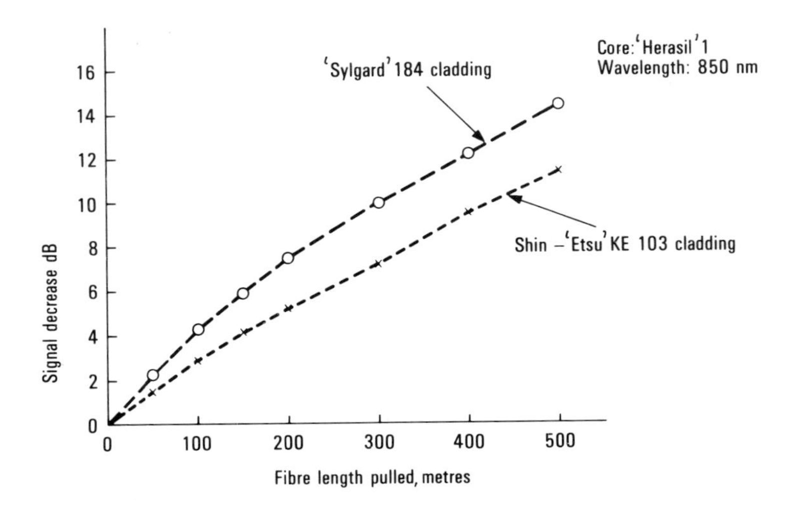

Figure 44. On-line monitoring plots for plastic clad silica fibres as a function of length

practice, i.e. in the range 10–20 ns/km. The effect on fibre loss as a function of length can be seen in the results of on-line measurements shown in Figure 44. This measurement, using the light generated by the hot zone where the fibre is drawn down from the silica rod, fills the acceptance aperture of the fibre, and over the first 200 m the loss, given by the slope of the curve, decreases as the lossier higher order modes are stripped out. The effect is more marked with the higher loss 'Sylgard' than the 'Shin-Etsu' silicone.

MECHANICAL AND ENVIRONMENTAL PROPERTIES

The mechanical performance of a fibre is of importance for cabling and installation purposes. Failure in fibres is by brittle fracture which arises through stress-induced cracking. A stress applied to a fibre will induce growth of any flaw above a given minimum size. The relationship between crack growth and applied stress has been predicted theoretically by Griffiths[44] in terms of the bond strength of the material. In well controlled processing gross bulk flaws are eliminated and the dominant flaw type is that arising on a surface due to mechanical damage or chemical attack. Application of polymer coatings to the fibre surface immediately after pulling minimizes such effects and gives rise to the tensile performance results described in Chapter IV, where strains at failure 7–8 per cent approach the theoretical predicted values on short lengths of fibre. Undoped silica surfaces offer a significant advantage in this respect since the Si–O bond energy is higher than that of most other components used for fibre making.

Fatigue effects due to prolonged stressing of fibres or to cycling of the stress levels have been shown to be much reduced for the routinely pack-

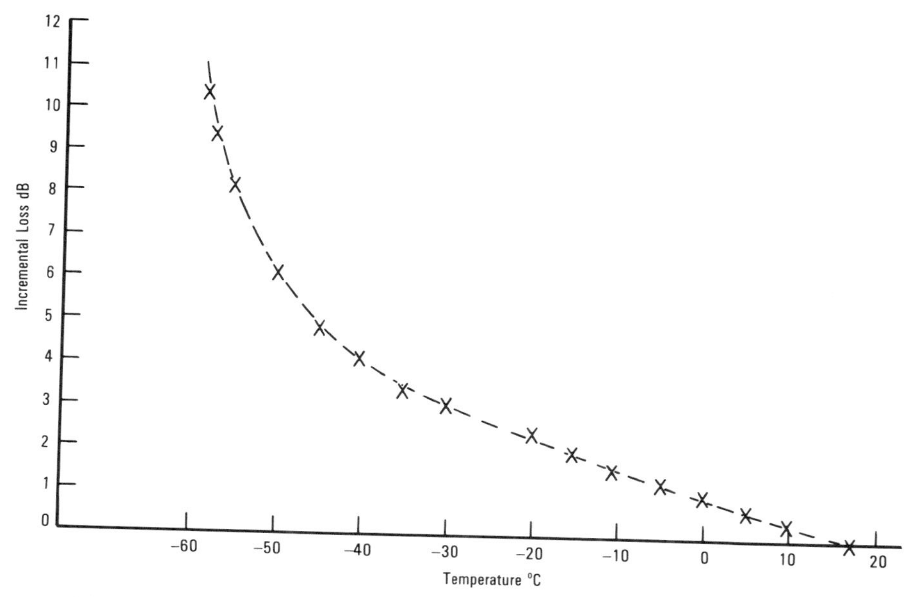

Figure 45. Low temperature incremental loss in plastic clad silica fibres

aged fibres, and occur at levels which are not likely to give problems in normal operating conditions.

The all-glass fibre structures are not sensitive to temperature over a wide range of values, but in the packaged condition greater sensitivity is observed owing to the forces applied to the fibre through differential thermal expansion effects within the package. Close attention to package design has allowed incremental values of no more than a few tenths of a dB/km to be achieved over all normal operating conditions.

The plastic clad silica fibre shows an additional temperature sensitivity in that the refractive index of the cladding has a much higher temperature coefficient than that of the core silica. At low temperatures the indices converge, producing a drop in numerical aperture and increased loss (Figure 45). Guidance is lost at temperatures where the indices become very close (<60°C for some silicones). At elevated temperatures the converse applies and signal levels increase due to increased NA of the fibre.

Nuclear radiation effects vary with the fibre type and the particular core composition. Friebele[77] has reported that germanium phosphosilicate cores show least short time sensitivity but that pure silica core fibres, such as plastic clad silica, show the fastest recovery.

CHAPTER IV

Fibre-optic Cables

INTRODUCTION

This chapter is concerned with the conversion of the fibres discussed in the previous chapter into practical cables which can be handled in the same way as conventional electrical transmission cables.

Visual comparison between multicore metallic cables and an experimental optical cable with roughly equivalent transmission capacities, shown in Figure 2 (Chapter I), highlights the small physical size and considerable flexibility of optical cables. However, the glass materials from which the optical cable is constructed are brittle, and are particularly susceptible to catastrophic damage during normal handling procedures owing to the small cross-sectional area of the fibres.

This was recognized at an early stage of development and steps were taken to improve the poor tensile strength and to protect fibres against external influences by applying a coating of plastic with a high elastic modulus.[78] This principle was extended to composite structures containing a multiplicity of coated fibres. More recently, as a result of the low optical losses which have been achieved in the basic fibres, considerable attention has been paid to minimizing additional losses which might be introduced during cable-making due to mechanical and environmental factors.[79–81] Considerable improvements have also been achieved in the strength of fibres by applying special surface treatments during the fibre drawing process.

Plastic coated fibres may be considered as the basic units in the design of practical cables, in the same way that insulated metallic wires are employed in electrically conducting cables. In fact, all subsequent operations in the construction of fibre optical cables are similar to those presently employed in conventional cable manufacture, although certain modifications are necessary owing to the inherent physical limitations of the fibres.

PERFORMANCE REQUIREMENTS

A performance specification is the starting point for all cable design and should cover not only the finished product but also its component parts to ensure that the manufacturing processes are practical. The specification will vary with the application and conditions of use, but can be compiled by selection from the principal features given in Table IV.

Table IV
Main factors in optical cable design

Optical	— number of fibres for required transmission capacity — attenuation at specified wavelength (normally 850 nm) — pulse dispersion — numerical aperture
Mechanical	— tensile properties — resistance to radial compression — flexing resistance — cold bend resistance — abrasion resistance — vibration resistance — environmental protection
Constructional	— material and dimensions of core, cladding, and coatings of fibres — sheath material and dimensions

CABLE COMPONENTS

A number of different cable designs have been proposed in the literature, but the essential component parts are all included in the following list: coated optical fibres, tensile reinforcement, radial reinforcement or cushioning, sheath and moisture barriers, insulated metallic conductors and screens, core wraps, and fillers.

The simplest basic structure is the single coated fibre, with or without a sheath, and this may have sufficient integrity without the need for further components. Enhanced mechanical properties can be obtained by providing additional components for tensile and radial reinforcement, and by including filling material over, under, or between the coated fibres. Increased signalling capability is obtained using groups of fibres, and control or power feeding by the inclusion of metallic conductors. Core wraps are used to maintain geometric uniformity in groups of fibres and conductors and to provide heat barriers during the application of plastic components (for example, sheathing) by thermoplastic extrusion. In practice, the most critical components are the coated fibres and the tensile reinforcement or strength members.

COATED FIBRE CHARACTERISTICS

The basic fibre material is either a multicomponent glass or silica, details of which have been given in earlier papers. From the point of view of cable-making we are primarily interested in two properties. First, an inherent degree of tensile strength and flexibility that will see it safely through the rigours of cable manufacture, installation, and use. Secondly, preservation of the fibre attenuation and dispersion throughout these processes, the most critical of which is the application of plastic coatings.

Immediately after drawing, bare fibres, particularly of silica, have an extremely high tensile strength, but this rapidly deteriorates under normal atmospheric conditions by as much as two orders of magnitude, and a typical bare silica fibre of 125 μm diameter (a commonly used size for optical transmission) will break at an average tension of about 5 N, with an elongation to break of 0.5–0.6 per cent (Figure 46). However, a major increase in both the load and the strain at break is obtained by applying a thin plastic coating in line with the fibre drawing process described in Chapter III. For example, values up to at least 100 N load and 8 per cent strain have been achieved by this method. These on-line or primary coated fibres may be used directly for further cable-making operations, but an additional or secondary coating with a greater wall thickness is commonly applied by extrusion to provide mechanical reinforcement.

Breakage of fibres under tension is believed to be caused by the propagation of microcracks, present in the fibre surface, across the fibre cross section. These microcracks vary in size and result in considerable variations in strength along a length of fibre. In practice the most significant breaking strength is that of the weakest point rather than the average, and statistically this is likely to be smaller the greater the length of fibre. Several series of measurements on short samples (30–80 cm) of bare fibre have been analysed[82,83] and extrapolated using Weibull failure probability statistics to determine the minimum strain at break in a 1-km length, such as might be

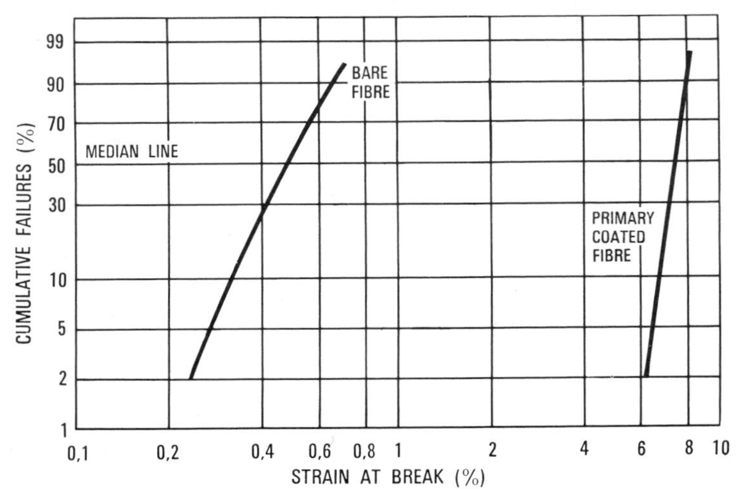

Figure 46. Plots showing the cumulative breaking strain for bare and primary coated fibres

employed in cable-making. With a median breaking strain of 0.5 per cent, the extrapolated value for 1 km was only about 0.1 per cent, which would be unacceptably low for a practical cable. Similar results on bare fibres have been obtained at Standard Telecommunication Laboratories, but it was found that the Weibull equation is not a sufficiently accurate representation for an extrapolation of this magnitude. However, cumulative failure plots on Weibull probability paper are a useful indication of the suitability of a fibre. This is illustrated by Figure 46 which shows typical plots of experimental data for a bare fibre and a good primary coated fibre. Note that a favourable plot is near vertical below a high median value. Tensile strength values follow a similar pattern but may be improved by the addition of a secondary coating, particularly at the important lower end of the plot.

SECONDARY COATINGS

Secondary coatings are relatively thick plastic coverings applied over primary coated fibres, their main purpose being to enhance the tensile strength and provide radial protection.

The tensile effect is based on the principle of load sharing between the fibre and the plastic coating. For this purpose the cross section of the plastic may be made up to 200 times that of the fibre to allow for its lower elastic modulus, which in simple extruded form is at least an order of magnitude lower than that of the fibre. Adequate mechanical protection is obtainable with nominal coating diameters in the range 0.8–1 mm.

A number of high modulus plastics have been used for secondary coating, including amorphous polyethylene terephthalate (polyester), polypropylene, and nylons. These can be applied to fibres by thermoplastic extrusion using a small conventional wire insulating line. The equipment employed by STL and Standard Telephones and Cables consists of a 38-mm plastic extruder together with a special pay-off and take-up equipment capable of processing fibres with low tensile strengths. The techniques of extrusion and for cooling the extruded product have to be closely controlled to avoid degradation of the optical transmission properties of the fibre due to stresses in the fibre caused by thermal contraction or morphological changes in the plastic. These stresses tend to cause columnar collapse of the fibre as the plastic is either cooled after extrusion or heated during the application of a sheath, producing many regions of small radius which radiate light from within the fibre. In the latter case stresses are generally a result of longitudinal alignment of the plastic molecules, which can be deliberately introduced by some methods of application in order to enhance the elastic modulus and improve the tensile properties of the fibre plastic composite.[84] However, there is a risk that contraction may occur during subsequent cable manufacturing processes, or even after installation, with adverse effects on the transmission properties of the enclosed fibre.

With any application technique it is possible to arrange for the coating to be either a tight fit on or a loose fit around the fibre. The latter construction has been proposed as a means of isolating the fibre from strain in the coating, but experiments have shown that a relatively small longitudinal contraction of the coating causes the fibre to adopt a helical configuration of short pitch thereby increasing the optical loss. In general, it is felt at STL that a coating which conforms to the fibre without exerting much pressure on it is the preferred form.

STRENGTH MEMBERS

In principle all the cable components add to its tensile strength. However, with limitations imposed on the strain by the optical fibres it is normally advantageous to include a component specifically to increase the effective tensile modulus without unduly stiffening the cable.

Tensile strength is of special importance in long cables (for example, cables installed in ducts) where the tension required for pulling in increases throughout the process as a function of the coefficient of friction between the cable sheath and adjacent surfaces. For straight ducts the tension increases linearly with the length, but for curves in ducts an additional exponential increase occurs which can substantially raise the pulling-in tension.[85] A figure of merit for tensile properties is provided by the ratio of the tension at a standard strain to the weight per unit length: this applies to individual components as well as to the completed cable. Preferred features of a strength member are therefore:

(a) high Young's modulus,
(b) strain at yield greater than the maximum designed cable strain,
(c) low weight per unit length, and
(d) flexibility, to minimize restriction of the bending capability of the cable.

Other features that may be relevant include friction against adjacent components, transverse hardness, and stability of properties over a range of temperatures including those encountered during cable manufacture and in service at subzero temperatures.

High modulus materials are inherently stiff in the solid form, but flexibility can be improved by employing a stranded or bunched assembly of units of smaller cross section, preferably with an outer coating of extruded plastic, helically applied tape, or a braid. Such a coating is particularly necessary if the strength member comes into contact with coated fibres since a resilient or smooth contact surface is required to avoid optical losses due to microbending, a phenomenon commonly observed in fibres subject to localized mechanical stress.[79,80]

Five main types of material have been proposed or employed for the construction of strength members on account of their high Young's high moduli: steel wires, plastic monofilaments, multiple textile fibres, glass fibres, and carbon fibres. Some significant features of these materials are summarized below:

(1) Steel wires: these have been widely used in conventional cables for armouring and longitudinal reinforcement. Various grades are available with tensile strengths to break ranging from 540 to nearly 3100 MN/m^2. All have the same Young's modulus (19.3×10^4 MN/m^2) and the choice is guided by the preference for a high strain at yield compatible with that of optical fibres. The main disadvantage of steel is its high specific gravity which substantially adds to the cable weight.

(2) Plastic monofilaments: although these are available commercially in several basic materials, STL have developed a specially processed polyester filament which combines a high elastic modulus (up to 1.6×10^4 MNm^{-2}) with good dimensional stability at elevated temperatures, and a smooth cylindrical surface. This type of strength member is of particular interest where low weight or absence of metals are prime requirements for the cable, but is not technically competitive with steel for cables to be installed in long ducts (500 to 1000 m).

(3) Textile fibres: commercial forms normally consist of assemblies of many small diameter fibres laid up in twisted or parallel configurations. Typical examples in conventional cables are polyamides (nylon) and polyethylene terephthalate ('Terylene', 'Dacron', etc), with elastic moduli which may be as high as 1.5×10^4 MN/m^2 for the individual fibres. Owing to the large number of individual fibres they are resilient in a transverse direction and are useful as cable fillers and binders as well as providing improved tensile properties in optical fibre cables, but are more bulky than monofilaments of equivalent strength. An exceptional member in this class which has been widely employed in optical cables is 'Kevlar', an aromatic polyamide. The individual fibres have the exceptionally high modulus (for an organic material) of up to 13×10^4 MN/m^2 which, coupled with its specific gravity of 1.45, gives it an effective strength-to-weight ratio nearly four times that of steel. Commercial forms of 'Kevlar' suitable for cable reinforcement consist of composites of large numbers of single filaments assembled by twisting, stranding, plaiting, etc. and/or resin bonding, and retain a high proportion of the single fibre modulus.

(4) Glass fibres: for some applications the optical fibres may supply sufficient tensile strength, but additional nonactive fibres can be used, generally in a manner similar to textile fibres, if higher strength is required. Elastic modulus is high, typically 9×10^4 MN/m^2, but the elongation to break may deteriorate to an unacceptably low value. Silica fibres are included under this heading, and when protected by on-line plastic coating have superior properties to those of multicomponent glasses and greater stability.

(5) Carbon fibres: this material has been successfully employed in rigid and semirigid plastic or metal composites, and has a modulus of up to 20×10^4 MN/m^2 in single filaments, but in STL's experience it is unsatisfactory as a flexible strength member in plastic encapsulated form.

Table V
Properties of strength member materials

Material	Specific gravity	Young's modulus (MN/m^2)	Tensile strength (MN/m^2)	Strain at break (%)	Normalized modulus-to-weight ratio
Steel wire	7.86	19.3×10^4	$5–30 \times 10^2$	25–2	1.0
Polyester monofilament	1.38	$1.4–1.6 \times 10^4$	$7–9 \times 10^2$	15–6	0.3
Nylon yarn	1.14	$0.4–0.8 \times 10^4$	$5–7 \times 10^2$	50–20	0.3
'Terylene' yarn	1.38	$1.2–1.5 \times 10^4$	$5–7 \times 10^2$	30–15	0.3
'Kevlar' 49 fib.	1.45	13×10^4	30×10^2	2	3.5
'Kevlar' 29 fib.	1.44	6×10^4	30×10^2	4	1.6
S-glass fib.	2.48	9×10^4	30×10^2	3	1.4
Fused Silica	2.65	7.5×10^4	$45–60 \times 10^2$	6–8	1.2

The relevant properties of the above materials, with the exception of carbon fibre, are summarized in Table V.

In practical cables the other components may add considerably to the weight but relatively little to the strength, and this most strongly affects the modulus-to-weight ratio when the strength member has a low specific gravity. The net effect is that the optimum modulus-to-weight is generally achieved with steel wires and 'Kevlar' 49, followed some way behind by glasses and 'Kevlar' 29, the remainder coming a poor third. The ultimate selection of material for the strength member depends upon the relative importance of cost, mechanical performance, and the acceptability of a metallic component.

CABLE DESIGN

Sufficient detail is now available to define the principal features of a practical cable design.

(1) Strength member that enables the cable to be held under a high tension at low strain (0.5–1.0 per cent) and with sufficient flexibility to allow bending round small radii.

(2) Plastic coated optical fibres of high intrinsic strength, arranged within the cable so that when bent or stretched the fibre strain remains within the limits set for the cable.

(3) Other components to protect the coated fibres from environmental influences which may cause deterioration of their optical performance.

The ideal position for either of the first two components to provide maximum flexibility with minimum longitudinal strain is at the neutral axis of the cable, and consequently two types of construction are possible with one component arranged around the other. Both systems have specific

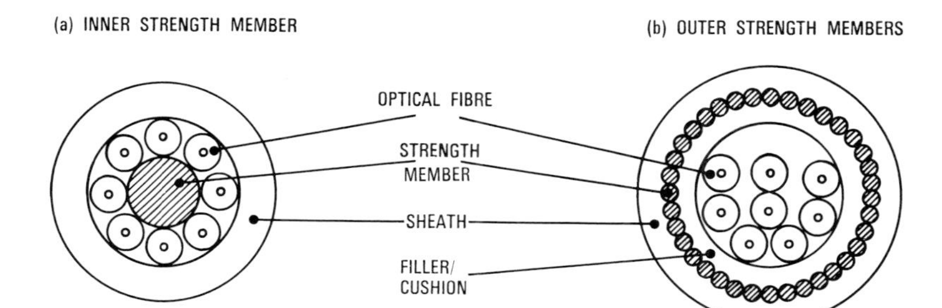

Figure 47. Typical constructions used for optical fibre cables. (a) Inner strength member. (b) Outer strength members

advantages. A centrally located strength member provides maximum flexibility, but if placed around the fibres may protect them from radial crushing forces. Both types of construction are shown in Figure 47, and a number of variations have been developed by manufacturers in the US, UK, and Japan.

We believe that it is important for fibres to be retained in a fixed geometrical configuration within the cable under all stress conditions, and that the fibres should be in line contact with adjacent components to avoid microbending effects. In the case of a single-fibre cable this is best achieved by locating the fibre at the cable axis and placing the other components around it, with or without a cushioning layer in between. However, in a multicore cable the optimum configuration for the fibre is a long helix wound in close contact with a central supporting member which may be the strength member or another coated fibre. The pitch of the helix must be long in relation to its diameter to minimize curvature of the fibre which causes optical loss. Fibres should then be secured in position by further layers of fibres, binding tapes, and so on. However, a number of designs have been proposed in which freedom from radial compressive forces has been achieved by allowing unrestricted radial movement within the cable structure, as shown in Figure 47(b).

The construction shown in Figure 47(a) is a multifibre design developed at STL early in 1972 which has been used extensively with only minor modifications since that time for several applications. It is simple, small, and lightweight, particularly if a nonmetallic strength member is employed, and has been constructed in lengths of over 1300 m without any significant change in the optical properties of the fibres. Further mechanical protection or tensile reinforcement can be provided if necessary. An additional option is the incorporation of cushioning layers close to the coated fibres to minimize concentrated mechanical stress, an important feature of a number of optical cables produced by ITT Electro-Optical Products Division in the US. A generalized structure incorporating all these features is shown in Figure 48.

The transmission capacity of optical fibres is such that most present requirements can be met by cables with a relatively small number of fibres,

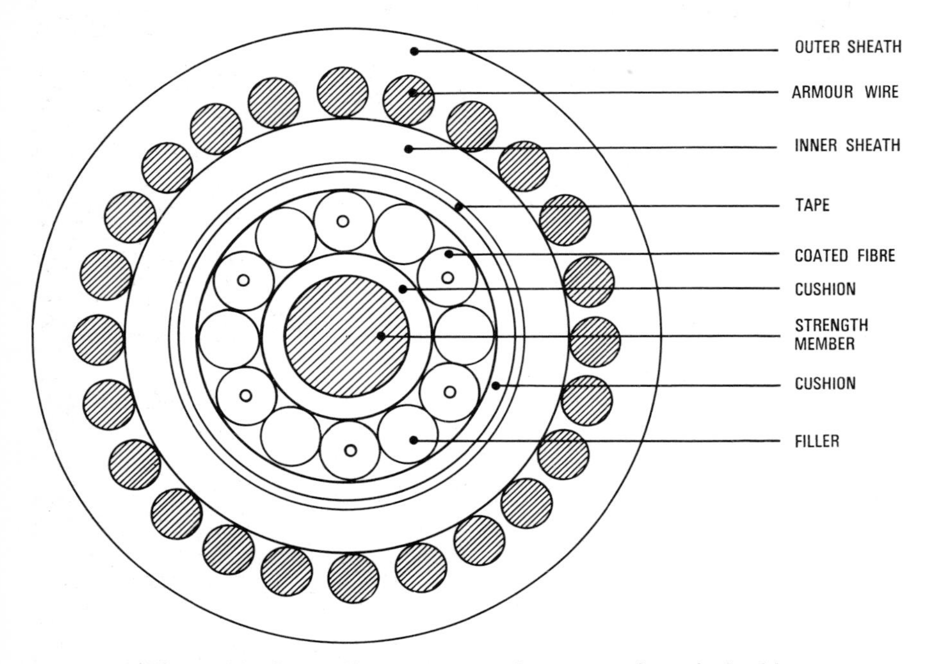

Figure 48. Generalized structure of a concentric optical cable

and the above construction in which they are arranged in a single layer is eminently suitable for up to 8–10 fibres. In the near future, however, cables with many more fibres are likely to be in demand and will require some design modifications, possibly involving multilayer structures or unit constructions. The need to control the fibre strain is likely to favour the latter using units of 4–8 fibres which are then stranded together to form high capacity cables. Since the units are inherently strong compared with individual fibres, manufacture can be undertaken with standard cabling equipment. To construct the units, however, it is advisable to employ well engineered purpose-built stranding machines, such as the one shown in Figure 49, if fibre breakage is to be eliminated.

The information capacity of single-mode fibres is so large that their use may well be concentrated in cables containing only one fibre. For many specialized short links multimode fibres may also be used singly. A very compact design with a secondary coated fibre on the neutral axis, stranded Kevlar 49 filaments around this and a final sheath has been made. Because of the very small dimensions no difficulties were experienced with the external high modulus strengthening filaments, while their stranding minimized the shift of the neutral axis on bends. An example is shown at the foot of Figure 50. This very compact cable has a working load over 1000 N and an outside diameter of about 2 mm. A 1-km length of single mode fibre has been incorporated in such a design with the performance shown in

Figure 49. Cable stranding machine

Figure 50. Examples of optical fibre cables

Figure 40. The attenuation increment after cabling was too small to measure. This type of cable may well offer greater information-carrying capacity than any other guiding medium.

While STL advocates the compact concentric type of cable design, and a number of companies (notably in Japan) have adopted the same principle, several companies have successfully produced nonsymmetric types[86,87] in which the fibres are located between and in line with two strength members. A novel construction employing fibres mounted on flat tapes and acting as their own strength members has also been proposed.[88] Future practical experience will determine the relative merits of these different cable designs, some of which are shown in Figure 50.

Figure 51. Horizontal tensile test bed

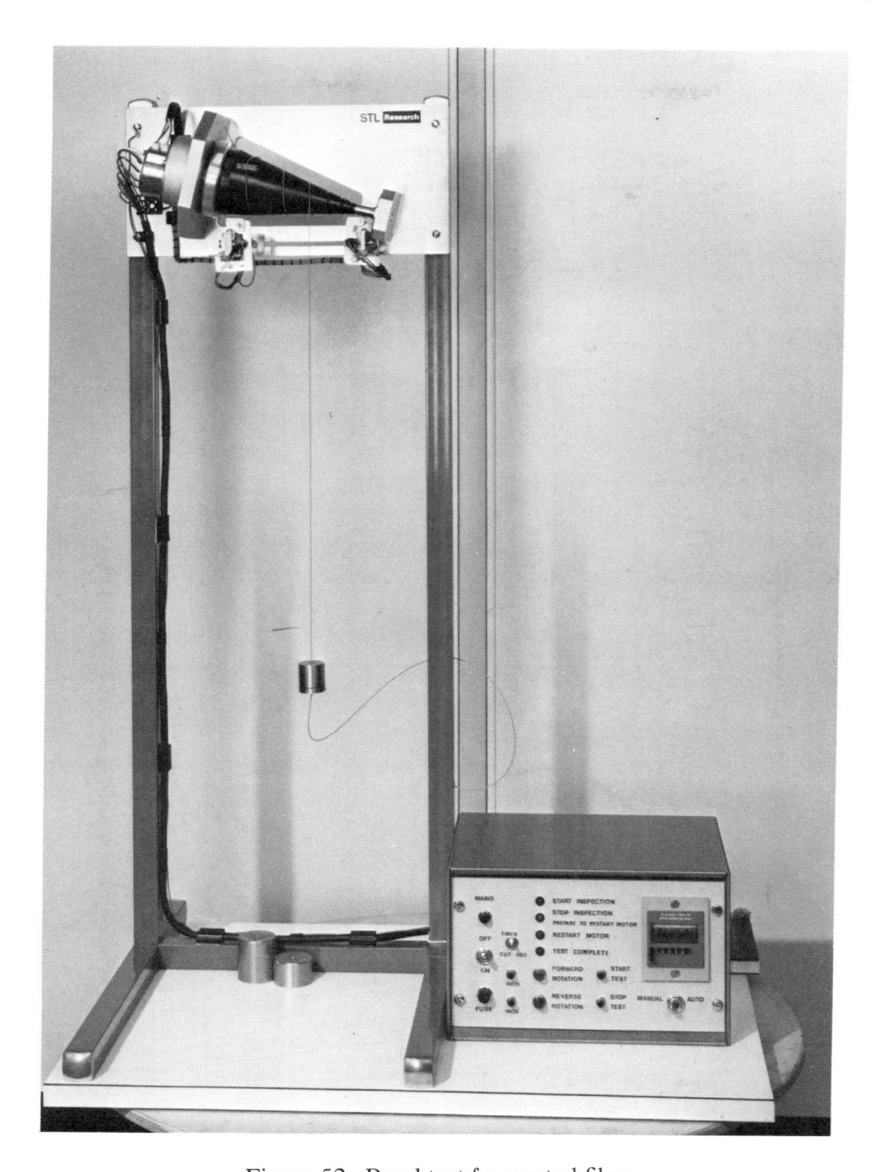

Figure 52. Bend test for coated fibres

CABLE TESTING

Close control of quality is an essential feature of optical cable manufacture, bearing in mind the catastrophic results of damage which may not be directly visible. Optical tests are described in Chapter VI, and only mechanical tests will be considered here.

Specific tests have been developed, applicable to coated fibres and strength members individually and also to finished cable. The properties of

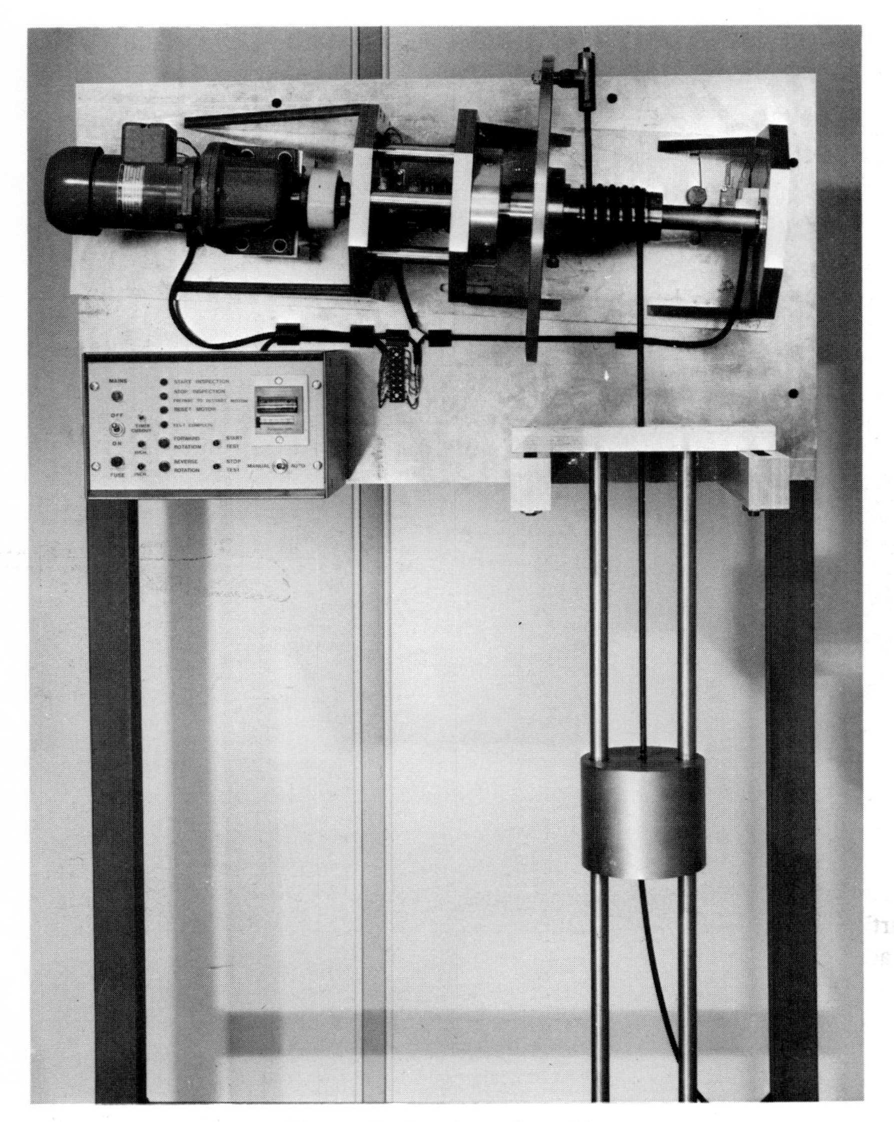

Figure 53. Bend test for cables

major significance include: (a) stress/strain characteristics; (b) bending; and (c) crushing. Tension versus strain curves are graphically recorded up to the point of failure, with the low strain rate of 0.02/min to avoid unduly optimistic results characteristic of 'snatch' tests. Coated fibres have been tested in short gauge lengths (about 50 cm) which can be accommodated in a conventional tensile testing machine (e.g. Instron), a minimum of fifty samples being used to obtain statistical data such as that shown in Figure 46.

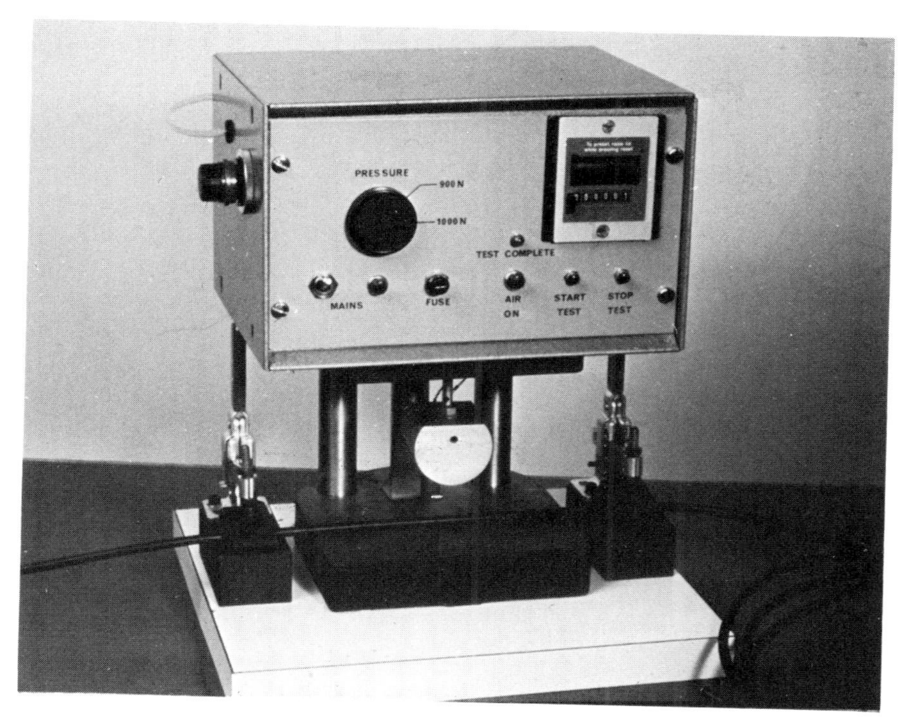

Figure 54. Crushing test for cables

Tension/strain curves for strength members and cables are more suitably obtained by using the purpose-built horizontal tensile tester seen in Figure 51, in which a gauge length of 5–12 m can be employed, giving accurate meaurements of strain in the range of practical importance, 0–1 per cent. A further useful cable test with this equipment is to cycle between two tensile load limits in order to detect mechanical fatigue effects. Special grips are necessary to avoid localized damage to the optical fibres in these tests.

Bending behaviour of coated fibres and cables is conveniently determined by winding on and off mandrels, as shown in Figures 52 and 53. The coated fibre is wound continuously into a groove in a conical mandrel under a steady tension in such a manner that the fibre will bend to a steadily reducing radius of curvature, and then unwound; the cycle is repeated a number of times with continuous observation of fibre continuity. A similar test is applied to cables with individual mandrels of fixed diameter related to the cable sheath diameter.

The crush resistance of cables is determined by application of a crushing force between flat metal plates, applied and removed over a number of cycles (see Figure 54). In a modification of this test the cable is laid on a flat plate and subjected to impulsive blows from a spherical or cylindrical ended weight, as shown in Figure 55.

Figure 55. Hammer test for cables

Figure 56. Cable samples after hammer test

In all of the above mechanical tests, fibre failure is determined by direct observation of visible light transmitted along them. Figure 56 shows some cables in which the fibres survived several hundred cycles of the hammer test after considerable damage to the rest of the structure, including breakage of the copper conductors incorporated into this design for repeater power feeding.

CHAPTER V

Optical Fibre Cable Connections

INTRODUCTION

Following the fundamental research leading to the availability of critical components, such as low loss fibre and long life sources, attention was concentrated on the engineering problems of efficiently coupling the fibres in cables. Methods of splicing fibres and fabricating connectors are described in this chapter.

The problem of coupling single strand fibres arises from the small active core area of the fibre cross section. Many civil and military systems use multimode fibres with core diameters ranging from 50 to several hundred microns. Future PTT applications may use fibres with core diameters down to 3 μm.

Two main types of joint are being developed:

(1) fibre splices, which are permanent joints in fibres, are analogous to soldered joints in an electrical system, and

(2) demountable connections for optical fibres, which are equivalent to electrical plugs and sockets.

The fibre alignment requirements for each type are similar and will be outlined before discussing specific splice and connector techniques.

FIBRE ALIGNMENT CONSIDERATIONS

One of the major problems associated with the connection of optical fibre arises from the need to align the cores with great precision. The absolute alignment accuracy necessary depends on the core diameter and type of the particular fibre and the maximum permitted joint attenuation.

Two categories of fibre are currently available. The first has a glass or silica core and a lower refractive index glass or silica cladding. The interface between the core and cladding can be either a step (step index) or a gradual (graded index) change in refractive index. The second category of fibre has a silica core, usually with a diameter of greater than 100 μm, and a low refractive index plastic cladding.

Most techniques for joining fibres rely on accurate geometry, in particular of the core and cladding diameters. The core/cladding concentricity is also very important in silica/silica fibres where alignment of the cores usually depends on alignment of the outside surfaces of the cladding.

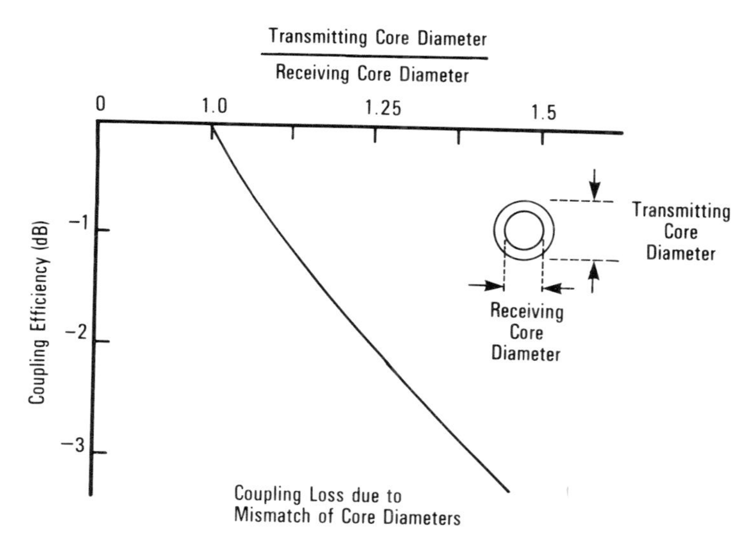

Figure 57. Coupling loss due to mismatch of core diameters

The effects of errors in fibre geometry and core alignment[89,90] on the efficiency of a connection are shown to a first approximation for a step index fibre in Figures 57–60.

The relative importance of each of these contributions to attenuation depends on the type of connection being made and the core size and numerical aperture of the fibre being used.

The improvements in fibre production technology which have occurred over the last few years have reduced the contribution of fibre geometry to connection loss to a low level. With the further improvements which can be expected, the fibre geometry's contribution to joint attenuation will decrease to a few tenths of one decibel which will only be significant where low loss splices with small core fibre are required.

In a demountable connector an important contribution to the loss is lateral misalignment, since a number of machining tolerances are usually involved. Figure 58 shows that for a step index fibre a 13 μm misalignment of two 80-μm cores contributes 1 dB to the connector loss. Axial separation of fibre ends induces a loss which is dependent upon the numerical aperture of the fibre, as shown in Figure 59. In a well designed connector or splice this separation should be small. Figure 59 also shows that if an index matching medium is used between the fibre ends the loss due to their separation is reduced. This occurs because, without index matching, refraction increases the angles of the rays as they emerge from the end of the fibre into air. Whilst index matching is normally used in a splice, its use in a demountable connector is rare owing to the problem of selecting a suitable index matching material which will not collect dust and dirt.

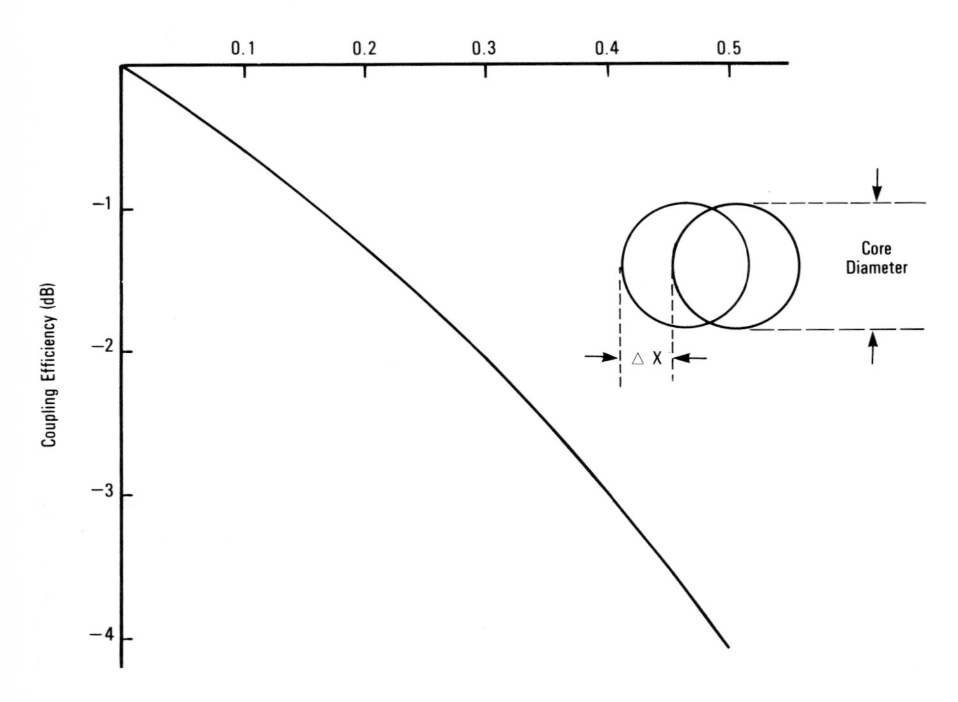

Figure 58. Coupling loss due to lateral misalignment of fibres

Another contribution to connection loss is due to Fresnel reflection at the end faces of the fibre. This loss is given by

$$\left(\frac{n_{\text{air}} - n_{\text{core}}}{n_{\text{air}} + n_{\text{core}}}\right)^2 \tag{19}$$

for each surface, n_{air} and n_{core} being the refractive indices of the gap and fibre core, respectively. This loss, which is about 0.35 dB for two silica–air interfaces, is virtually eliminated by the use of index matching.

Angular misalignment[90] of two butting fibre cores also contributes to the connector loss since some of the light incident at the receiving fibre core is not within its acceptance angle. This is shown for two different numerical aperture fibres in Figure 60. The effect of using index matching material in the gap is also shown.

These misalignments contribute in varying degrees to the loss between two butt jointed fibres. The losses present in another type of connection technique using expanded beams will be discussed below in the section describing lens terminations.

Poor preparation[91] of, subsequent damage to, or contamination of the fibre end can also contribute to optical attenuation in a connection.

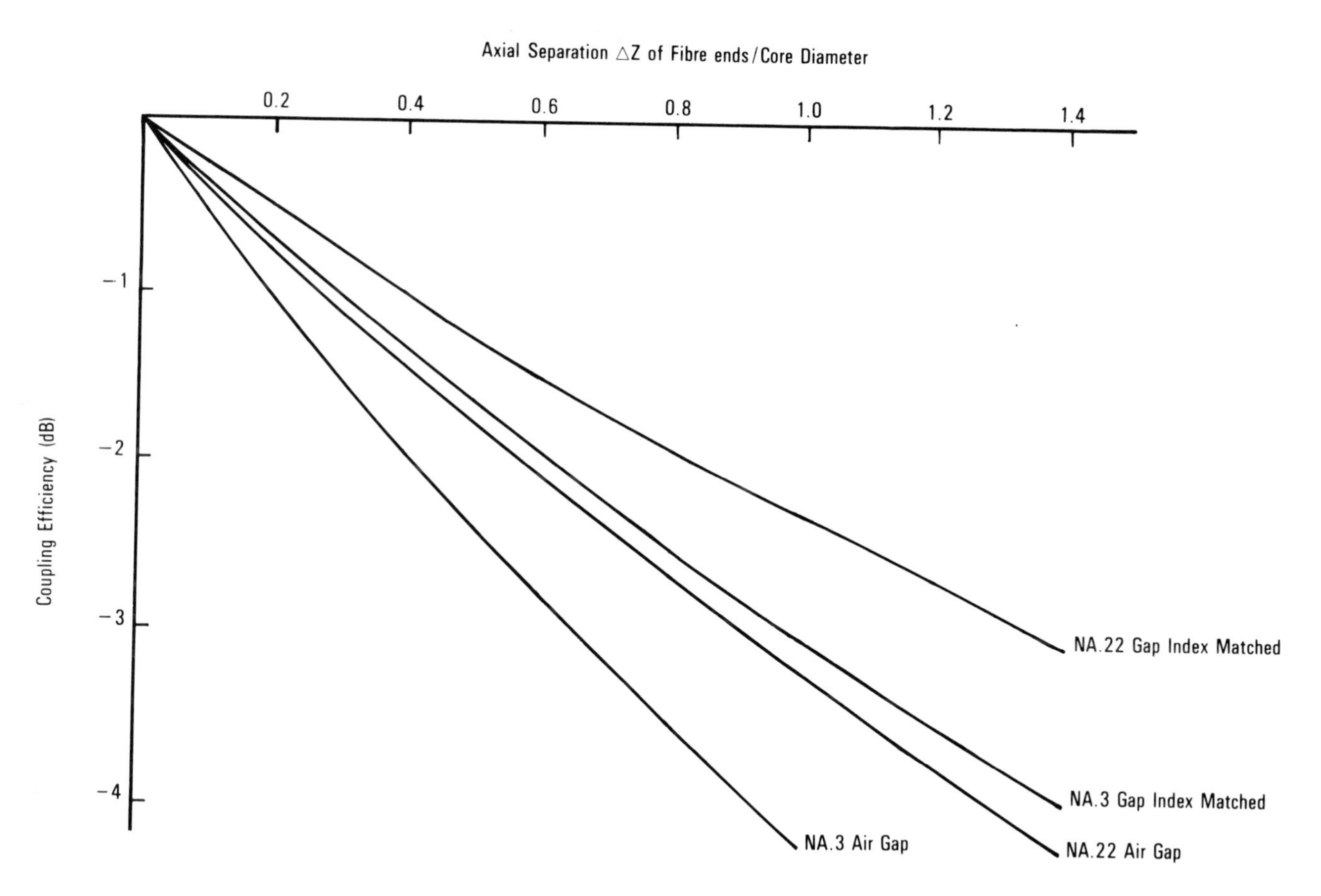

Figure 59. Coupling loss due to gap between fibre ends

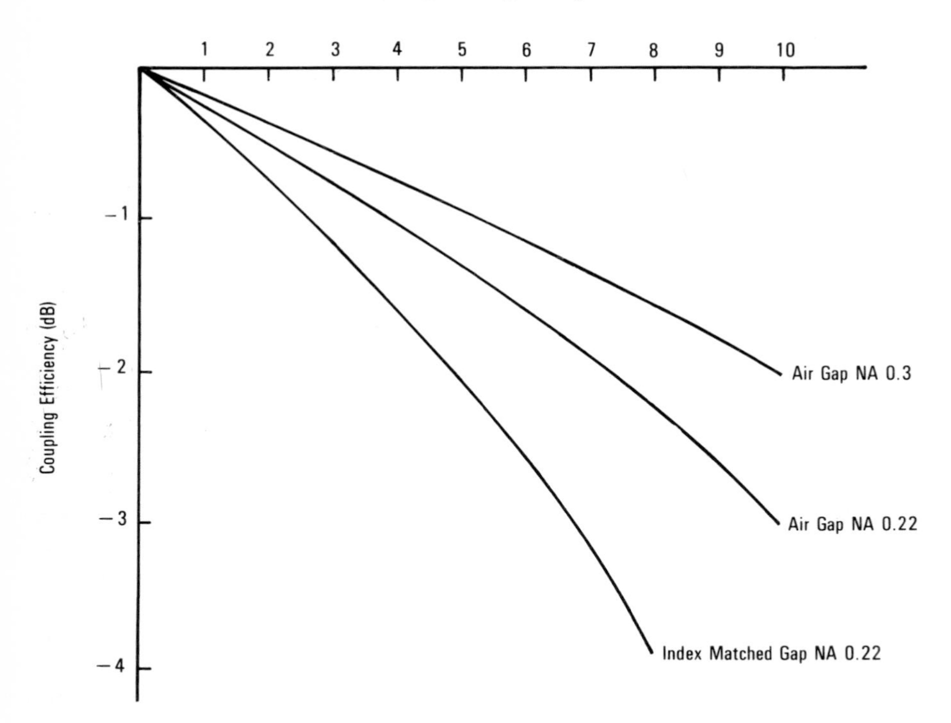

Figure 60. Coupling loss due to angular misalignment

OPTICAL FIBRE SPLICES

A number of techniques for splicing optical fibres have been described in the literature, including fusion splices, [92–95] V-groove splices,[96–98] and sleeve splices.[89–101] Sleeve splices have obvious mechanical advantages for single fibre field splices but have the disadvantages of 'slop', due to the clearance required to insert the fibres in the sleeve. Fusion splices are also attractive but they require precise external alignment of the fibres to be joined. Both sleeve and fusion splices have been developed at STL and are described in this chapter.

The STL sleeve splice minimizes the clearance between the fibre and the tube by collapsing a glass sleeve around one fibre end to form an accurate socket into which the second fibre is cemented.

The technique involves the use of a glass sleeve with a lower softening temperature than the fibres to be joined. Use is made of the fact that when a glass tube is heated to its softening point, surface tension forces tend to act to collapse the tube into a solid rod. Figure 61(a) shows a partially collapsed glass sleeve formed by local heating of the sleeve. In the basic fibre-to-fibre splice, a glass sleeve is collapsed around one prepared fibre end to form an

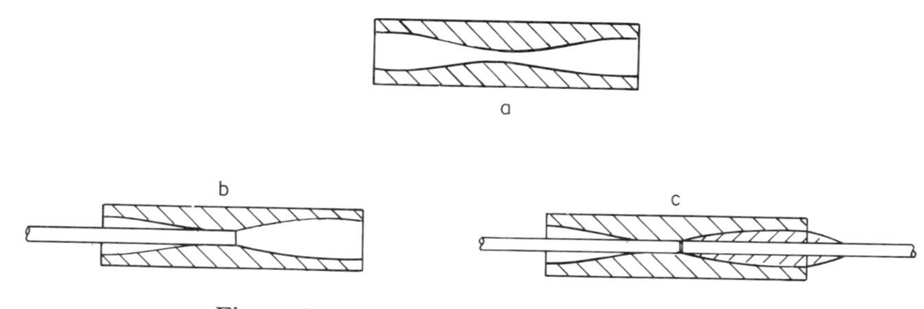

Figure 61. Collapsed sleeve splicing technique

accurately fitted socket into which the second prepared fibre is cemented. This is shown in Figure 61(b). The collapsing process is well controlled, but should one fibre have a larger diameter than the other, for example, as is likely to be encountered when jointing different cable lengths, then the sleeve can be collapsed onto the larger of the two fibres. The collapsing is continued until the sleeve lips over the fibre end to form a socket which is a close fit to the second (smaller) fibre. In this way a better alignment of the two fibre ends can be achieved than would be possible with a close fitting sleeve or V-groove.

A small portable jig has been developed to join two plastic coated fibres. This is shown in Figure 62. The two fibres to be joined are stripped of their protective coating over a length of ~15 mm and their ends prepared by cleaving. They are then placed in the prealigned Vee grooves (1). These are

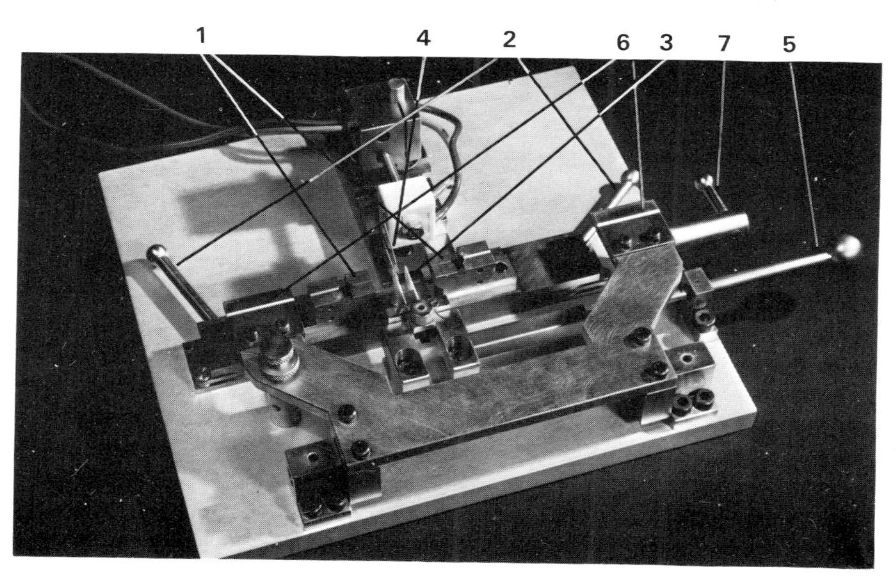

Figure 62. Optical fibre splicing jig

mounted on a co-linear dovetail slide and can be moved longitudinally by the levers (2). The glass sleeve (typically pyrex for silica fibre and approximately 10-mm long) is placed on a pair of notched alumina supports (3) which are prealigned with the V-groove axis so that the fibres can be automatically inserted into the tube which has a bore typically 30 per cent oversize on the fibre diameter. An open loop heater (4) is also mounted on a slide and can be moved into a preset position to collapse the sleeve. The process is observed with a binocular inspection microscope. Once the sleeve has been collapsed onto one fibre, cement is run into the back, by capillary action, to provide mechanical strength since there is a high stress concentration at the bond line between sleeve and fibre. The second fibre is then inserted into the tube together with an optical cement which both holds it in place and provides index matching between the fibre ends. During the cementing operation the alumina supports can be dropped clear by means of the push rod (5). After the cement has cured, the plastic coated fibres are clamped to the outer Vee grooves (6) and the splice unclamped from the inner Vee grooves so that the completed splice can be raised off the jig by a cam (7) and a stainless steel ferrule slipped over the splice. The ferrule is then bonded to the plastic coating and the completed splice removed from the jig. Figure 63 shows a completed splice on a bare silica fibre and in its finished form with a ferrule bonded to the plastic coating.

The jig has been mounted in a small portable glove box with viewing optics for cable jointing in manholes. This is shown in Figure 64. This jig was used early in 1977 for the installation of the ITT 140 Mbit/s Hitchin–Stevenage Demonstration link described in Chapter XI. Space is provided at the right of the jig to accommodate the base of a pigtail splicing enclosure into which the two cables to be joined were potted. For the demonstration link a modified British Post Office Sleeve 31A was used, the major modification being the addition of a drum onto which approximately 1 m of spare fibre could be wrapped to allow for subsequent rejointing. A completed splice minus the sleeve end cap is shown in Figure 65. The splices are supported on two rods which are screwed into the base after splicing has been completed.

Typical attenuations for this technique of 0.2–0.3 dB have been measured using 100 m + 100 m of 30-μm core, 100-μm o.d. graded index silica fibre

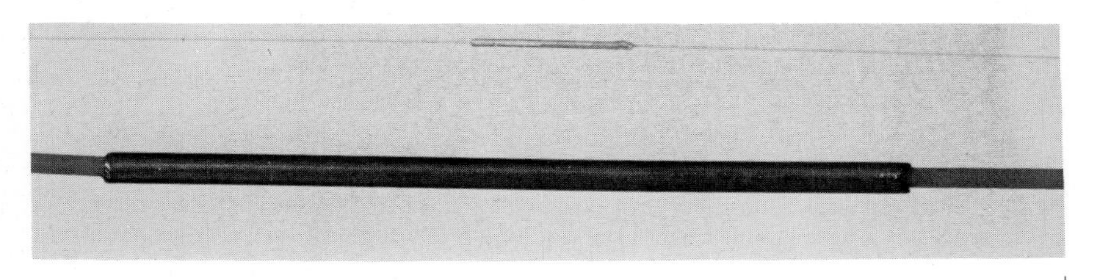

Figure 63. Completed splice on bare fibre (top) and with protective ferrule on plastic coated fibre

Figure 64. Portable splicing equipment

on a break and make basis. In the field demonstration average losses of 0.5 dB were obtained. The additional losses were due to the jointing of lengths of cable containing fibres produced from a number of preforms. The 140 Mbit/s demonstration has shown that this technique using simple easily operated jigging, offers a viable low loss field cable splice.

The fusion splice is under development at a number of companies including STC/STL. This type of splice does not have the inherent alignment

Figure 65. Spliced fibre in enclosure for use in manholes

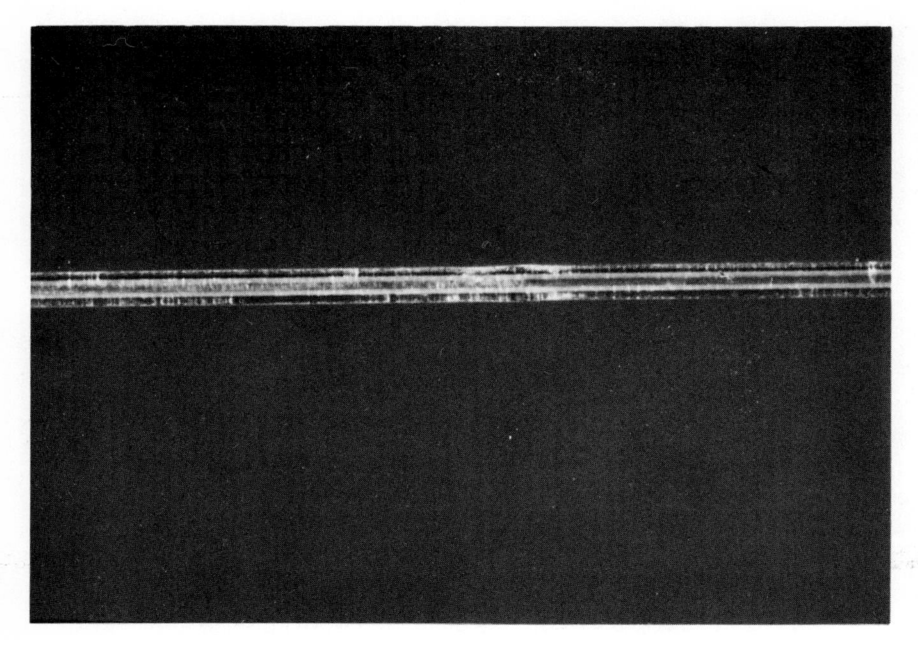

Figure 66. Fusion splice

which is associated with the collapsed tube splice but does eliminate the need for index matching adhesives (see Figure 66).

Fusion splice techniques have been developed at STL and STC for both glass and silica fibres. The fibres are aligned visually using micromanipulator movements. The aligned fibres are then fused together using a suitable heat source. An open loop electric heater similar to that shown in Figure 62 is used to fuse glass fibres; however, the higher softening temperature of silica necessitates the use of an oxy-hydrogen microtorch or an electric arc for VDS or PCS fibres. Typical losses of less than 0.5 dB are obtained for a 30-μm VDS fibre. For 350-μm PCS, losses of less than 0.1 dB have been measured. Figure 66 shows a fusion splice in a 50-μm core, 100-μm o.d. VDS fibre. The protection needed for the fusion splice will depend on the type of system in which it is used. For PTT-type applications the Post Office 31A sleeve has been used.

DEMOUNTABLE OPTICAL FIBRE CONNECTORS

Demountable connectors for optical fibre cables need to perform several functions. Their primary function is, of course, efficiently and repeatably to couple light from one optical fibre to another. If a connector is to operate satisfactorily it must also protect the fibre ends from damage which may occur due to handling, tensile load on the cable, or environmental factors such as moisture and dust, whilst still allowing rapid connection and disconnection when required. For optimum performance in hostile environments the cable and connector must be considered as an integral unit.

An optical fibre connector can usefully be considered in three parts.

(1) Fibre terminations which protect and locate the fibre end.

(2) Alignment guides which position the pair of fibre terminations for optimum coupling.

(3) Connector shells which protect the optical contacts from the environment, hold the alignment guides and fibre terminations in place, and terminate the cable sheath and strain member.

There are two major categories of demountable optical connector. The first is the butt joint[102–104] in which the prepared ends are close to each other and are aligned so that their fibre axes coincide. The second major category uses the expanded beam technique[105,106] for joining fibres. In this approach the size of the transmitted beam is increased by one-half of a connector and this expanded beam is reduced again to a size compatible with the core of the receiving fibre by the second half of the connector. This expansion can be achieved by tapering the fibre or by using lenses.

FIBRE TERMINATIONS

Fibre terminations must transform a fragile bare fibre into a component that is sufficiently robust to survive repeated matings and that is suitable for incorporating into the shell of an optical connector. The termination must provide suitable reference surfaces so that when two are aligned in a suitable guide the optical power transfer is repeatedly high.

Butt Joint Terminations

In a butt joint termination the fibre is usually protected by a metal ferrule which also accurately locates the fibre end.[107] One such ferrule, developed by ITT, is the jewelled ferrule shown in Figure 67. This is manufactured with an accurate outside diameter and a 1-mm counter bore at one end. A standard watch jewel, with a hole size closely matching the diameter of the fibre to be used, is press fitted into the counter bore giving a fibre size hole which is accurately concentric with the outside diameter of the ferrule. The close diameter and concentricity tolerances of the fibre hole are achieved more easily and cheaply by drilling a 1-mm hole and using a watch jewel than by drilling a small hole directly into the end of the metal ferrule. The jewel approach also allows accurate sizing of the jewel hole to match the fibre by holding a stock of inexpensive standard jewels with a range of sizes rather than manufacturing many different ferrules.

Measurements of the accuracy of several hundreds of these ferrules have shown that the outside diameter of the ferrule is held within $\pm 2.5\ \mu m$ and

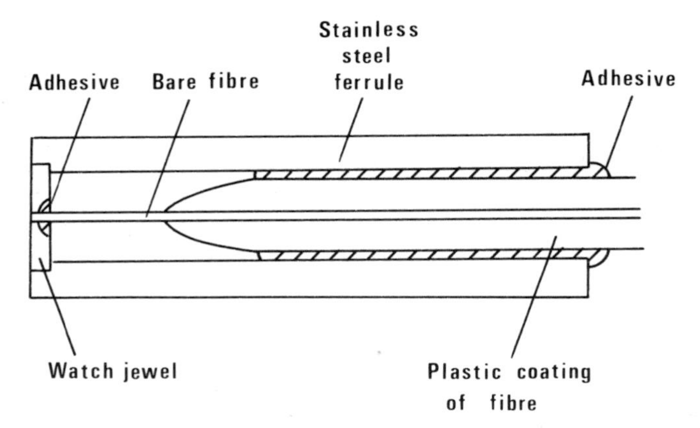

Figure 67. Watch jewel connector ferrule

the standard deviation of the concentricity of fibre hole to ferrule outside diameter is 3 μm.

To terminate a fibre in a jewel ferrule the plastic coating is removed over a short length from the end, and the fibre is fed into the ferrule filled with a suitable adhesive and through the close fitting jewel hole at the front. The fibre is then polished back flush with the ferrule end. Typical concentricity errors between the fibre core and the outside diameter of the ferrule are 2–6 μm.

This termination technique is available for both chemical vapour deposited (CVD) silica and glass fibres. It is currently being developed to accomodate plastic clad silica (PCS) fibre and to allow simple field terminations to be made.

Expanded Beam Terminations

When a prepared fibre end is fixed at the focus of a lens, a collimated beam, with a diameter greater than the fibre core diameter, emerges from the lens. When two of these terminations are aligned an optical connector is produced, as shown in Figure 68. The fibre must be positioned at the focus of

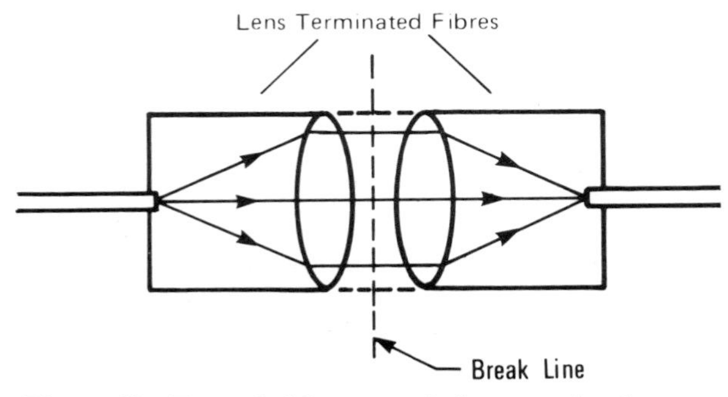

Figure 68. Expanded beam optical connection between two fibres

the lens with the same accuracy as two fibres in a butt joint since the receiving fibre is in effect forming a butt joint with the image of the transmitting fibre. At first sight this may appear to forfeit the advantages of this technique; however, since the accurate positioning and fixing of the fibre in the lens termination unit is only required once, it would normally be done in a factory. Making and breaking of the connection occurs, as shown in Figure 68, between the two collimating lenses. The required connection tolerances are reduced since the increased beam diameter allows greater lateral misalignment of the expanded beam terminations than directly butting fibres. Owing to the collimation of the beam a small separation of the terminations can be tolerated without significantly increasing the attenuation. Figure 69 shows the effect of such lateral displacement and axial separation on the coupling efficiency of the connector.

The increased beam diameter reduces the effect of dust on the connector attenuation. The separation of the terminations minimizes the risk of permanent damage arising from grit scratching or chipping the optical surfaces when the connector is inadvertently coupled in a dirty condition. These considerations make this an important termination technique for rugged connectors.

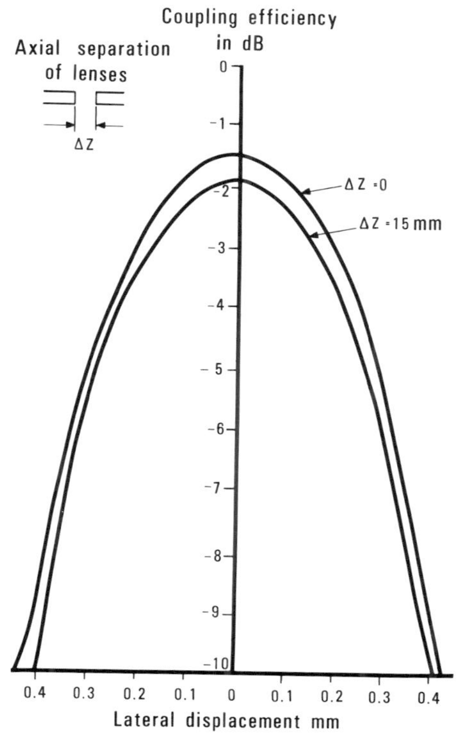

Figure 69. Coupling efficiency of expanded beam terminations

The angular alignment tolerance between expanded beam terminations is tighter than between butt jointed fibres. However, this should not be too difficult to maintain in practice, provided the beam expansion is not too great. A beam diameter of 300 μm will significantly ease the lateral alignment requirements without imposing impractical tolerances on the angular alignment.

The beam expansion can be produced by a number of techniques including glass and plastic lenses, graded index lenses, and tapered sections of fibres.

ALIGNMENT GUIDES

The method of aligning two terminated fibres in the shell will depend on the termination technique chosen.

X-Shaped Jewel Guide

For accurate alignment of two fibres it is desirable to reduce the effect of ferrule machining tolerances by using the fibre itself as a reference surface.

The accuracy to which watch jewels can be produced, their large length-to-hole-diameter ratio of more than 5 to 1, and the highly polished inside surface of the hole, means that a reasonably accurate alignment of two fibres can be achieved in the hole of a long jewel of the type shown in Figure 70.

The female portion of the connector is formed by a fibre with a cleaved end, fixed centrally into the parallel part of the jewel hole, as shown in Figure 70. The male half of the connector has a fibre approximately centred in, and proud of, the end of its ferrule. The proud portion of the fibre is protected by a silicon rubber cone which also helps the initial alignment of the fibre within the jewel hole of the female half of the connector.

Some alignment accuracy is lost owing to the few microns difference that is required between the jewel hole and the fibre outside diameter to achieve a sliding fit of the fibre into the hole. Good alignment can, however, be obtained using these accurate parts which are relatively inexpensive since they are manufactured in very large quantities for the watch-making industry.

Attenuations of 1.4 dB have been obtained in a laboratory environment with poor geometry 85-μm diameter fibre. This performance would however deteriorate fairly quickly in a less benign environment due to the likelihood of dirt becoming entrapped in the alignment hole and damaging the fibre.

Precision Bore Guide

Robust connectors[108] rely on good optical coupling between two fibre terminations and require very accurate alignment of the ferrule tips. Techniques which offer the required alignment accuracy include the precision bore, helical spring, and V-groove and spring guides.

PLASTIC COATED FIBRE
FERRULE
ADHESIVE
BARE FIBRE
X-JEWEL
SILICONE RUBBER CONE
BARE FIBRE
FERRULE
PLASTIC COATED FIBRE

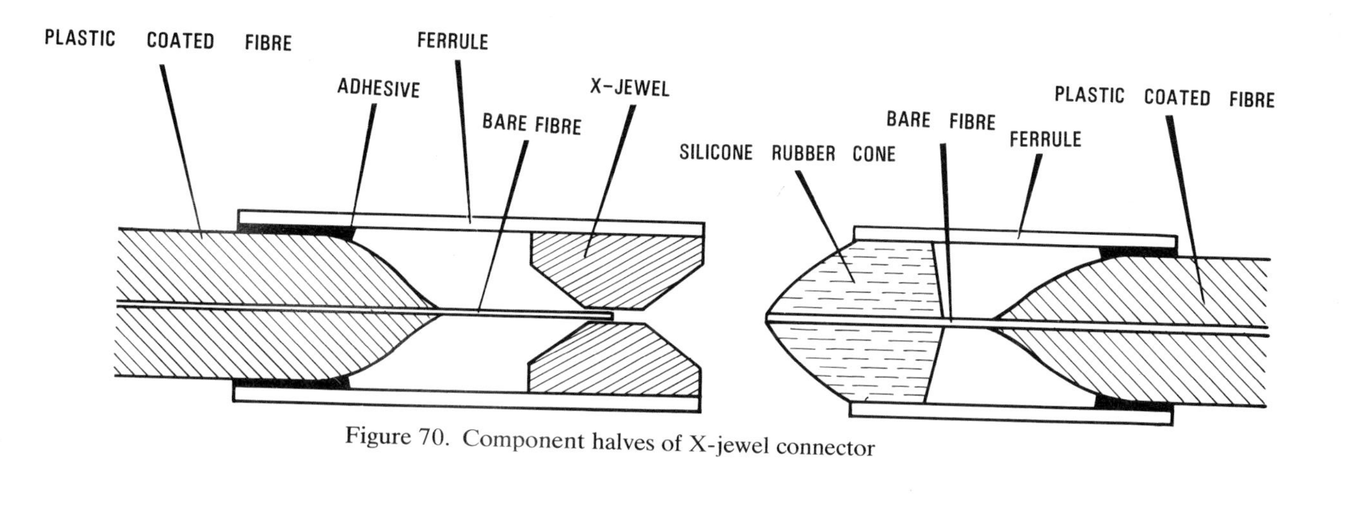

Figure 70. Component halves of X-jewel connector

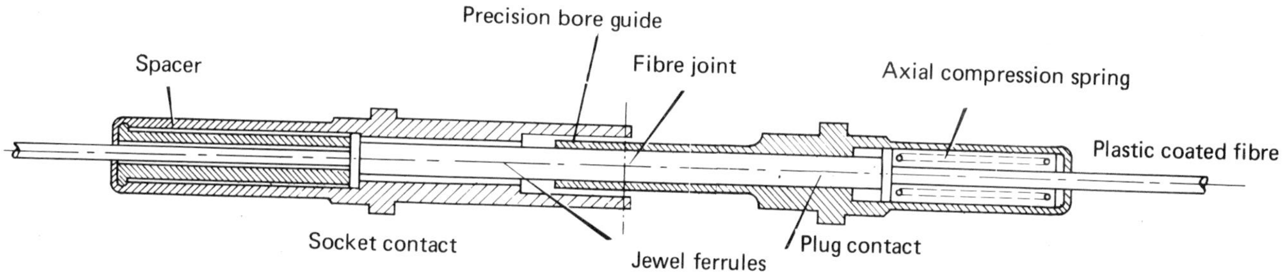

Precision bore guide
Spacer
Fibre joint
Axial compression spring
Plastic coated fibre
Socket contact
Jewel ferrules
Plug contact

Figure 71. Precision bore optical contact assembly

Figure 72. Eight-Way PVX optical connector

The precision bore guide consists of an accurately machined tube which is a close sliding fit to the precision termination ferrule. The two jewelled ferrules butt into the centre of the tube and are held together by compression springs. This technique, shown in Figure 71, is used in a range of connectors which are currently being produced by ITT in the UK. This range includes the multiway optical version of the Cannon PVX Connector shown in Figure 72. With 85-μm core high NA glass fibres the attenuation of these connectors ranges from 1.5 to 2.5 dB.

Helical Spring Guide

One source of misalignment in the precision bore guide results from the need for a sliding fit tolerance between the ferrule and the bore. A technique, being developed, which eliminates this tolerance is the helical spring guide. This guide is a close wound helix of narrow spring strip with a relaxed inner diameter slightly smaller than the outer diameter of the ferrule. When the spring is slightly unwound the inner diameter increases and the two jewelled ferrules are inserted and butted together in the centre. As the spring is allowed to relax it winds down pulling the two ferrules together and aligning and gripping their tips.

Figure 73 shows a 'demountable splice' and the helical spring which it uses to align two jewelled ferrules. The connector is called a demountable splice since it does not provide the protection normally required from a

Figure 73. Helical spring demountable splice

connector. The demountable splice does however provide a compact method of joining prepared ends in a benign environment, such as within equipments, where occasional disconnection of maintenance joints may be required.

Measurements of the attenuation of the helical spring guide have shown the expected decrease in attenuation compared with the precision bore. Losses of 0.7–1.6 dB have been measured with 85-μm diameter high NA glass fibre and 0.5–0.6 dB with 85-μm diameter low NA silica fibre of better geometry. Both sets of measurements were without the use of index matching between the ferrules.

A prototype bayonet action connector using the helical spring alignment guide has been designed and is shown in Figure 74.

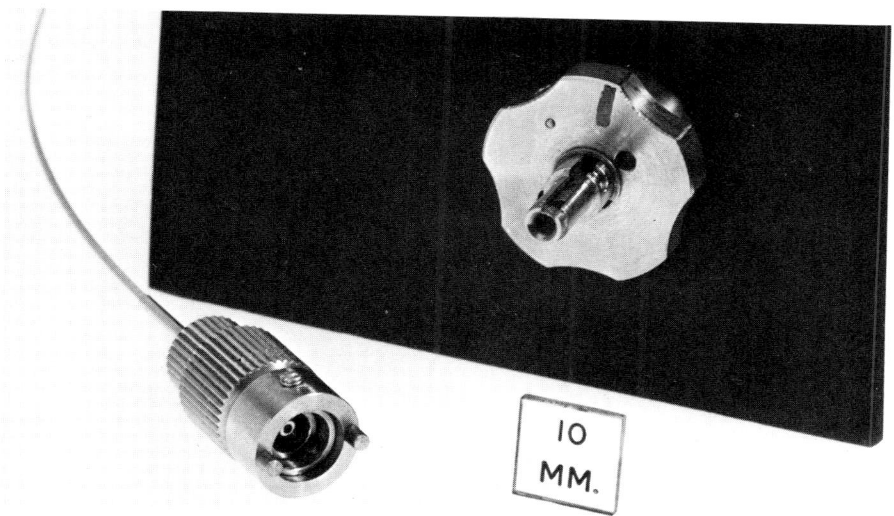

Figure 74. Bayonet action helical spring connector

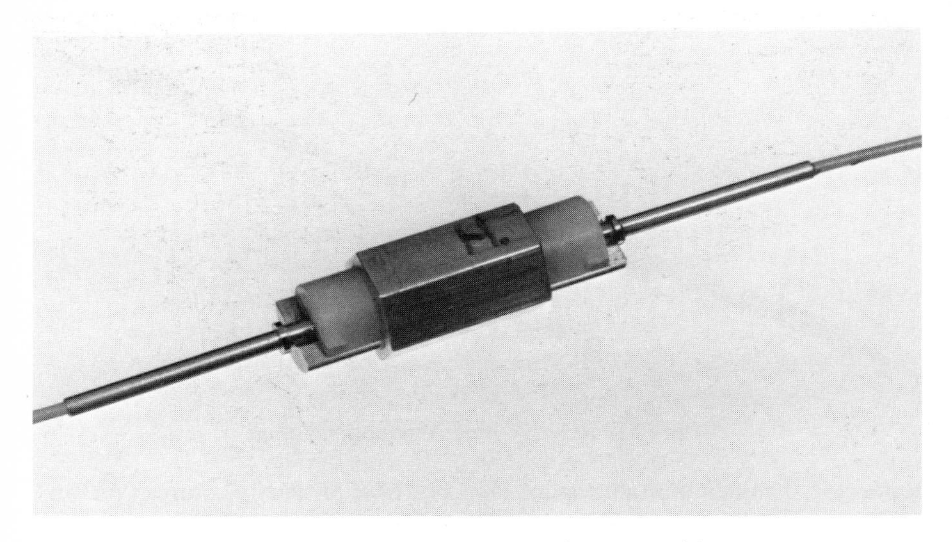

Figure 75. Compact V-groove alignment guide.

Vee Groove Alignment

Another technique for aligning terminated fibres uses a V-groove to laterally position two butting jewelled ferrules. The ferrules are held in the V-groove by springs. A compact version of this alignment guide which achieves optical attenuations of about 1 dB using 50-μm core diameter fibre is shown in Figure 75.

For certain applications, particularly military, hermaphroditic connectors are required. Where an even number of fibres are used this can be achieved by sharing the male and female contacts between each connector half. If an odd number of contacts are required hermaphroditic optical contacts must be used. Figure 76 shows such a connector contact based on the V-groove guide. Each termination can be seen resting in its V-groove. When the connector halves are mated the axially sprung ferrules are butted together and then aligned by being sandwiched between the two V-grooves. The optical loss of this alignment technique is comparable with the V-groove and spring.

Adjustable Connectors

For a very low loss connection of small core fibres and particularly for single mode waveguides, an adjustable connector is necessary to remove the effect of ferrule and alignment guide machining tolerances.[108–111]

One such connector is currently in use as a device-to-fibre coupler on the 140 Mbit/s demonstration link. In this connector the male part contains a short fibre tail from the laser to the connector face. The female half of the

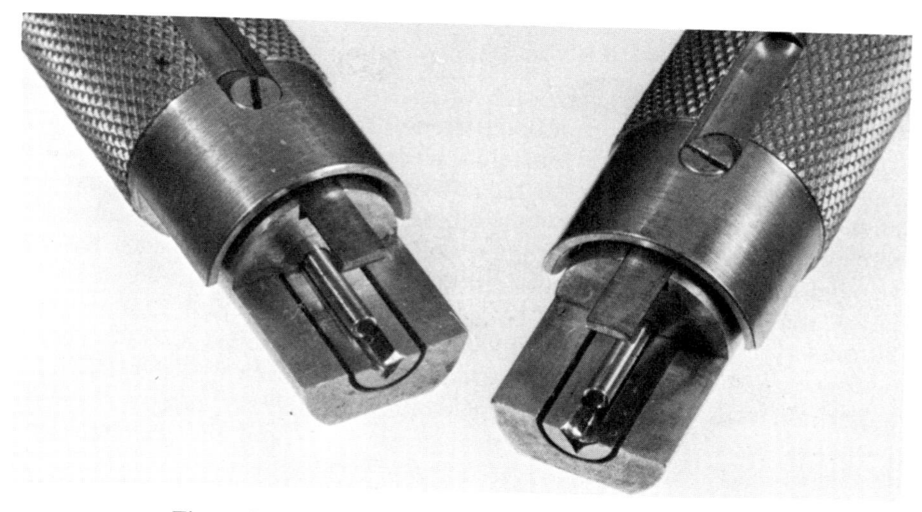

Figure 76. Single way hermaphroditic connector

connector is shown in Figure 77. This half contains a fibre, terminated in a jewelled ferrule, which can be adjusted by screws acting on a pivoted ferrule holder in order to maximize the power launched into the fibre. The connector halves are aligned by a V-form block and V-grooves and locked in position by a spring. These connectors enable attenuation governed primarily by the factors in Figure 57 and equation (19) to be achieved.

Figure 77. Adjustable connector

OVERALL CONNECTOR DESIGN

A number of the requirements for an optical connector shell are common with their electrical counterparts. Where the action for making and breaking a pair of optical contacts is push-pull, electrical shells can often be used to house optical contacts. This has the advantage of eliminating a large portion of the shell development and tooling costs and consequently this increases the number of shell configurations available. For example, ITT has developed a range of demountable optical connectors which are based on the ITT Cannon PVX Pattern 602 electrical connector which is a multipin bayonet-type coupling developed for use in current military aircraft. Typical performance and environmental requirements are given in Table VI.

The connectors ensure correct polarization of shell and receptacle by the 5 key/keyway orientations. The shell is the male and the receptacle the female connector (Figure 72).

Coupling can be made by a simple push action to couple and a pull for uncoupling; however, most military connectors are either threaded or bayonet couplings. Rotation of the coupling nut engages the shell and receptacle and in turn the pin and socket contacts are also engaged.

The final rotation of the coupling nut engages the shell with receptacle to fully compress the peripheral and inter-facial moisture seals, and also allows the bayonet pegs to drop into the lock position of the cam track.

The connector has a closed entry hard socket insulator to ensure positive mating of any misaligned pin contacts. This is accomplished by the lead-in chamfer of the hard insulator guiding the tip of the pin into the socket contact, as shown in the right-hand half of Figure 72.

The basic configuration of a simple butt joint pair of optical contacts is shown in Figure 71. The outer dimensions of the optical contact halves are exactly the same as their electrical analogues. Internally, a close fitting tube locates two identical fibre terminations in radial alignment. Individual contacts can easily be removed from the connector shell using a simple tool.

Table VI
PVX pattern 602 connector performance and environmental requirements

Temperature range	208–473 K
Low air pressure severity	20×10^2 N/m^2
Vibration severity	10–2000 Hz 1.5 mm at 196 m/s^2, duration 10 h, including 1 h at 208 K and 3 h at 473 K
Shock severity	490 m/s^2 for 11 ms
Acceleration severity	981 m/s^2
Durability	500 cycles engagement/ disengagement
Resistance to sand and dust	with particles <152 μm in an air stream 4.76 m^3/min

This construction gives flexibility and versatility in civil and military applications, with a performance adequate to meet the severe mechanical and environmental specification requirements.

Cable Termination

Current designs of optical fibre cable can withstand very high tensile loads which must be transmitted across a connector without affecting the optical power transfer. It is therefore necessary to design suitable cable/connector interfaces. For benign applications this need be little more than a simple clamp on the strain member and a grommet seal between the outer sheath of the cable and the connector shell. For more demanding environments more sophisticated termination of the cable strength member is necessary and a moisture sealing moulding between the cable and connector hardware is required to provide a cable bend restriction as well as an environmental seal.

CONCLUSION

When considering optical connections it is necessary to take into account not only the optical alignment but also the protective package or shell. In an operational system it is the consistency of the attenuation and convenience of use, over a period of time in a real environment, which are important, rather than the lowest optical attenuation which can be achieved in ideal conditions. Techniques are now becoming available which fulfil these requirements for both permanent splices and demountable connections.

CHAPTER VI

Measurement Techniques

INTRODUCTION

As optical fibre waveguides have progressed from the original theoretical proposals into prototype practical systems of the type described in Chapter XI, so the measurement equipment and techniques for evaluation of fibre and cable performance have been developed and expanded to meet the changing requirements. In particular, now that fibre attenuation has been reduced to a commercially viable level, the early preoccupation with this characteristic has given way to a wider study of a number of parameters concerned with transmission quality.

Much of the measurement work discussed in this chapter is of a diagnostic nature, primarily concerned with feeding back information to assist in the development of the transmission medium. For example, attenuation measurements are made over a wide spectral range since this gives some insight into loss mechanisms and identifies offending impurities. Diagnostic measurements being carried out at STL fall into five main groups:

(1) fibre attenuation;
(2) fibre (and bulk) absorption;
(3) fibre scatter loss;
(4) pulse dispersion; and
(5) fibre refractive index profile.

Fibres also need to be assessed in terms of system performance. Here, the parameters to be measured and the conditions of measurement should be of practical significance to the system designer. For a long haul, wideband, civilian optical communications system the primary design requirement is to maximize the distance between repeaters. This distance is limited by the signal attenuation and distortion which increase with fibre length. The overall attenuation includes not only the inherent scattering and absorption loss of the fibre, as discussed in Chapters II and III, but also the losses incurred when coupling optical power from the source to the fibre and those due to jointing (splicing). Signal distortion arises mainly from pulse dispersion or bandwidth limitation which can be caused, for example, by a multipath effect in a multimode fibre due to the differing propagation velocities of the various modes. Therefore, attenuation and dispersion as a function of length are the key system parameters, together with physical characteristics of the fibre waveguide such as acceptance angle (numerical aperture, NA), core diameter, circularity, and concentricity, which affect the efficiency of launching and jointing. Unfortunately these parameters cannot be completely separated, and the length dependence is not predetermined, so the choice of measurement procedure is not straightforward.

In addition to laboratory system measurements there is a need for the assessment of cables in the field. Equipment developed at STL and used to measure pulse dispersion and attenuation of cables installed in British Post Office ducts as part of the 140 Mbit/s optical communications system discussed in Chapter XI will be described.

'On-line' attenuation measurement techniques have also been introduced into the fibre pulling and extrusion coating processes. These techniques were developed to provide an immediate feedback so that optimum processing conditions could rapidly be established in order to reduce the load on off-line measurements and to form the basis for quality control in manufacture.

GENERAL ASPECTS OF MEASUREMENTS ON MULTIMODE FIBRES

Multimode fibres with a graded refractive index profile can have a low enough dispersion to be acceptable for long haul systems, although generally it is high enough to be a critical constraint in system design. Compared with single mode fibres, they have advantages as regards launching and jointing but present some complications in the measurement or even specification of the essential parameters.

The dominant contributor to pulse dispersion is the range of propagating velocities in the modes, and in a graded index fibre this range of velocities can be much less than for a step index fibre.

Measurement complications arise because attenuation and dispersion per unit length depend on the distribution of energy between modes. Higher loss modes die out first and the resulting change in mode distribution leads to a variation in the attenuation and dispersion per unit length (along the length of the fibre). A further factor is mode conversion, i.e. energy launched into one mode being coupled to others, either abruptly at an imperfection or in a gradual manner due to minor variations along the fibre length. This can be caused by cabling, probably due to microbends which are known to cause mode coupling.

These considerations show that results will depend considerably on the launching conditions and on the length of the fibre being measured. Consider, for example, the measurement of acceptance angle. It can be shown that for a short fibre the light launched within say $\pm 15°$ is propagated, so that a large proportion of light from a semiconductor laser can be accepted. If, however, the high order modes (steeper angles) are preferentially attenuated, with negligible mode conversion, then only light launched within say $\pm 5°$ may contribute significantly to the received signal from a long fibre, so that the effective acceptance angle would be length dependent. Further problems are presented by the leaky modes (see below) and by possible mode conversion at joints.

Fortunately, for a fibre with a constant mode conversion coefficient along its length, the distribution of modes tends towards an equilibrium.[112] For fibres which have settled to this dynamic equilibrium, attenuation is proportional to the length while pulse disperion tends to become proportional to the square root of the length. Launching an equilibrium distribution of modes would therefore simplify interpretation of the measurements, especially when the results are being applied to the design of a long haul system in which equilibrium mode distribution is likely to be established.

However, launching an equilibrium distribution is extremely difficult to do in practice. A number of methods have been suggested,[113,114] in particular the use of a scrambler on the fibre close to the launch end to create very high mode conversion in a short length. If sufficient mode conversion results then a distribution will be created that will remain unchanged from then on. Although in the case of step index fibres scramblers have been shown to produce an equilibrium distribution,[114] we have not been able to show that the same is true for graded index fibres.

OPTICAL MEASUREMENTS IN THE LABORATORY

Fibre Attenuation Measurements

Basically, two types of attenuation measurement are conducted: spectral and single wavelength (spot measurement). Spectral measurements are useful for showing up the absorption peaks of metallic impurities, water peaks, and scattering peaks.[28] The spectral loss measurement equipment uses a monochromator and white light source, as shown in Figure 78, and has a wavelength range of 500–1800 nm with a detector change at 1100 nm from silicon to lead sulphide. Alternative long wavelength detectors that are

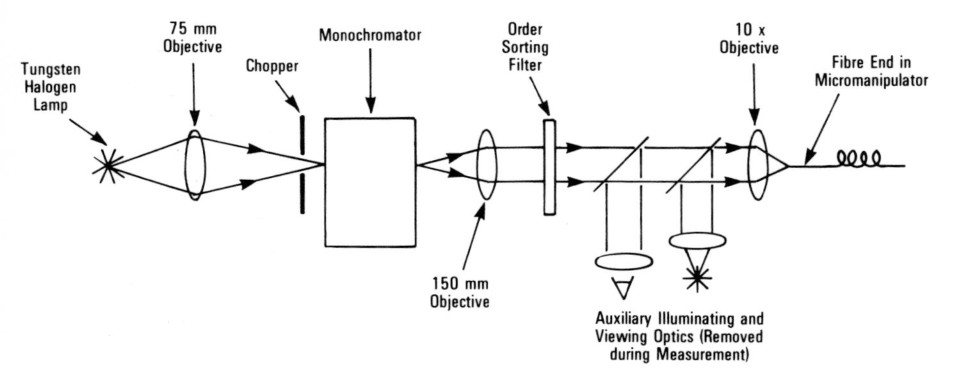

Figure 78. Optical arrangement for the measurement of spectral attenuation

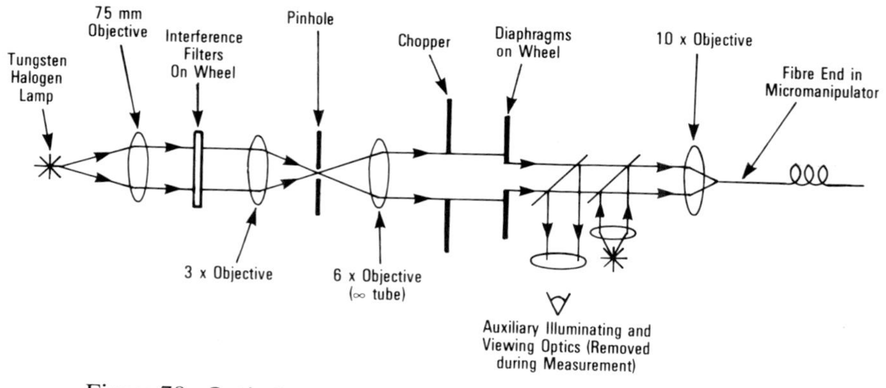

Figure 79. Optical arrangement for spot attenuation measurements

worthy of consideration are cooled germanium (700–1800 nm) and cooled indium antimonide (1000–5500 nm). Typical spectral loss curves are found in Chapter III.

Spot measurements made at one wavelength are quicker and at present are carried out at 850 nm since the majority of applications are at or very near to this wavelength. The optical arrangement for spot measurements is shown in Figure 79. Figure 80 shows the electronic equipment used for both spectral and spot attenuation measurements.

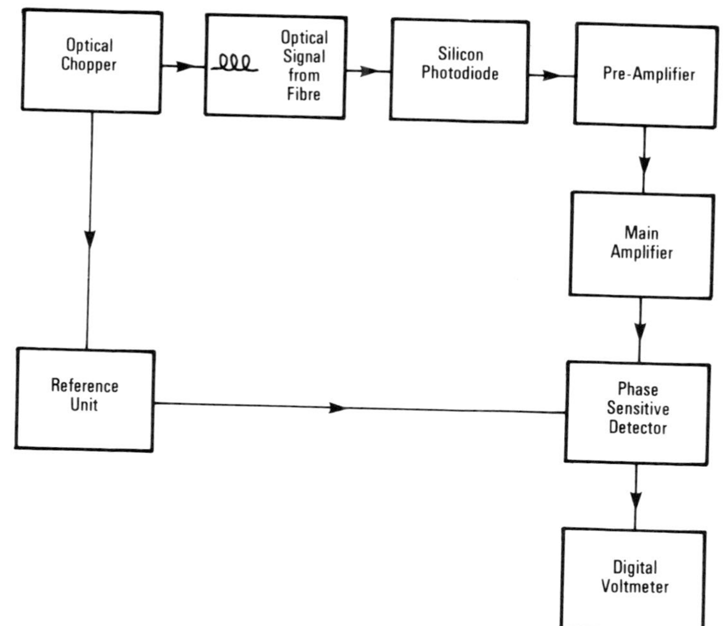

Figure 80. Block schematic of the electronic equipment used for both spectral and spot attenuation measurements

Spectral measurements are nearly always conducted on the bare or primary coated fibre and since the curve shape is unlikely to change during subsequent extrusion coating or cabling, spot measurements, if necessary, are used for all the later stages of the cabling process. The reason is that any loss increase due to subsequent coating and cabling is entirely radiative and for multimode fibres almost wavelength independent.

Both techniques use a single beam approach in which the light transmitted by a fibre, usually several kilometres in length, is measured. The fibre is then cut 3 m from the input end and the signal from this short length is measured. The ratio of this signal to the received signal from the total section is used to work out the fibre loss in the length removed according to the formula

$$A = \frac{10 \log_{10} V_2/V_1}{L_1 - L_2} \quad \text{dB/km} \tag{20}$$

where V_1 is the signal at the original fibre length L_1, in volts; V_2 is the signal at the cut back fibre length, L_2 (usually 3 m) in volts; and L_1 and L_2 are in kilometres.

A typical numerical example would be

$$V_2 = 6.0 \text{ V}; \qquad V_1 = 3.0 \text{ V}$$

$$L_1 = 1 \text{ km}; \; L_2 = 0.003 \text{ km}$$

$$A = \frac{10 \log_{10} 6/3}{0.997} = 3.0 \text{ dB/km}$$

The dynamic range of measurement of the monochromator set up at 850 nm is 25 dB for 30-μm and 29 dB for 50-μm core graded index $GeO_2 : SiO_2$ fibre. For the spot wavelength equipment, again at 850 nm, it is 39 dB for 30-μm and 43 dB for 50-μm fibre. Equipment using a GaAlAs laser as the light source, typically has a dynamic range of the order of 60 dB.

The accuracy of the attenuation measurements is limited mainly by the repeatability one can obtain (in a reasonable time) of the output signal from the fibre. The total uncertainty in the loss is roughly $0.2/\Delta L$ dB/km, where ΔL is in km; for example, ± 0.2 dB/km for a 1-km length.

As previously mentioned, for multimode fibres the measured loss using this technique depends upon the launch conditions. In the spot measurement technique the launch beam numerical aperture can be varied by a series of diaphragms (Figure 79) from 0.04 to 0.24 so that the loss at 850 nm as a function of launch numerical aperture can be determined.

For a graded multimode fibre with a parabolic profile it can be shown that when launching all modes equally, 25 per cent of the light is launched into leaky modes.[115] The theoretical attenuation of these modes has been calculated.[116] For example, for fibres of 30-μm and 50-μm core diameter with an NA of 0.2 and a parabolic profile, the leaky mode power in the fibre at

$\lambda = 0.85\ \mu m$ is as follows, respectively

input end: 25%; 25%
at 3 m: 3.6%; 7%
at 200 m: 2.5%; 5.5%
at 1 km: 2.1%; 4.5%

Thus, when measuring 50-μm core diameter fibres between 3 m and 1 km, the attenuation will be pessimistic by 0.13 dB/km because of leaky mode attenuation. It seems likely therefore that bound modes rather than leaky modes cause the main variations in the loss results obtained using various launch conditions.

Since it is difficult to achieve equilibrium launching (and for other than long haul systems it may not be the best choice anyway), our general approach with respect to systems attenuation measurements has been to provide the system designer with pessimistic, or the 'at worst', figures by launching with a spot size larger than the fibre core and at a larger NA than the fibre so as to excite all possible modes of propagation in the fibre under test. Typically the spot size is 95 μm and the launch NA 0.24 ± 0.01. Once the system source and launch conditions have been specified then the fibres can be measured under the system operating conditions if required.

For example, in the case of the 140 Mbit/s field demonstration the fibres were first measured under 'at worst' or overfilled conditions and later measured under the system operating conditions (using the overlay equipment described later in this chapter). Losses obtained from the two sets of measurements are compared in Table VII. It can be seen that the overfilled loss is larger than the systems loss but that the difference reduces as the length of fibre measured increases and is of the order 0.5 dB/km for a 1-km length. This is as expected owing to the high order modes launched in the overfilled case being preferentially attenuated. Thus, in this case of a 30-μm core fibre the overfilled measurement, although slightly pessimistic, is a good guide to the system designer as to how the fibre will perform. However, the current trend in fibre design is to increase the core diameter from 30 μm to 50 or 60 μm in order to ease splicing problems and the overfilled measurement may not be the right choice for these larger core fibres. In fact, for a long haul system, using sections of 60-μm fibre spliced together, there is evidence to suggest that loss measurements avoiding the excitation of high

Table VII

Fibre length (in m)	Overfilled loss (in dB/km)	GaAlAs loss (in dB/km)	Difference (in dB/km)
720	4.5	3.9	0.6
1750	4.2	3.9	0.3
2270	3.8	3.8	0.0

order modes would be of most relevance to the system designer.[117] In summary it is clear that for meaningful measurements for the system designer, the measurement conditions must be carefully chosen with due consideration given to both the fibre parameters and the system application.

As mentioned previously, measurements are made on the fibre at most stages of its manufacture into a cable. To measure any change due to each stage it is important that the fibre is measured under the same conditions. Thus, bare or primary coated fibres wound on drums are cooled to reduce microbending loss;[80] this is straightforward since the aluminium drum has a much larger coefficient of thermal expansion than glass or silica fibre. Extrusion coated fibres are measured loosely in boxes. The input or launch end of the fibre is usually prepared by a fibre breaking machine.[91]

Fibre (and Bulk) Absorption Measurements

There is a need to separate the scatter and absorption components of the loss from the fibre spectral attenuation curve as a further means of loss diagnosis, as has been done, for example, in Figure 81.

The technique used to measure absorption loss is similar to that developed by White and Midwinter.[118] The temperature rise due to energy absorbed from a powerful laser beam which is launched into the fibre is measured with a thermocouple. Figures 82 and 83 show details of the calorimeter used for measuring fibres.

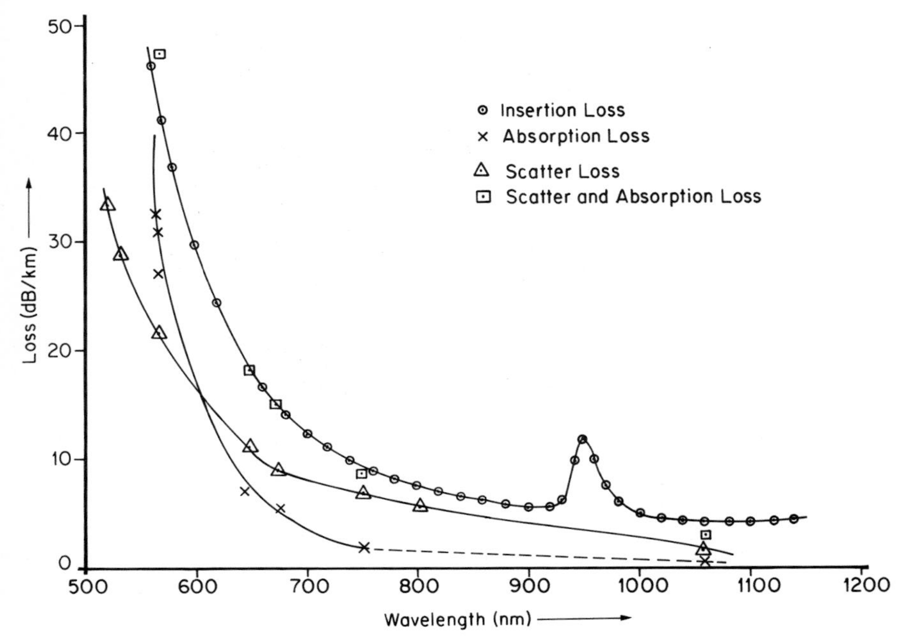

Figure 81. Typical curves of scatter and absorption components of fibre loss

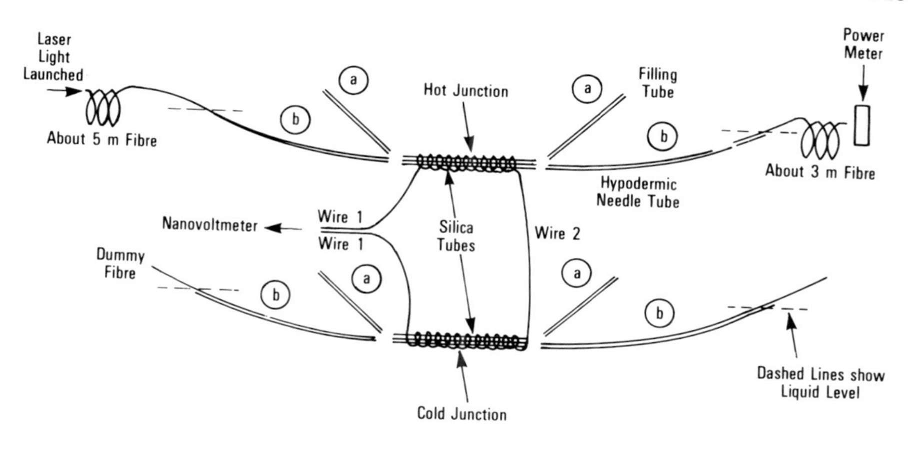

Figure 82. Schematic of a calorimeter used for fibre absorption measurements

Figure 83. A calorimeter for fibre absorption measurements

A Krypton ion laser and a Nd:YAG laser are used as sources to provide measurements at nine different wavelengths in the visible range up to 800 nm and also one at 1060 nm.

The fibre end is passed down a hypodermic needle tube (Figure 82) which accurately locates with a fine bore silica tube around which a 25-μm diameter thermocouple is wound. The silica tube can be filled with methanol via the filling tubes to obtain good thermal contact between the fibre and the silica tube. A dummy fibre in an identical arm is used as a reference.

For both fibre and bulk absorption requirements (where another calorimeter is used) the temperature rise is monitored at the centre and the ends are thermally clamped to remove end effects (in the bulk loss case these are caused by surface losses on the sample).

The theory is complex[118] but the fibre absorption, *N*, can be determined from the heating or cooling curve of the fibre under test (Figure 84), the

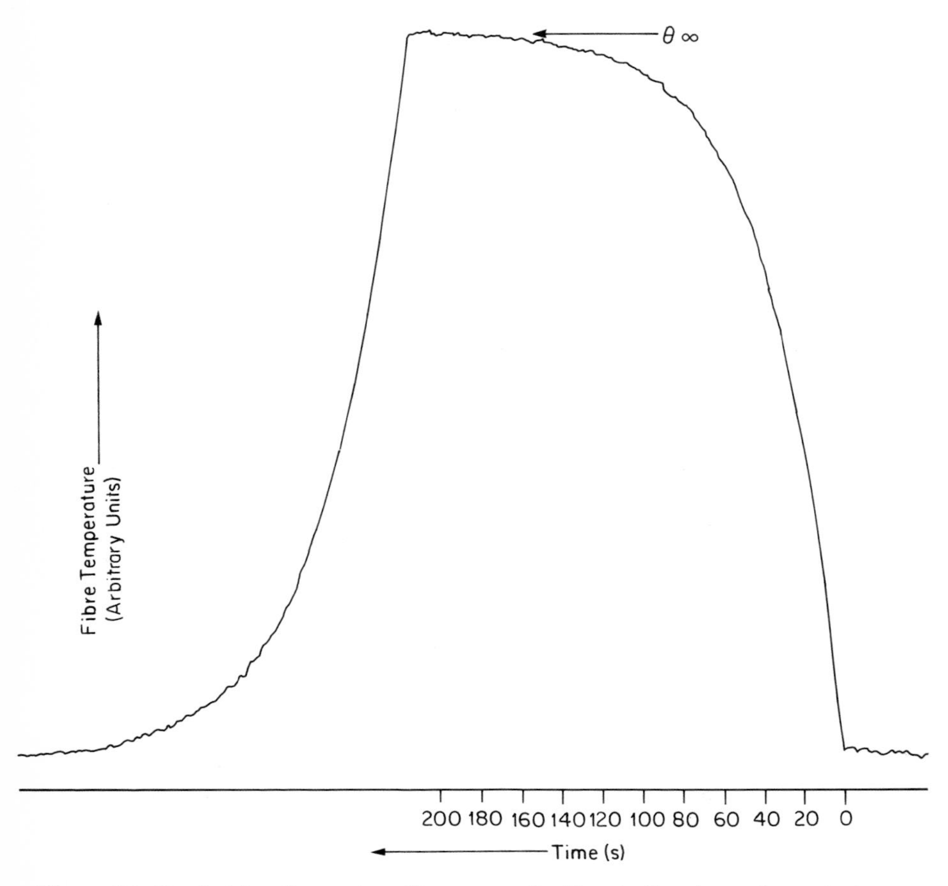

Figure 84. Typical heating and cooling curve of a fibre in the absorption calorimeter

optical power propagated, and the thermal capacity of the methanol filled silica tubes, according to the formula

$$N = \frac{K}{P_0} \frac{\theta_\infty}{T_{11}} \quad \text{dB/km} \tag{21}$$

where K is proportional to the thermal capacity per unit length of the silica tube and methanol in the tube, P_0 is the power propagating in the fibre, θ_∞ is the maximum temperature rise that the fibre attains, and T_{11} is a time constant obtained from the curve of $(\theta_\infty - \theta_t)$ plotted on a logarithmic scale against the time t (Figure 85), where θ_t is the temperature of the fibre at time t. We have

$$\frac{1}{T_{11}} = \frac{\ln(\theta_{t1}) - \ln(\theta_{t2})}{t_2 - t_1}$$

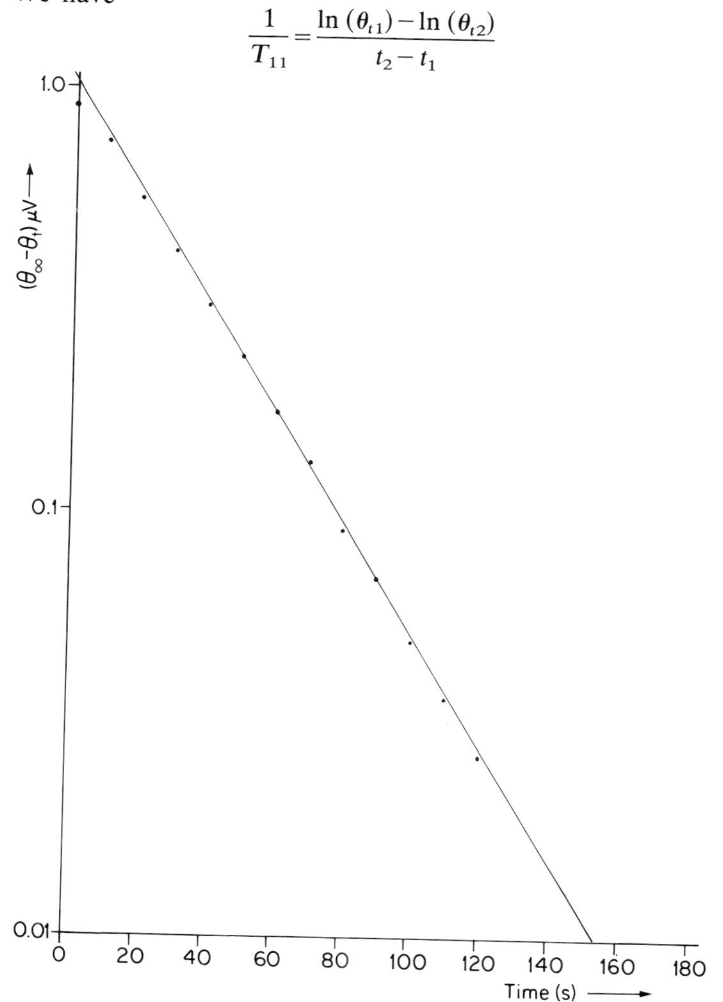

Figure 85. Plot of $(\theta_\infty - \theta_t)$ versus t to determine T_{11} and fibre absorption

The constant K can be obtained by calculating the thermal capacity of the methanol filled silica tube or from calibrating the calorimeter using a known heat source in the form of a known current passing through a fine wire replacing the fibre. Calibration yields a value for the thermal capacity within 6 per cent of that calculated from the masses and the specific heats.

T_{11} is a constant for the calorimeter inversely proportional to its rate of heat loss. It is generally determined using a fibre of high absorption which gives a high temperature rise, thus minimizing experimental errors. Once T_{11} is calculated, the absorption loss of better fibres can be determined from their corresponding maximum temperature rise θ_∞.

Absorption measurements as low as 0.57 ± 0.11 dB/km at 1060 nm have been measured with the equipment. In this case the optical power propagated was 250 mW and the fibre was passed three times through the calorimeter to triple the heating effect.

By way of numerical example we illustrate the case where K is 202.9×10^2 J/deg. The heating and cooling curve of a high absorption fibre (17.4 dB/km) is shown in Figure 84. Figure 85 shows the corresponding curve of $(\theta_\infty - \theta_t)$ against time. From Figure 85 we have

$$\frac{1}{T_{11}} = \frac{\ln \theta_{20} - \ln \theta_{120}}{120 - 20}$$

Temperatures are measured in terms of the thermocouple output in μV and so we have for the case quoted above

$$\frac{1}{T_{11}} = \frac{\ln(0.57) - \ln(0.028)}{100}$$

giving $T_{11} = 33.2$ s.

Substituting into equation (21) we have

$$N = 611.2 \frac{\theta_\infty}{P_0} \quad \text{dB/km}$$

In a typical example a fibre was passed through the calorimeter three times and the temperature rose to a maximum of 0.00077 deg °C (or thermocouple voltage of 0.033 μV). The optical propagated power at 752 nm was 83 nW and the fibre absorption is given by

$$N = \frac{611.2 \times 0.00077}{3 \times 83 \times 10^{-3}} = 1.9 \text{ dB/km}$$

Fibre Scatter Loss Measurement

Figure 86 shows the scatter loss integrating sphere designed to measure scatter loss over a short path length. However, the fibre can be drawn through to study how the scatter loss changes along the length of the fibre and to isolate discrete scattering centres for study. Krypton ion and

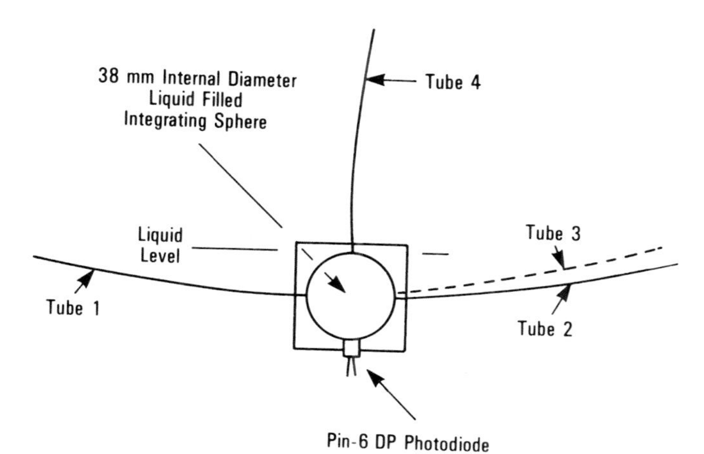

Figure 86. Measurement of scatter loss using an integrating sphere

Nd : YAG lasers are again used as sources so that a spectral measurement can be built up in a similar way to the absorption results.

The cavity is machined from brass and a thin coating of aluminium is evaporated onto the inside prior to a coating of high reflectance white paint. The aluminium prevents leaching and formation of coloured copper salts by chemical interaction with the white paint. It also serves as a back layer to the paint to increase the reflectivity.

Carbon tetrachloride is used as the index matching liquid in the sphere for silica fibres because it gives a good index match at the wavelengths in question, has a low optical loss, and is compatible with the paint.

The fibre being measured passes once across the cavity tubes 1 and 2 (Figure 86) which are accurately aligned opposite each other. Tube 3 is used for measuring the light from the fibre end and tube 4 enables air to escape when filling or topping up the cavity via tube 2 and also allows the efficiency of the sphere to be measured.

The scatter loss N_s is given by

$$N_s = 10 \log_{10}\left(\frac{S_1}{S_1 - S_3}\right) \text{ per } \Delta L \text{ dB/km} \tag{22}$$

where S_1 is the propagating signal at the cell (usually measured in volts) and S_3 is the scattered signal in volts from the short length of fibre ΔL.

Since $S_3 \ll S_1$ the logarithm may be expanded and we have

$$N_s = 4.343 S_3/S_1 \times 1/\Delta L \text{ dB/km}$$

Owing to the large dynamic range and good linearity of the photodiode, neutral density filters are not required for the measurement.

The repeatability of scatter measurements using this technique is about 5 per cent while systematic errors may be up to 10 per cent. This means that the shape of the scatter loss curve is accurate to 5 per cent but the curve may be shifted up or down by 10 per cent.

To measure the contribution of Rayleigh scattering loss alone, care is taken to ensure that only low order modes are launched at the input end (this is possible using the Krypton ion and Nd:YAG laser sources).

The length of fibre between the launch end and scatter cell is approximately 5 m in a loose coil, and approximately 10 m between the cell and the output end. When measuring the scattered signal the output end is immersed in an index matching liquid to prevent back reflected light being picked up by the cell.

For the scatter sphere described, ΔL is 3.71 cm and we have

$$N_s = 1.170 \times 10^5 \, S_3/S_1$$

Since a small percentage (0.4 per cent) of the total scattered radiation from the fibre falls directly on the photodiode a small correction has to be made to this formula which gives

$$N_s = 1.143 \times 10^5 \times \frac{S_3}{S_1} \text{ dB/km}$$

e.g. for a fibre at 752.5 nm

$$S_1 = 6.67 \; V, \quad \text{at a gain of } 10^5$$

$$S_3 = 2.83 \; V, \quad \text{at a gain of } 10^9$$

$$N_s = 1.143 \times 10^5 \times \frac{2.83}{6.67} \times \frac{10^5}{10^9}$$

$$= 4.85 \text{ dB/km}$$

Fibre Refractive Index Profile Measurements

The refractive index profile of a graded multimode fibre may be described by the equation

$$n_r = n_0[1 - 2\Delta(r/a)^\alpha]^{1/2} \qquad (23)$$

where

n_r = refractive index at radius r,
n_0 = refractive index at $r = 0$,
$\Delta = (n_0 - n_2)/n_0$,
n_2 = cladding refractive index,
a = radius of the fibre core, and
α = exponent which depends upon the refractive index profile.

The technique used for routine measurements at this and many other establishments is, at the time of writing, the near-field technique.[119] Figure 87 shows a typical curve obtained by this technique on a graded index fibre

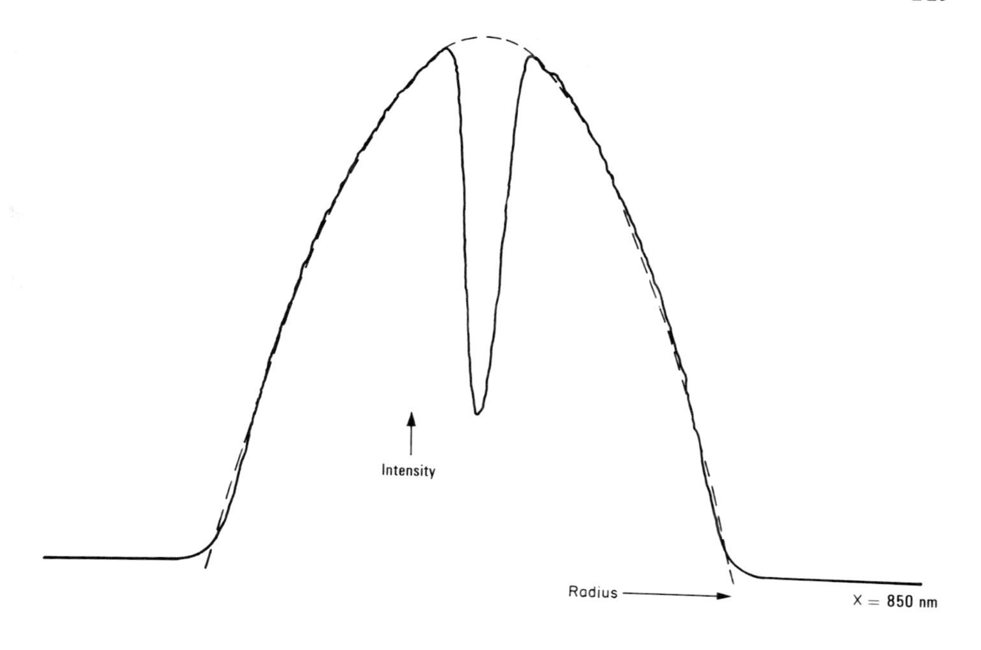

Figure 87. Typical plot of near-field intensity distribution

of 50-μm core diameter using the arrangement shown in Figure 88. The dashed line shows the best fit of equation (23) to the experimental curve.

For the result shown, $\alpha = 1.94$ at the measurement wavelength of 850 nm. The abscissa and ordinate axes are allowed to float to obtain the best overall fit.

Figure 88 shows the optical arrangement. Essentially the technique consists of illuminating the input end of the fibre by a Lambertian source and scanning the output end near-field intensity distribution. The sample length is typically a few metres. The source is a tungsten ribbon filament lamp which enables us to fulfil the Lambertian source requirement.

The near and far field of the launch spot have been carefully measured. The near field is uniform to within 5 per cent over the spot diameter of 70 μm. The far field is uniform to within 5 per cent over a numerical aperture greater than 0.3.

A fibre in an XYZ micromanipulator in conjunction with an extended red response photomultiplier is used to detect the intensity distribution. The 850-nm interference filter of 13-nm bandwidth is sited in front of the photomultiplier tube. The photomultiplier may, if desired, be cooled to dry ice temperature.

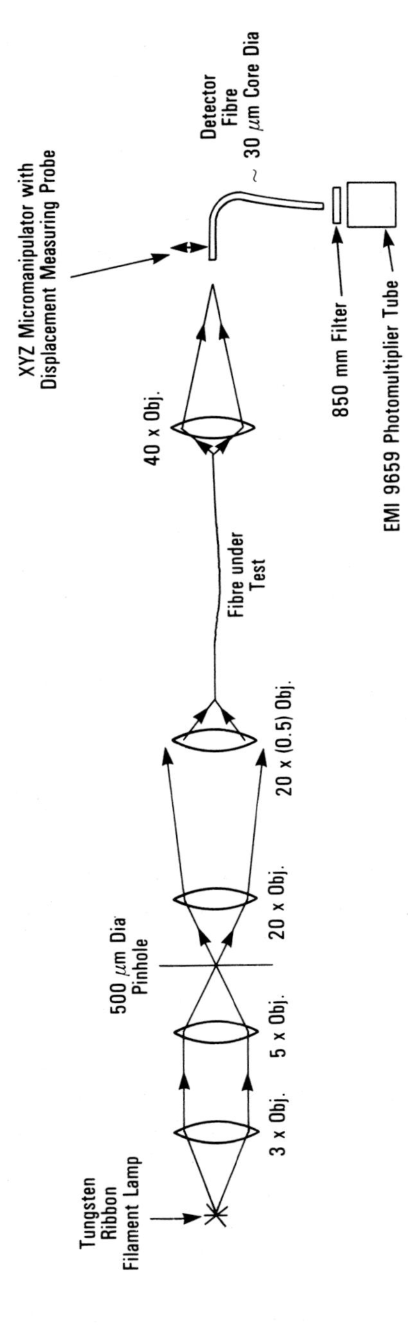

Figure 88. Optical arrangement for measurement of refractive index profile by near-field method

Payne *et al.*[116] have shown that if all modes are equally excited on the input end and the bound modes suffer negligible loss, then

$$\frac{I(r)}{I(0)}=\frac{n_r^2-n_2^2}{n_0^2-n_2^2}C(r, z) \tag{24}$$

where $I(r)$ is the intensity at radius r on the fibre output core, and $C(r, z)$ is the leaky mode correction factor[116] which is dependent upon the sample length z.

From equations (23) and (24) we have

$$I(r)=I(0)[1-(r/a)^\alpha]C(r, z) \tag{25}$$

For the curve fitting in Figure 87 we have assumed $C(r, z)=1$. This implies that leaky modes are assumed to be completely attenuated.

For a circular core fibre with no microbending loss or diameter variations $C(r, z)$ may be calculated.[116] This correction factor does itself depend upon the α value but a good approximation is obtained to $C(r, z)$ for many fibres by putting $\alpha=2$. For a circular fibre of NA 0.2, diameter 50 μm, length 1 m, and wavelength 850 nm the $C(r, z)$ calculated reduces the α value of a parabolic curve from 2.0 to 1.4.

However, a slight ellipticity in the fibre core affects the attenuation of the leaky modes and hence $C(r, z)$ significantly.[120] For a given minimum value of leaky mode attenuation, say 10 dB/m, which would enable us to ignore all leaky modes in the near-field measurement, the ellipticity required to produce this loss is dependent upon α. For $\alpha=2$ an ellipticity of only 0.03 per cent is required while for $\alpha=1.95$ and 2.05, 0.8 per cent is required. These values of core ellipticity would be difficult to measure.

The implication of this is that if one is very close to the optimum parabolic profile, leaky modes may be ignored, but not necessarily otherwise.

By squeezing the fibre one could deliberately introduce ellipticity in order to remove the leaky modes; however, high order bound modes would almost certainly be attenuated as well in the process.

The above arguments suggest that it is reasonable to put $C(r, z)$ equal to unity for some fibres, but not for all. This uncertainty is a disadvantage of the near-field index profile measurement technique. The resolution of the technique is limited by two effects: first the finite V value of the fibre; for the fibre in Figure 87 $V=37$ at 850 nm. Since there are $V^2/4$ bound modes in a parabolic fibre[115] this number is 1369. The finiteness of the number of modes propagating produces an amplitude ripple on the near-field distribution.[121]

The second limitation on the resolution is due to the distortion of the near-field intensity profile by the limited local numerical aperture of the fibre. The resolution of the intensity profile is dependent upon the quality of the image produced by the 40× lens. This depends upon λ/NA, where NA is normally the lens numerical aperture but in this case is the local fibre NA since it is the smaller of the two quantities.

As the local NA of the fibre varies across the core diameter, the effect is most pronounced near the core/cladding interface.[122]

Set against these disadvantages the technique has the following advantages: it is extremely easy to set up and carry out routine measurements. Since the measured intensity depends upon n^2, it is very sensitive.

The reflection technique[123,124] for measuring the refractive index profiles of fibres has also been investigated in this laboratory. Basically this consists of scanning a small spot of light ($<1\ \mu$m diameter) across the end of a fibre and measuring the intensity profile of the back reflected light. The reflectivity depends upon the Fresnel reflection coefficient.

At normal incidence the reflectivity $R = [(n-1)/(n+1)]^2$

$$\frac{\Delta n}{n} = \frac{\Delta R}{R} \frac{\sqrt{R}}{1-R} \tag{26}$$

For $n = 1.5$, $R = 0.04$, $\sqrt{R}/(1-R) = 0.208$. Thus, for a 1 per cent change in n one gets a 5 per cent change in R. For small Δ the reflected intensity profile varies linearly with the refractive index profile.

The problems with this technique are principally the difficulty of accurately measuring a very small change with a large background signal and the sensitivity of the technique to end preparation. The latter problem limited the usefulness of the technique to that of measuring Δ. Furthermore, the technique produces inaccurate results for fibres which are affected by atmospheric attack.[125]

Stewart has proposed a technique using the refracted rays for fibre profile measurement.[126] Figure 89 shows the basic optical arrangement. Light from

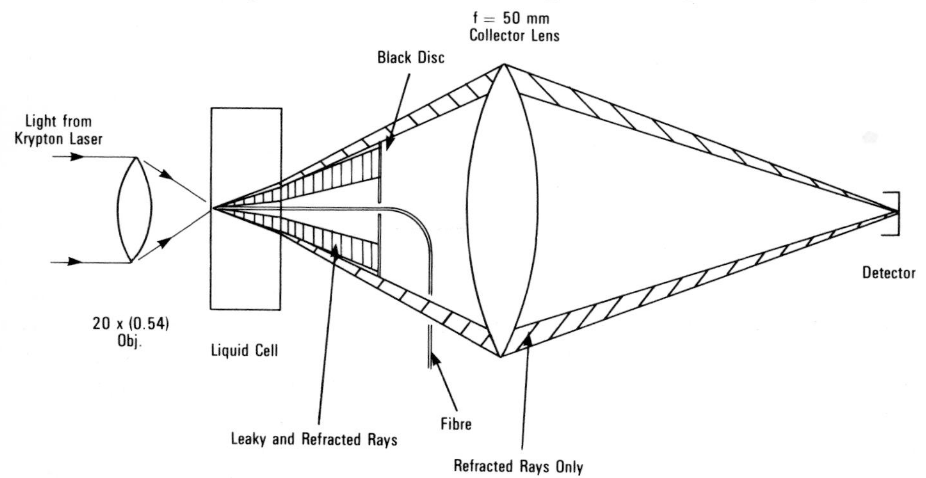

Figure 89. Basic optical arrangement for measurement of refractive index profile by refracted rays technique

a Krypton laser is focused to a small spot (approx 1 μm diameter) onto the end of the fibre being measured. The fibre passes through a hypodermic needle tube into a liquid cell the refractive index of which is slightly higher than that of the fibre cladding. A black disc is located so that leaky rays are not measured by the detector. As the small spot of light is scanned across the fibre end an intensity profile is obtained which is related to the refractive index profile.

A further technique worthy of mention is the interference microscope technique.[127] This can yield accurate values for α but requires careful preparation of thin slices of the fibre. The equipment necessary is very costly and has to be computer controlled.

DISPERSION MEASUREMENTS

With the attainment of very low attenuation in optical fibres, the limitations imposed by dispersive mechanisms in long distance optical fibre communication systems have become increasingly important and consequently measurements providing a fuller characterization of fibre dispersion are essential. The dispersion measurements at STL may be roughly divided into two classes and these will be dealt with separately following the brief description of the dispersive mechanisms in fibres given below.

Dispersion in Optical Fibres

In optical materials the phenomenon of dispersion is usually connected with the variation of the material refractive index with the frequency of the incident electromagnetic radiation or, more generally, the variation with frequency of the propagation constant of the electromagnetic radiation in the material. In optical fibre, however, dispersion, whilst embracing the above definition, is used to describe all mechanisms in the fibre which produce a distortion at the output of signals launched into the fibre. The three principal causes of dispersion in optical fibre are described below.

Inter-Modal Dispersion

One of the results of the solutions of Maxwell's electromagnetic equations in round optical fibres shows that only certain modes, each having its own velocity of propagation, are allowed which correspond to discretely defined angles of incidence of rays in the fibre. When a short duration optical pulse is launched into the fibre, depending on the spatial and angular distribution of power in the pulse, some or all of the allowed modes in the fibre are excited and propagation in these modes takes place. However, owing to the range of different group velocities of these excited modes, there is a spread in the arrival times of power transmitted in the individual modes which

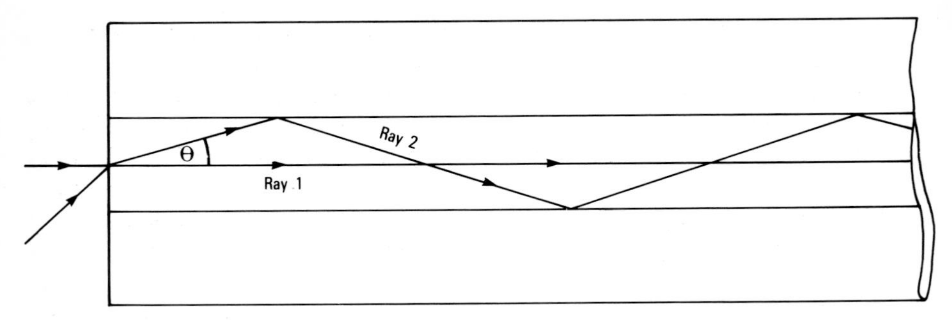

Figure 90. Direct and limiting rays in a step index fibre

results in a wider, dispersed pulse at the output. This is known as inter-modal dispersion or group delay dispersion, and in multimode fibres is the dominant dispersion mechanism. For a step index fibre this contribution may be calculated approximately using geometrical optics. In Figure 90 ray 1 is the ray incident normally on the fibre end and ray 2 is the limiting ray which propagates by total internal reflection at an angle (θ) just below the critical angle. The total length traversed in a fibre of length L is given by $L/(\cos\theta)$ and the delay difference ($\Delta\tau$) between rays 1 and 2 is given by

$$\Delta\tau = \frac{nL}{c}\left(\frac{1}{\cos\theta} - 1\right) \tag{27}$$

Thus, for $n = 1.45$ and $\theta = 8°$ (corresponding to a relative index difference of 1 per cent) $\Delta\tau = 47$ ns/km. Dispersion as high as this is rarely met in practice owing to the effects of mode mixing and mode filtering in the fibres. Mode mixing, in which the launched power in each mode is continuously transferred to higher and lower order modes, results in the modes propagating at some average velocity. Mode filtering is a special case of mode mixing in which the power in the higher order modes is mixed to radiative modes thereby reducing the power in the slower high order modes with the result that most of the power is transmitted in the lower order modes having a lower spread in their group delays. With the additional effect of mode mixing in typical STL step index fibres having a core diameter of 30 μm and a relative refractive index of 1 per cent, the measured dispersion is 7 ns/km. Another consequence of mode mixing is an improvement of the dependence of inter-modal dispersion on the fibre length. For a multimode step index fibre having negligible mode mixing the length dependence of pulse spreading is linear, as predicted by equation (27). However, when there are imperfections or bends along the fibre length mode mixing occurs and it may be shown that after a certain distance, known as the coupling length, a steady state modal power distribution is set up with the result that pulse spreading follows a square root dependence with length.[112] For fibre lengths less than the coupling length, the pulse spreading lies between the root dependence and the linear dependence. The coupling length may be less than 1 m in lossy fibres but exceeds several kilometres in fibres having low microbending loss and an attenuation of a few dB.

As described earlier, the presence of mode mixing gives rise to improved inter-modal dispersion in fibres; however, the minimum dispersion in step index fibres with strong coupling is relatively high and is achieved at the expense of a reduced power throughput as a result of high mode conversion or deliberate launching of low order modes. Inter-modal dispersion may be greatly reduced, however, without associated power losses or reduced launching efficiency by arranging for the energy transmitted in the higher order modes to travel faster than the lower order modes such that the increased velocity compensates the longer distance traversed by the higher order modes to travel faster than the lower order modes such that the simultaneously. Since the modal group velocity is an inverse function of refractive index this may be achieved in practice by the fabrication of fibres having a gradual decrease in refractive index from the centre of the core to the core/cladding interface. Theoretical studies[128,19,129] of the impulse response of fibres having a graded refractive index have shown that very low inter-modal dispersion may be obtained with the class of refractive index profiles given by

$$n_r = n_0\left(1 - 2\Delta\left(\frac{r}{a}\right)^{\alpha}\right)^{1/2}, \qquad r \leqslant a \tag{28}$$

$$n_r = n_0(1 - 2\Delta)^{1/2}, \qquad r > a \tag{29}$$

Where n_0 is the refractive index at the centre of the core, n_r is the refractive index at radius r, a is the core radius, and α the index gradient. Gloge and Marcatili[19] have shown that for fibres having a small relative refractive index ($\Delta \ll 1$) and in which other dispersive mechanisms may be neglected, the maximum delay difference (τ_{max}) per kilometre of fibre between the fastest and slowest modes for an optimized profile, is given by

$$\tau_{max} = \frac{n_0 \Delta^2}{8C} \tag{30}$$

Thus, for $n_0 = 1.45$ and $\Delta = 0.01$ the delay spread is 0.060 ns/km which is three orders of magnitude smaller than an equivalent step index fibre. The value of α in equation (28) for such fibres is given by $\alpha = 2(1 - \Delta) = 1.98$ for $\Delta = 0.01$. Other factors determine the actual optimum α value such as the influence of material dispersion[130] and mode coupling[131] and, more importantly, the profile dispersion which describes the variation in mode group velocities with wavelength. A common finding in the theoretical studies of propagation in a multimode graded index fibre is that the actual pulse spreading is highly sensitive to small departures of the α value from the optimum which puts tight restraints on the control of fibre making processes. In a practical multimode fibre the intermodal dispersion is much higher than that predicted by equation (30) possibly as a result of slight variations in α

along the fibre or due to the presence of the refractive index perturbation at the centre of the fibre core which arises as a result of out-diffusion of the dopants during the tube collapse stage of the fibre making process. Calculations by Khular et al.[132] have shown that for a Gaussian-shaped dip in a parabolic index fibre having a $1/e^2$ width to core radius ratio of 0.06 the intermodal dispersion is a factor of three larger than that in a fibre with no dip. In spite of the discrepancy between theoretically predicted dispersion and results in practical fibres, intermodal dispersions much less than 1 ns/km are now commonplace and profiles are being developed to reduce intermodal disperion to negligible values in 50μm core fibres.

Material Dispersion

As the inter-modal contribution to pulse spreading approaches the theoretical minimum as a result of improved control on fibre fabrication processes, the influence of material dispersion on the net pulse spreading in fibres is increasingly important. Furthermore, in single mode fibres in which there is no inter-modal dispersion, material dispersion is the dominant fibre bandwidth limiting mechanism. Material dispersion, which is also known as wavelength dispersion, arises as a result of the refractive index variation of the fibre with the wavelength of the signal source. Practical sources used in optical fibre transmission systems have a finite spectral linewidth and since the fibre refractive index is different for different frequency components of the source there is a finite range of group velocities which results in a spreading of a pulse propagating in the fibre. The pulse spread ($\Delta\tau_{\mathrm{m}}$) due to material dispersion is proportional to the second derivative of the refractive index (n) with respect to wavelength (λ) and is given by[133]

$$\Delta\tau = -\frac{\lambda}{c}\frac{\mathrm{d}^2 n}{\mathrm{d}\lambda^2}\Delta\lambda L \tag{31}$$

The $\mathrm{d}^2n/\mathrm{d}\lambda^2$ for silica has been plotted against λ by Miller[134] and for a GaAlAs laser having a linewidth of 4 nm centred at 900 nm the contribution of material dispersion is 0.35 ns/km, for an LED this is an order of magnitude higher at 3.5 ns/km. The former result is about five times greater than the theoretical limit of intermodal dispersion and for maximum bandwidth capability in fibres it is important to reduce the material dispersion contribution. One attractive way of achieving this is by working at a longer wavelength. Gambling *et al.*[42] have shown that the second derivative of the refractive index and hence the material dispersion in equation (31) falls to zero at a wavelength in the region of 1.2–1.4 μm and then increases with wavelength but with a change of sign (i.e. longer wavelengths are delayed with respect to shorter wavelengths). Thus, for a finite linewidth source the material dispersion contribution[135] would be negligible in this wavelength range. For graded index fibres the exact wavelength at which the material is minimized is determined by the dopant used and its concentration in the core since, with the exception of phosphorous pentoxide, the addition of dopant slightly alters the host material refractive index.[136] For example, the refractive index at 1.2 μm in pure silica is 1.447 and with 7 mol.% GeO_2 doped silica it is 1.464.[137] As shown by equation (31) the pulse spreading due to material dispersion is always a linear function of fibre length and is unaffected by mode mixing effects.

Waveguide Dispersion

If it were possible to eliminate inter-modal dispersion and material dispersion completely, a pulse launched into such a fibre would still exhibit a broadening as a result of a third dispersive mechanism known as waveguide dispersion. The magnitude of the group delay difference due to this effect ($\Delta\tau$) which arises as a result of the wavelength dependence of the group velocity in each mode, may be calculated from the relation[133]

$$\Delta\tau = -\frac{1}{\lambda c} V(\Delta n) \frac{d^2(Vb)}{dV^2} \tag{32}$$

where $\Delta n = n_1 - n_2$, the refractive index difference b is a normalized propagation constant, and V is the normalized frequency which for a fibre of core radius a and core and cladding refractive indices n_1 and n_2, respectively, is given by

$$V = \frac{2\pi a}{\lambda}(n_1^2 - n_2^2)^{1/2} \tag{33}$$

The second derivative in equation (32) has been plotted as a function of V for the HE_{11} mode by Gloge[133] which allows an estimate of $\Delta\tau$ for a single mode fibre to be made. For single mode operation V is less than 2.4048.[18] Thus, for a core diameter of 3 μm and a cladding refractive index of 1.4518 (pure silica at 900 nm) the refractive index difference at 900 nm from equation (33) is 0.018. Substituting this, and the value for $d^2(Vb)/dV^2$ of 0.1 at $V = 2.4048$ into equation (32) gives a group delay difference due to waveguide dispersion of 6.6 ps/nm km. Thus, in most cases this contribution may be neglected but at longer wavelengths where material dispersion is low the waveguide dispersion is of the same order.[138] However, since the material dispersion falls to zero and then changes sign after about 1.3 μm, by working at a slightly longer wavelength than 1.3 μm cancellation of the waveguide dispersion with material dispersion may be obtained.

Dispersion Measurements at STL

The two classes of dispersion measurements at STL are those concerned with an accurate specification of the fibre in terms of a component in a communication system and the more diagnostic measurements in which the actual dispersive mechanisms in fibres are investigated. The latter measurements are aimed at providing, in conjunction with other measurements such as refractive index profile and attenuation measurements, the relevant information for improving the fibre characteristics. The division between the two measurements is not distinct, however, since the results of some measurements are useful both for systems design and for providing insight into the quality of the fibre. For example, dependence of the dispersion on the square root of the fibre length enables the systems designer to calculate the minimum repeater spacing required and allows predictions of the emerging pulse shape at the output to be made. (The dispersion is proportional to the square root of length in fibres which are fully mode mixed and the impulse response for such a fibre, tends to a Gaussian shape.) It also gives the fibre designer an indication of the extent of irregularities at the core/cladding interface and inhomogeneities in the fibre.

System Measurements

As previously discussed, graded index fibres with the optimum refractive index profile have the minimum spread in group delays of the various launched modes, and slight departures from this optimum result in large variations in the group delay spread. Moreover, for a fibre possessing a large variation in group delays amongst the launched modes, the actual pulse width emerging from the fibre is principally determined by the actual modes launched into the fibre. In the absence of mode mixing in fibres, the modal power distribution propagating in the fibre is determined by the physical characteristics of the excitation source and its position in relationship to the core. Therefore in order to make accurate measurements of the dispersion of a fibre for use in a particular system it is a fundamental requirement that the measurement is made under conditions as near identical to the system as possible. If the measurements are not made under similar system conditions, accurate predictions of the fibre performance in the system cannot be made. The largest source of discrepancy occurs when the modal power distribution launched into the fibre or detected at the output of the fibre is not the same as that in the system. Additional sources of discrepancy arise if other system conditions are not met in the measurement of fibre dispersion. Table VIII shows the type of system conditions which must be adhered to in laboratory or environmental dispersion measurements to enable accurate predictions of system performance to be made.

A brief description of the sources of discrepancy which occur if the measurement conditions are different is also given. Other effects such as modal noise and mode mixing are also very important and may depend on one or more of the above system conditions.

Another important consideration in dispersion measurements relevant to system applications is the format of the results obtained, which may be in either the time domain or frequency domain. Time domain measurements

Table VIII

System condition	Source of discrepancy for measurement condition other than system
(a) Optical source, e.g. sawn cavity laser c.w. laser, LED (b) Source spot size (c) Launch conditions, e.g. butt joint, microlens fibre tail	Different modal power distribution launched.
(d) Detector receiving conditions, e.g. butt joint, microlens fibre tail	Different modal power distribution received.
(e) Source output wavelength	Inter modal dispersion is a function of wavelength (profile dispersion).
(f) Source output linewidth	Material disperson directly proportional to linewidth.
(g) Source drive conditions	Presence of chirping gives an apparent increased bandwidth.

are made by measuring directly the temporal spread of a pulse on transmission through the fibre whereas in the frequency domain the response of the fibre is measured over a broad frequency range. Both these modes of fibre dispersion measurement are made at STL and are described separately in the next section.

Time Domain Measurements

Time domain measurements are aimed at determining the impulse response of the fibre. The experimental arrangement for the time domain measurements is shown in Figure 91. The GaAlAs laser source(s) produces a train of pulses each having a duration of 0.4 ns measured at the half-power power points and which are launched into the fibre by means of a one-to-one optical imaging arrangement consisting of two back-to-back microscope objectives having numerical apertures larger than the test fibre. This arrangement simulates the practical butt joint method of launching, commonly used in systems. At the output end of the fibre a similar optical arrangement is used to collect the transmitted pulse and focus it on to a silicon avalanche photodiode detector (D). The output from the detector is then displayed on a sampling oscilloscope. The two-position beam splitters (B) mounted between the launching and receiving optics serve a dual purpose. First, by

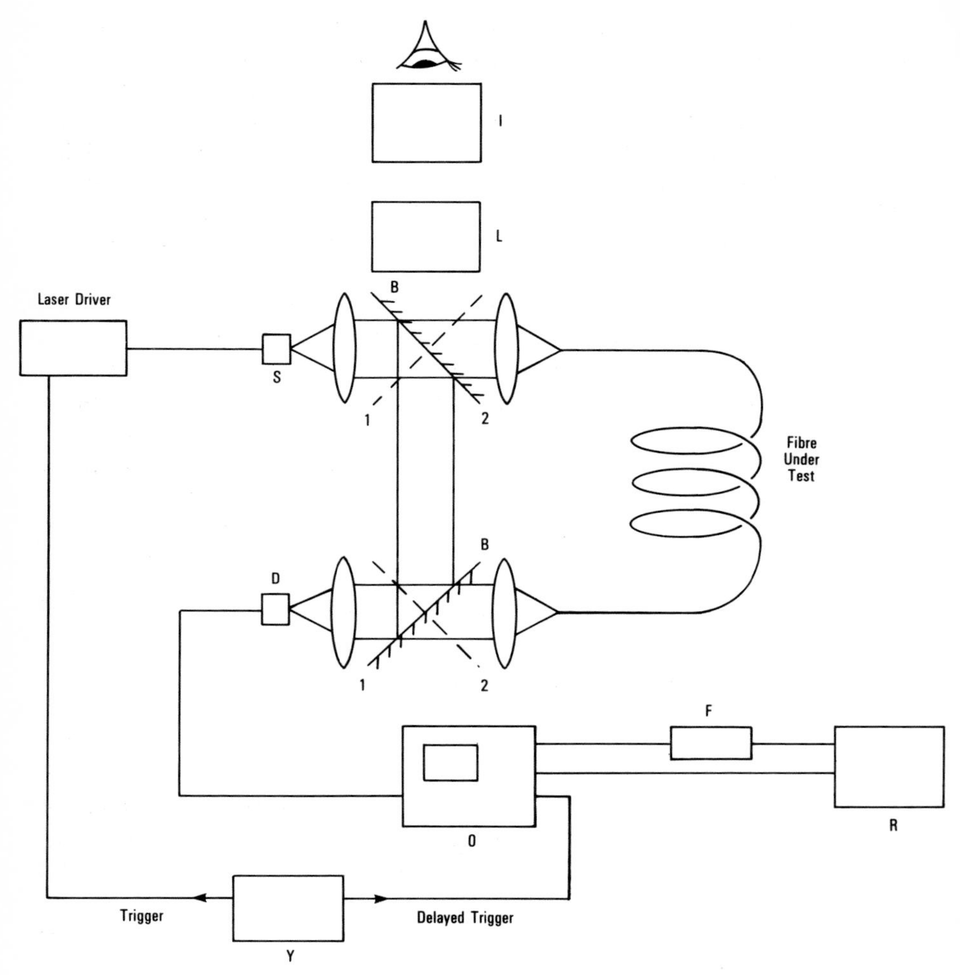

Figure 91. Block diagram of time domain dispersion measurement

means of the illuminator unit (L) and the image converter (I) and with the beam-splitter in position 1 the laser source is accurately focused. Then, by turning the beam-splitter to position 2, the fibre end may be inspected and the laser spot adjusted in position on the fibre core. Both laser and fibre end are supported on micropositioning devices for accurate alignment. Similarly, at the receiving end the image of the transmitted spot may be focused and positioned by adjustment of the micropositioning devices on to the active area of the detector. Secondly, with the beam-splitter at the launching end at position 2 and that at the receiving end in position 1, the input pulse to the fibre may be focused on the detector and the fibre impulse response determined from the measured width of the two displayed pulses. Time domain measurements are performed with either a sawn cavity laser or c.w.

laser depending on the system requirements. The c.w. laser operates at a pulse repetition rate of 10 MHz and the dispersed pulse may be observed on the sampling oscilloscope by triggering directly from a suitable point in the laser drive circuit. Owing to the duty cycle limitations in the sawn cavity laser the pulse repetition rate in this case is 25 kHz and since there is a long delay (5 μs for 1 km of fibre) between launching and receiving the laser pulse a similar delay is required before triggering the oscilloscope. This is achieved by means of a digital delay line which produces an adjustable trigger delay covering the range 0–100 μs in steps of 0.01 μs. For an improved sensitivity in these measurements the oscilloscope is switched to a slow scan and the signal x and y outputs are connected via a low pass filter to an xy recorder.

The impulse response of a linear system is defined as the time response at the output to an infinitesimally narrow pulse at the input. Since the direct pulse in these measurements has a finite width as a result of the finite rise-time of the laser, detector, and oscilloscope, the measured output pulse ($h_2(t)$) is not the fibre-impulse response, but is the convolution of the input pulse ($h_1(t)$) with the impulse response of the fibre ($h_F(t)$). This may be stated as

$$h_2(t) = \int_{-\infty}^{\infty} h_F(t-x)h_1(x)\,dx \tag{34}$$

where x is a dummy variable. This is often written

$$h_2(t) = h_F(t) \otimes h_1(t) \tag{35}$$

where $\otimes$ denotes convolution.

In order to extract $h_F(t)$ from (35) a lengthy mathematical process of deconvolution is necessary. However, if the shape of $h_1(t)$ and $h_2(t)$ are Gaussian or close approximations to Gaussian, equation (35) is given by the simple relation

$$T_2^2 = T_F^2 + T_1^2 \tag{36}$$

and the fibre dispersion over a length L of fibre is given by

$$\frac{T_F}{L} = \frac{(T_2^2 - T_1^2)^{1/2}}{L} \tag{37}$$

where T_1, T_2, and T_F are, respectively, the pulse widths at the input and output and the width of the fibre impulse response measured at some convenient fraction of the pulse amplitudes. Thus, for a Gaussian input pulse of 0.4 ns and a Gaussian output pulse of 1 ns measured at the half-amplitude height over a fibre length of 1.6 km the impulse response of the fibre is 0.92 ns and the dispersion is 0.57 ns/km. Equation (36) must be applied with caution, however, since large errors may occur if the output pulse is non-Gaussian. A typical launched and received pulse for an STL GeO_2 doped silica graded index fibre is shown in Figure 92.

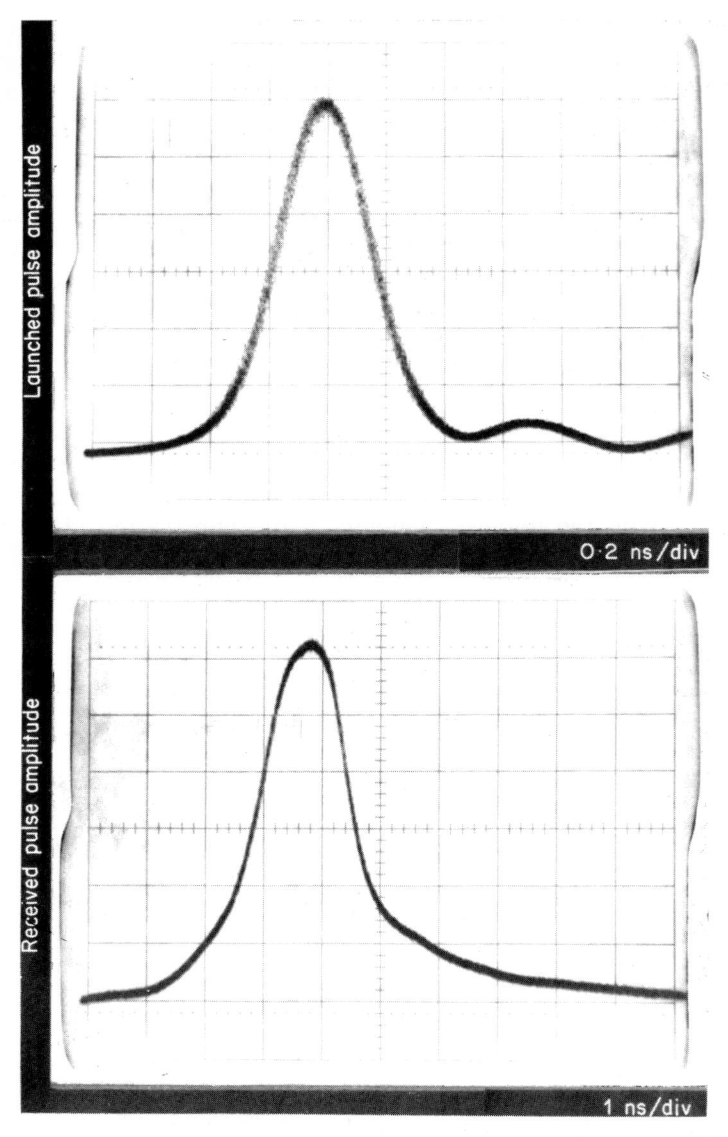

Figure 92. Launched and transmitted pulses in a graded index optical fibre

Since it is not possible to simulate all the conditions under which a particular fibre will be used in a system (for example, on long fibres the measurements in the laboratory are performed with the fibre wound on a drum whereas in the system the fibre is laid out along the communication route) it is necessary to perform measurements in the 'field'. Time domain dispersion measurements have been performed under such conditions as part of the 140 Mbit/s optical link field trials described later in this chapter.

Frequency Domain Measurements

Frequency domain measurements are performed by measuring the absolute response of the fibre over a wide range of frequencies. A block diagram of the two methods of frequency domain measurement used at STL are shown in Figure 93(a) and (b).

The optics and alignment procedures for both these methods are identical to the time domain arrangement described previously. Furthermore, method

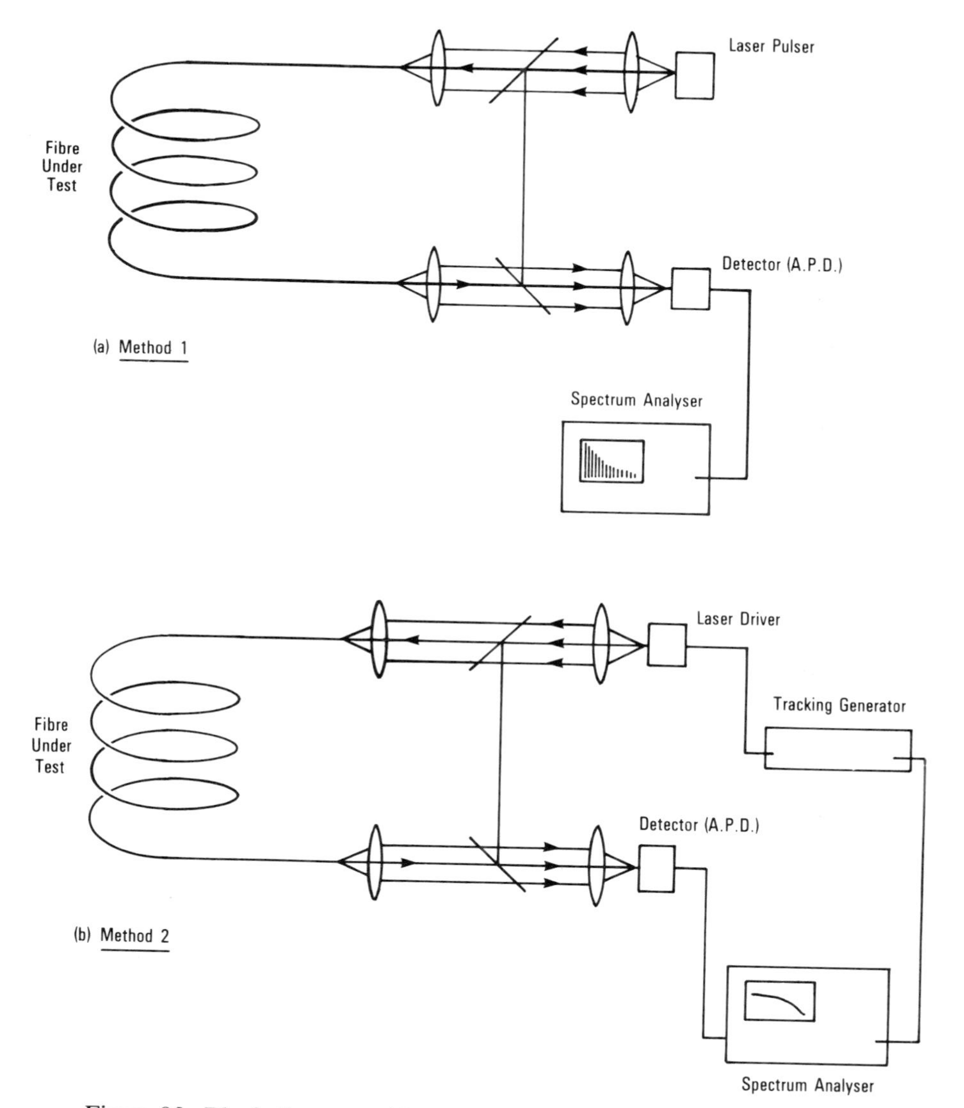

Figure 93. Block diagrams of frequency domain dispersion measurements

1 [Figure 93(a)] employs the same pulsed source as the time domain method but the sampling oscilloscope is now replaced by a spectrum analyser which displays the Fourier frequency components of the pulse separated by the pulse repetition rate, which in this case is 10 MHz.

An alternative method of frequency domain measurement is shown in Figure 93(b). Here, a tracking generator delivers a sinusoidal signal which is repetitively swept from 0.5 to 1300 MHz and is used to drive the GaAlAs laser source and simultaneously trigger the spectrum analyser. The display in this case is a continuous curve of response against frequency. Again in both methods in order to obtain the true response of the fibre allowance has to be made for the finite response of the laser and detector. This is simpler than the deconvolution process required for the time domain measurements since by the convolution theorem, if two functions are convoluted to produce a third function, the Fourier transform of the third function is equal to the product of the individual Fourier transforms of the other two. Thus, by taking the Fourier transforms of the functions in equation (35) we may write

$$H_2(w) = H_F(w)H_1(w) \tag{38}$$

where $H_2(w)$, $H_1(w)$, and $H_F(w)$ are the Fourier transforms of $h_2(t)$, $h_1(t)$, and $h_F(t)$, respectively. Furthermore, $H_2(w)$ and $H_1(w)$ represent the transmitted response of the fibre and the response of the laser and detector, respectively, from which $H_F(w)$ may be readily obtained.

The response of a typical STL fibre is shown in Figure 94 where both methods have been used. The slight discrepancies may be attributed to the

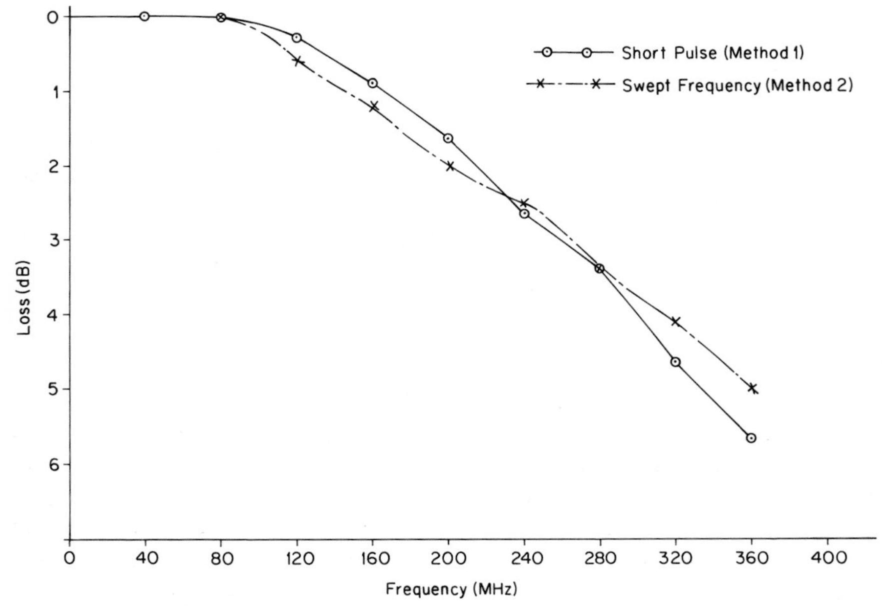

Figure 94. Baseband response of a graded index fibre

fact that different lasers driven under different conditions were used in each measurement which may have resulted in a different launched modal intensity distribution in the fibre.

In addition to these two experimental methods, the frequency domain information may be obtained by computation of the discrete Fourier transform of the pulses obtained in the time domain measurements. The Fourier series of a function $h(t)$ of period T is given by

$$h(t) = a_0 + \sum_{n=1}^{\infty} a_n \cos(nwt) + \sum_{n=1}^{\infty} b_n \sin(nwt) \tag{39}$$

where w is the fundamental angular frequency given by:

$$w = \frac{2\pi}{T}$$

Any harmonic n in equation (39) may combine as

$$a_n \cos(nwt) + b_n \sin(nwt) = c_n \cos(nwt + \phi_n) \tag{40}$$

where $c_n = (a_n^2 + b_n^2)^{1/2}$ and the phase angle $\phi_n = \tan^{-1}(b_n/a_n)$.

Computations of c_n and ϕ_n are readily performed on a digital computer and Figure 95 shows the plots obtained with the Hewlett Packard 9825 mini-computer for the fibre whose experimentally determined frequency response is shown in Figure 94. One of the advantages of computing the frequency response from time domain results is that the phase information may be obtained which gives the system designer insight into the type of distortions expected in the system.

Diagnostic Measurements

These measurements are aimed at providing more detailed information on the dispersive properties of fibres and, apart from routine measurements previously described, include length dependence, the determination of material dispersion, and dispersion measurements relating to the refractive index profile.

Length Dependence

As described in the earlier sections, the largest contribution to pulse spreading in practical multimode fibres is the inter-modal dispersion. Furthermore, for fibres having geometry and refractive index profile variations along their length, mode conversion is induced which alters the length dependence of pulse spreading in the fibre. In order to assess the extent of such variations in the fibre, length dependence measurements are performed. The simplest method of achieving this is by measuring the pulse spreading (or frequency response) on a long length of fibre and then successively

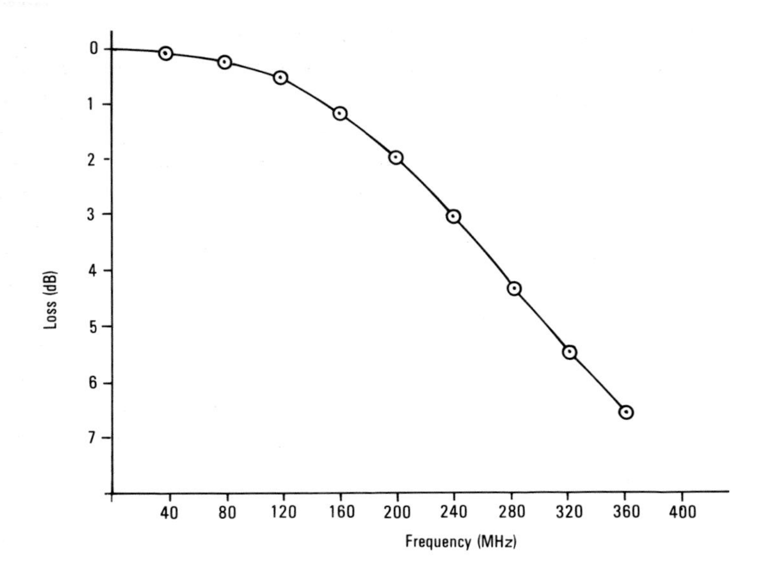

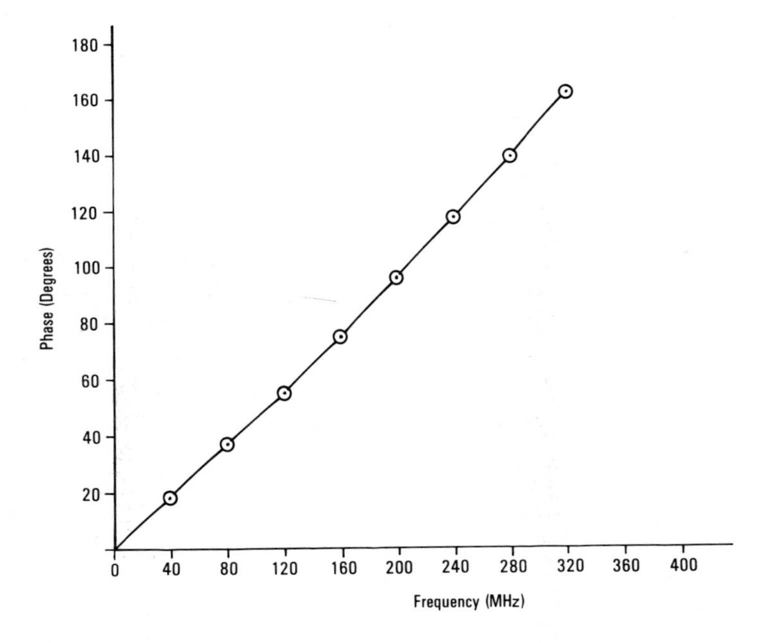

Figure 95. Computed frequency and phase responses of a graded index fibre

removing short lengths and repeating the process. The disadvantage of this method is that it is time-consuming and destructive. A more convenient and nondestructive way of investigating length dependence is the shuttle pulse technique demonstrated by Cohen.[139,140] This method is essentially the same experimental arrangement as that for time domain measurements except that partially transmitting mirrors are contacted on the fibre ends such that the laser pulse shuttles back and forth in the fibre. Thus, for a fibre of length L successive reflections at the output end corresponding to transmission lengths of L, $3L$, $5L$, etc. are detected. In order to pick out a particular pulse on the sampling oscilloscope, the oscilloscope is triggered after the same period of delay by means of an adjustable digital delay time. Using this technique Cohen *et al.*[140] were able to simulate a transmission length of 6.4 km in a fibre of 1.28 km corresponding to a pulse making five traversals of the fibre length. Using a shorter length of 106 m, however, up to ten traversals were possible before the signal-to-noise ratio was below a tolerable level. Similar measurements have been performed at STL but with an additional modification which allows measurements of the pulse widths at distances corresponding to transmission lengths of $2L$, $4L$, $6L$, etc. Referring to Figure 91 with the beam-splitter at the launch end in position 1 and at the receiving end in position 1, a path is arranged for pulses which have been reflected back towards the launch end. Results with this arrangement obtained on an initial length of 200 m of multimode graded index fibre are shown in Figure 96 where the slope is 1.0 suggesting that little mode coupling occurs along this length. The main drawback of this method is that mode conversion may occur at the mirrors, if the mirrors and fibre ends are not in good contact and if the fibre ends are not perpendicular to the mirrors.

Material Dispersion

Material dispersion may be determined by measuring the relative delay between two short laser pulses of different wavelengths.[141,142] For multimode fibres the accuracy of this method is limited by the extent of inter-modal dispersion. However, in single mode fibres material dispersion is the dominant pulse broadening mechanism and the broadening observed at one wavelength allows an accurate estimate of the material effect at that wavelength to be made, provided the input pulse is sufficiently narrow. With the experimental arrangement of Figure 91 pulse broadening measurements were performed on a 1.078-km length of single mode fibre. The linewidth of the laser in these measurements was 3 nm centred at 844 nm and the laser pulse width 0.30 ns. The deconvolved pulse spreading in this fibre was 0.28 ns giving a value of material dispersion at 844 of 85.8 ps/nm·km. We may compare this with that predicted by equation (31). At 844 nm the value of $\lambda^2 d^2n/d\lambda^2$ is ~ 0.02 for pure silica giving a value of material dispersion of 80.0 ps/nm·km. The difference in the calculated and measured values may be attributed to the finite waveguide dispersion and the fact that the core is GeO_2 doped silica which slightly alter the value of $\lambda^2 d^2n/d\lambda^2$.

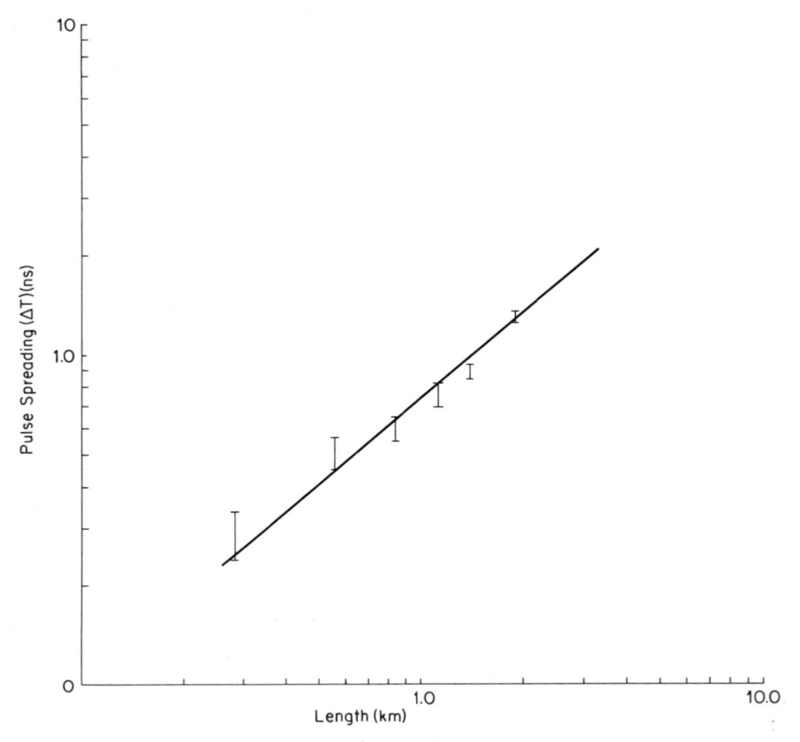

Figure 96. Shuttle pulse measurements on a graded index fibre of initial length 280 m

Profile Related Dispersion Effects

It has been shown by Olshansky and Keck[130] that the optimum refractive index gradient in a multimode graded index fibre is a function of wavelength owing to the variation in refractive index with wavelength of the materials used in the core. Furthermore, for fibres having refractive index profiles described by equations (18) and (19) if the α value is greater than the optimum, higher order modes travel slower than lower order modes whereas the converse applies for α values less than the optimum. Therefore as shown by Cohen[143] by measuring the relative delays of higher order modes relative to lower order modes at a particular wavelength it is possible to determine whether the fibre profile is above or below the optimum for that wavelength. The main problem with this technique lies in finding a suitable means of selective detection of the different modes. In step index fibres this may easily be achieved since there is a simple correspondence between the angle of propagation of a ray relative to the fibre axis and the angle launched or received. Thus, by inserting irises or annular masks in the collimated beam in the launch or receiving optics of Figure 91, a particular range of modes may be excited or received. In graded index fibre the angle of the rays

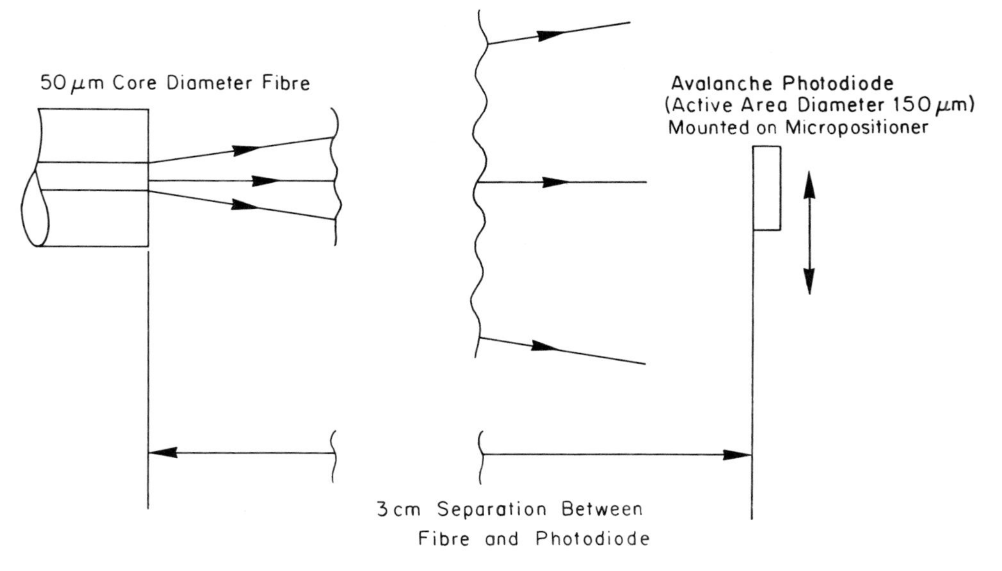

Figure 97. Mode selection by scanning in the far field

relative to the fibre axis vary along the length of the fibre and the simple correspondence between the launched or received rays does not exist. However, the limiting angle of lower order modes is less than that of higher modes, therefore by blocking off the low angles at the output sufficient mode separation is achieved. Also, by blocking off the high angles a mixture of low and high order modes is received; however, most of the emergent energy at low angles is from the low order modes. With this technique, by observation of the emerging pulse shapes and pulse widths, Cohen was able to determine whether the higher order modes arrived before or after the lower order modes thereby assessing the departure of the fibre α value from the optimum. Similar conclusions regarding the α values in STL fibres have been made in some initial measurements using both Cohen's technique and the method of mode separation at the output shown in Figure 97. Here the optics at the receiving end in Figure 91 are removed and the avalanche photodiode is scanned across the far field of the emerging light from the fibre. Although the sensitivity is reduced somewhat, measurements on fibre lengths of up to 1.6 km have been made using this technique.

ON-LINE MEASUREMENTS

On-line attenuation measurement techniques have been introduced to both the fibre pulling[144] and extrusion coating processes. These techniques have provided the following:

(i) an immediate feedback to the fibre-maker or coater so that optimum processing conditions can rapidly be established;

(ii) an alternative measurement to the two-point spot measurement previously described (see Fibre Attenuation measurements), thus reducing the measurement load;

(iii) further loss diagnostic information not previously available, in particular in the fibre pulling case a complete attenuation versus length profile is obtained; and

(iv) a basis for quality control in manufacture.

On-Line Attenuation Measurement During Fibre Pulling

The measurement system is shown in Figure 98. Part of the light produced in the preform/fibre white-hot zone is guided into the core of the fibre. The light signal from the fibre end is collected by a photodetector which is mounted on the fibre take-up equipment. To locate the fibre in the photodetector housing it is necessary to stop pulling after a few metres, position the fibre, and then restart without breaking the fibre. This can readily be done with present silica fibre pulling techniques—both CVD and plastic clad.

As the fibre is pulled the detector rotates with the take-up drum and the electric signal is fed via a slip-ring arrangement to a d.c. amplifier with variable gain. The amplifier output is then displayed on a chart recorder.

A bandpass interference filter, centred at 850 nm of 40 nm bandwidth and fully suppressed at other wavelengths, is located in front of the photodetector so that the system only monitors attenuation in the wavelength range of prime interest for optical communication. The output signal can be observed on the chart recorder and the attenuation either calculated from the change in signal output plotted against the number of metres of fibre pulled, or

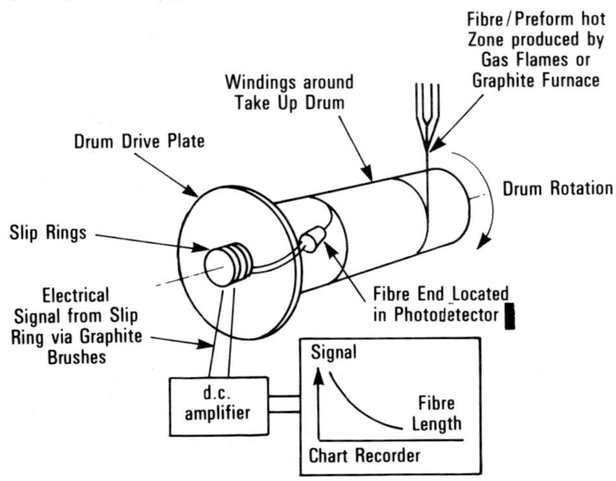

Figure 98. System for on-line attenuation measurement during fibre pulling

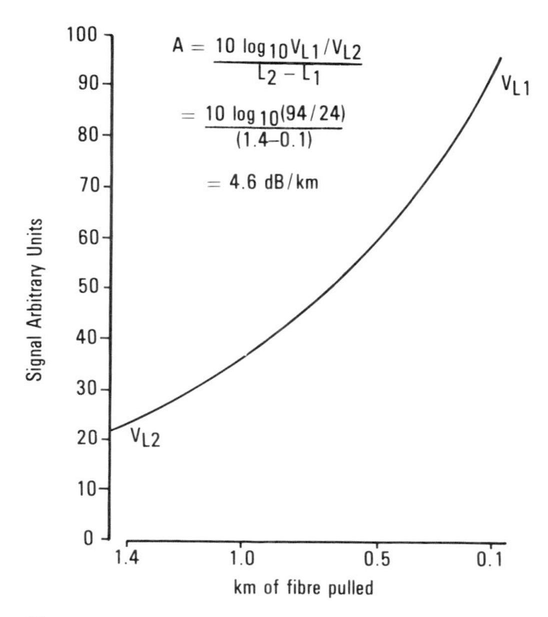

Figure 99. Typical plot of signal versus fibre length during fibre pulling

directly displayed on a digital read-out by operating on the signal output and the fibre length with a logarithmic amplifier and the accompanying logic circuitry.

Figure 99 shows a typical plot of signal versus fibre length, together with the corresponding calculation of attenuation. The present system can measure an overall fibre attenuation of 30 dB, and this could be increased by use of more sophisticated electronics, the typical short length signal power reaching the detector via the filter being of the order of 300 nW.

A complete attenuation/length profile of a fibre can better be seen by plotting the output signal on a logarithmic scale against the length of fibre pulled. A normal practical example is shown by curve (a) in Figure 100 where the signal has been plotted every 100 m. The attenuation of a fibre is given by equation (20).

A straight-line plot corresponds to constant fibre attenuation and, apart from the first few tens of metres, the attenuation of fibre (a) can be seen to be constant over the whole 2.2 km.

The apparent higher attenuation over the first few tens of metres is expected and is due to high order modes reaching the detector at the start of the fibre pull. These high order modes are preferentially attenuated as the fibre becomes longer and so indicate a higher initial loss. Thus, a two-point measurement obtained by cutting back to a short length can give a higher measured loss than one obtained by cutting back to a length where modal equilibrium has been established in the fibre.

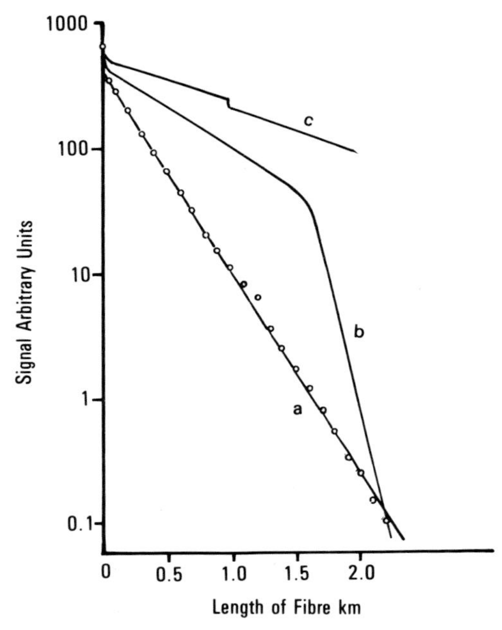

Figure 100. Signal plotted on logarithmic scale versus fibre length

Cladding light can also contribute to the apparent higher attenuation but this is very quickly stripped out by microbending caused by winding the bare fibre on the drum. It should be noted that it is standard practice now to in-line dip coat all fibres, but there is some 10–20 m of bare fibre pulled before coating commences.

Curve (b) in Figure 100 demonstrates one of the advantages of this measurement method as a continuous assessment of a fibre pull. In this hypothetical case there is a drastic increase in attenuation beyond 1.5 km. If the pulling conditions had been the same throughout, the fibre-maker knows that a processing problem exists which may be traced back to the preform. A normal two-point measurement of the two fibres (a) and (b) would, in fact, have given the same fibre loss and no indication of nonuniformity. This method will also identify the position of any step faults in the fibre as they occur so that they can, if necessary, be removed with the minimum loss of useful fibre. This is demonstrated by the hypothetical curve (c) in Figure 100.

A large number of fibres have now been monitored in this way and the results obtained agree to better than 1 dB/km with corresponding off-line differential measurements. These differential measurements were carried out by overfilling the numerical aperture of the fibre (typically 0.2) and cutting back to about 3 m. Figure 101 shows a histogram of the comparison of results of on-line and off-line measurements for 30-μm core graded index CVD fibre, and includes results on bare fibres and in-line dip coated fibres,

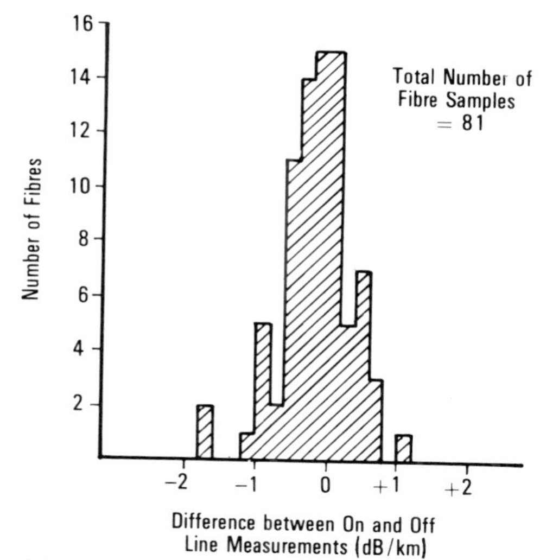

Figure 101. Comparison of on-line attenuation monitor during fibre pulling and two-point measurements

manufactured by using both gas flames and a graphite furnace as the heating method.

Thus, a method of continuously monitoring the attenuation of an optical fibre during the pulling stage of its manufacture has been developed which gives useful information for loss diagnostic purposes and eliminates the need for subsequent two-point measurements if a complete spectral attenuation curve is not required.

On-Line Attenuation Monitor During Extrusion Coating

The measurement system is shown in Figure 102. Light is focused into a prepared end of the fibre to be coated which has been accurately positioned along the axis of rotation of the payoff reel. A 20-W tungsten halogen lamp is used as the light source. The other end of the fibre is fed through the extruder and after an initial extrusion setting-up period the coated fibre is cut and the end is prepared and located along the axis of the take-up reel in close proximity to an interference filter in front of a large area silicon photodetector. The interference filter is an 850-nm bandpass filter of 40-nm bandwidth.

Once the initial setting-up period is completed the fibre will be coated all the time whilst the second fibre end is being prepared. Therefore, provision has to be made for the temporary accumulation of about 20 m of coated fibre before it is wound on the take-up reel. This can simply be a large box. Of course, if immediate reeling up is not required then the coated fibre can be collected in loose coils in a box and the fibre end can be directly located in the detector.

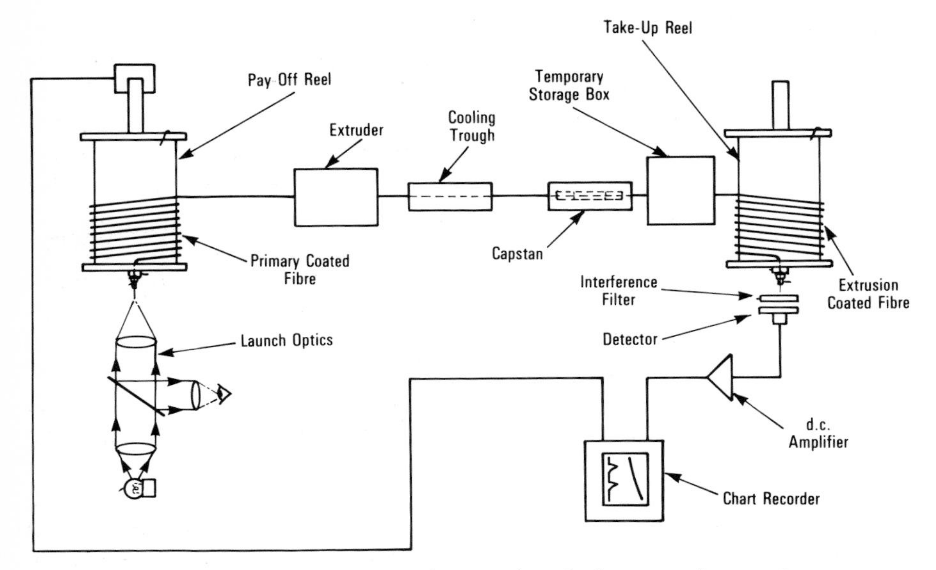

Figure 102. On-line attenuation monitor during extrusion coating

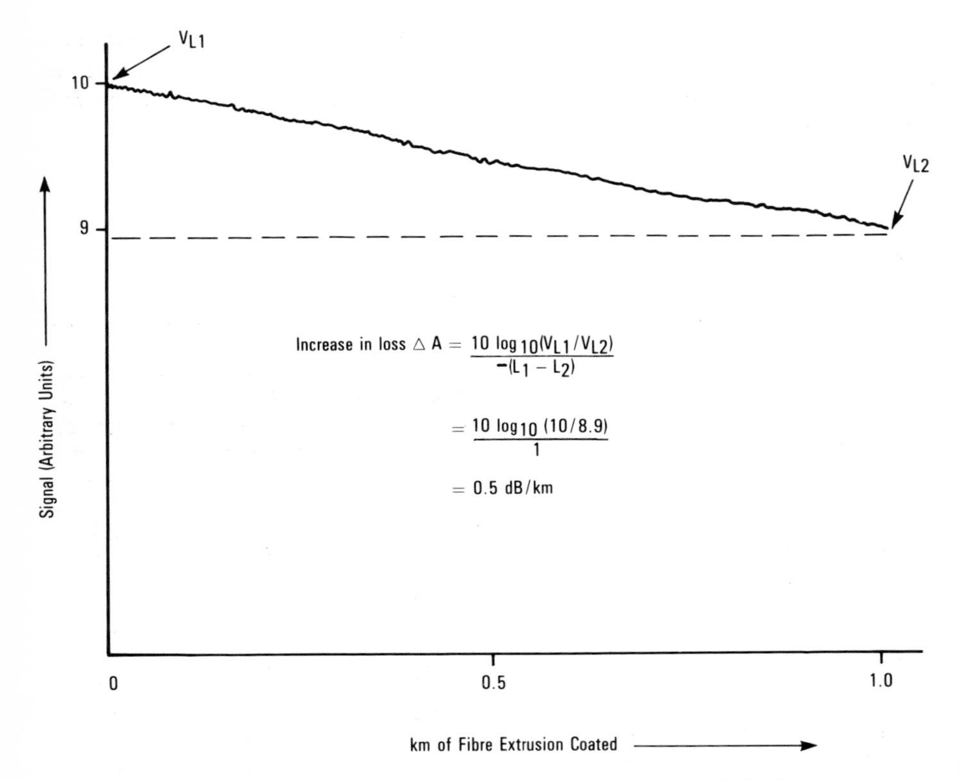

Figure 103. Example plot of signal versus length of fibre coated during extrusion

After detection the signal is amplified and displayed on a chart recorder. A typical plot of signal versus length of fibre coated is shown in Figure 103 together with the corresponding calculation of increase in fibre loss due to extrusion. It should be emphasized that this technique determines any increase or decrease in loss due to extrusion coating but it does not provide an absolute measurement of coated fibre loss. However, the two on-line techniques together allow the absolute loss of the primary coated fibre to be measured and the absolute loss of the extrusion fibre to be deduced. From Table IX it can be seen that the predicted final loss (A+B) and the actual final loss (C), as determined by the normal two-point measurement, overfilling the numerical aperture of the fibre and cutting back to 3 m, are in excellent agreement and certainly within the quoted accuracy of the on-line attenuation measurement during pulling of ±1 dB/km. So these two on-line techniques eliminate the need for any two-point measurements at a single wavelength until the fibre has been cabled.

Figure 104 demonstrates the usefulness of the technique to feed back rapidly to the fibre-coater the effect on the fibre loss of varying certain process parameters. In Figure 104 different cooling taps in the extrusion cooling trough have been switched off at points A, B, and C and the effect on loss is quickly and clearly defined.

Helped by the type of feedback produced by this measuring technique, extrusion coating has considerably improved since the results shown in Table IX were obtained and the mean increment in fibre loss on extrusion coating is now better than 0.2 dB/km.

Table IX

	Primary coated fibre	Extrusion coated			
Length in m.	loss A (in dB/km)	On-line increase in loss B (in dB/km)	Predicted final loss (A+B) (in dB/km)	Off-line final loss C (in dB/km)	Difference (A+B)−C (in dB/km)
1000	4.8	0.5	5.3	6.1	−0.8
1305	3.7	0.2	3.9	4.2	−0.3
1186	5.7	0.1	5.8	5.7	+0.1
1374	4.4	1.3	5.7	5.8	−0.1
1135	5.4	1.2	6.6	6.4	+0.2
1449	4.3	0.2	4.5	4.1	+0.4
492	5.1	3.4	8.5	8.5	0.0
864	4.2	0.3	4.5	4.1	+0.4
1278	4.9	0.3	5.2	5.0	+0.2
918	4.8	1.1	5.9	5.5	+0.4
883	5.4	0.1	5.5	5.4	+0.1

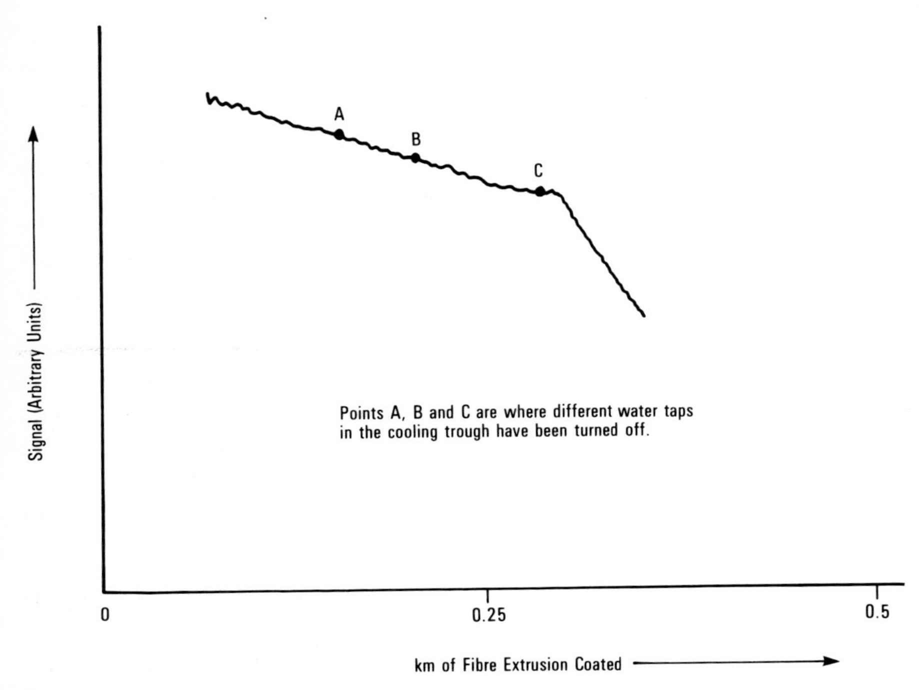

Figure 104. Plot showing the effect on fibre loss of turning off cooling taps during extrusion coating

FIELD MEASUREMENT OF FIBRE OPTIC CABLES

In addition to laboratory measurements there is a growing need to determine the attenuation and pulse dispersion of fibre optic cables in the field. Equipment described below was specifically developed and used to assess cables installed in British Post Office ducts as part of the 140 Mbit/s optical field demonstration.[145] The main equipment design criteria and the measurement requirements would be common to most practical installations.

The 140 Mbit/s field demonstration consists of a total 9 km of cable installed in standard British Post Office ducts between telephone exchanges in Hitchin and Stevenage. It is made up in three sections, repeaters being positioned at the junctions of the sections. The first and third sections each contain three spliced cables and the middle section four, the total length of each section being about 3 km. Measurements of attenuation and pulse dispersion in the time domain were required and performed on each of the ten cables before and after installation and on the jointed cables within the three sections after each splice joint had been completed. The cable comprises three optical silica fibres of 30-μm core and graded index, two to be used in the system and one spare, one filler and four copper wires for power to the repeaters and for engineer ordering.

For much of the work the equipment was housed in two small vans of 0.9 tons capacity stationed at either end of the cable under test, but it had to be sufficiently portable to allow work to be carried out in positions where vehicle access is not convenient. Communication between the two working points was by field telephone, using two of the four copper cores in the cable. Power for the equipment was supplied by 250-W petrol generators, two being required at the launch end and one at the receive end. Each van was equipped with the necessary hardware to gain access to the cables in the ducts, complying with BPO safety regulations. This hardware included items such as a water pump, gas detectors, manhole guards, pedestrian barriers, road signs, and a tent.

The attenuation and dispersion measurements were performed separately on each fibre in the cable and in the first method evaluated, different light sources and detectors were used for the two types of measurement. However, at a later stage an equipment was successfully introduced in which a common source was used for both operations switching between different source drive circuits. This latter method will be referred to as the 'overlay' measurement.

The first field measurement method is shown in Figures 105 and 106. It can be seen that the attenuation part of this equipment is basically the same as the laboratory spot attenuation arrangement previously described (see section on Fibre Attenuation Measurements). As in the laboratory the measurement of attenuation was carried out using the cut-back technique, in which the optical power from the full length of the cabled fibre is compared with that radiating from a short length of about 3 m under identical optimized launch conditions. This optical arrangement produced a launched spot size of 95 μm on the prepared 30-μm core fibre input end with a numerical aperture of 0.24, greater than that of the fibre (typically 0.2). At the other end of the cable the fibre end, after preparation using the 'scratch and pull technique', was placed in close contact with a large area silicon photodiode. Using the arrangement shown schematically in Figure 106, path 1, a signal proportional to the light intensity was displayed on a digital voltmeter. Further ends were prepared until three ends were obtained with a signal repeatability of better than 1 per cent. With similar detection equipment at the launch end, the same procedure was carried out for the short length measurement, taking care to strip out all the cladding light. After each day of measurements the two sets of detection equipment were cross-calibrated to obtain a sensitivity ratio. This ratio changed typically by only 0.2 per cent per day. The dynamic range of this field attenuation equipment, using an 850 ± 5 nm interference filter, is 30 dB. This is less than the corresponding laboratory measurement since phase sensitive detection is not used. It was thought that a means of feeding a reference signal from one end of the cable to the other would not always be available and so simple a.c. detection was employed. The error in these field measurements is estimated to be ± 0.4 dB/km for a 1-km cable length compared with 0.2 dB/km for the

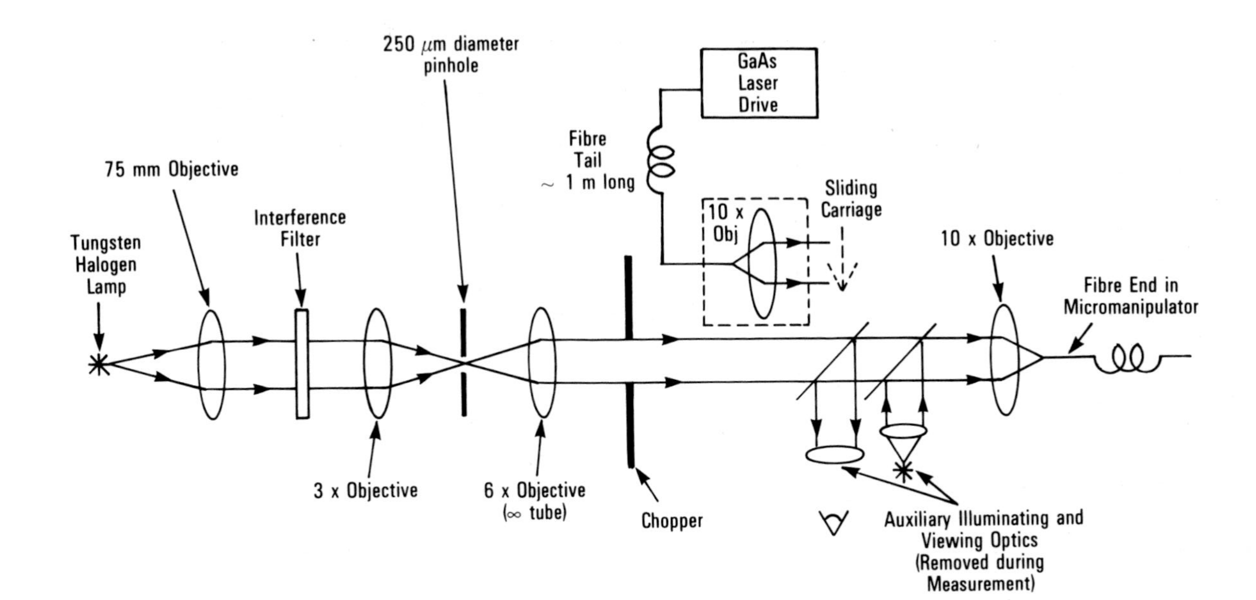

Figure 105. Optics of the field measurement launch equipment

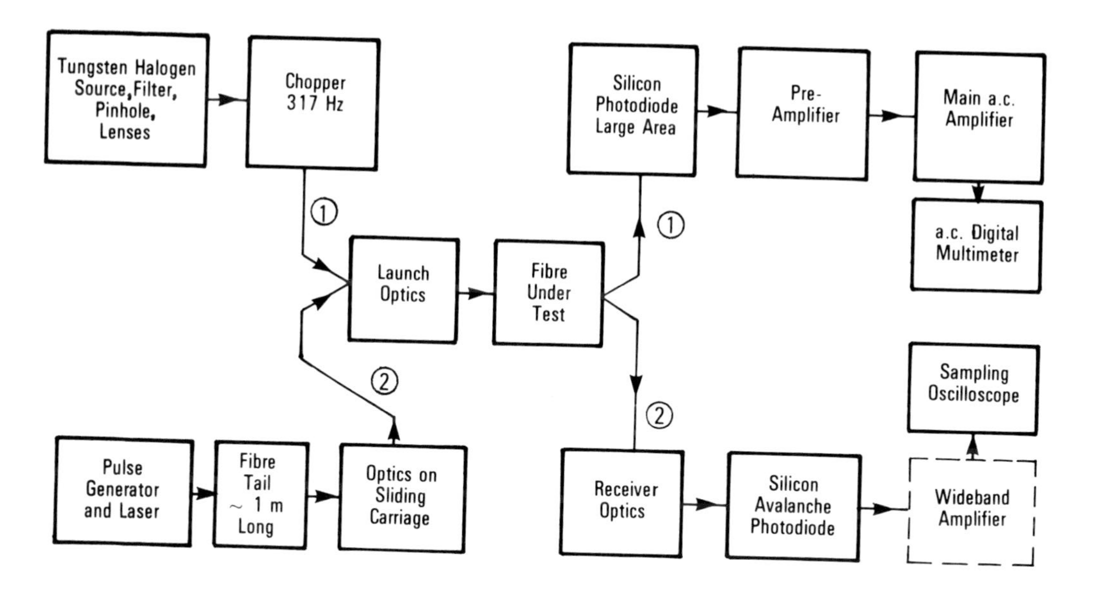

Figure 106. Block diagram of the field measurement equipment

laboratory measurement. This increase in error is mainly due to the larger ambient temperature fluctuations encountered in the field which leads to sensitivity variations.

Dispersion measurements in the time domain were performed by launching a narrow pulse into the fibre and, after detection, observing the output pulse on a sampling oscilloscope. The optical source consisted of a c.w. GaAlAs laser driven by fast switching step-recovery diodes, providing a train of symmetrical pulses having a width of 0.35 ns and a repetition rate of 13 MHz. At the output face of the laser a 1-m length of coated fibre was butted, the numerical aperture and core diameter being similar to that of the fibres under test. The other end of the short fibre tail was terminated in the male half of a demountable connector which was mounted rigidly at the focus of a X10 microscope objective. This latter assembly was placed in a pre-set position by means of a sliding carriage and stop. (See Figures 105 and 107). By using one-to-one launching optics, measurements were made under conditions similar to those of the actual system. At the receiving end the transmitted output was focused onto a high speed silicon avalanche photodiode by a similar one-to-one optical arrangement and the output from the photodiode was displayed and photographed on a sampling oscilloscope (see Figure 106, path 2). For measurements over the whole 3-km spliced sections, a wideband amplifier was included after the photodiode to provide

Figure 107. Detail of launch equipment on sliding carriage

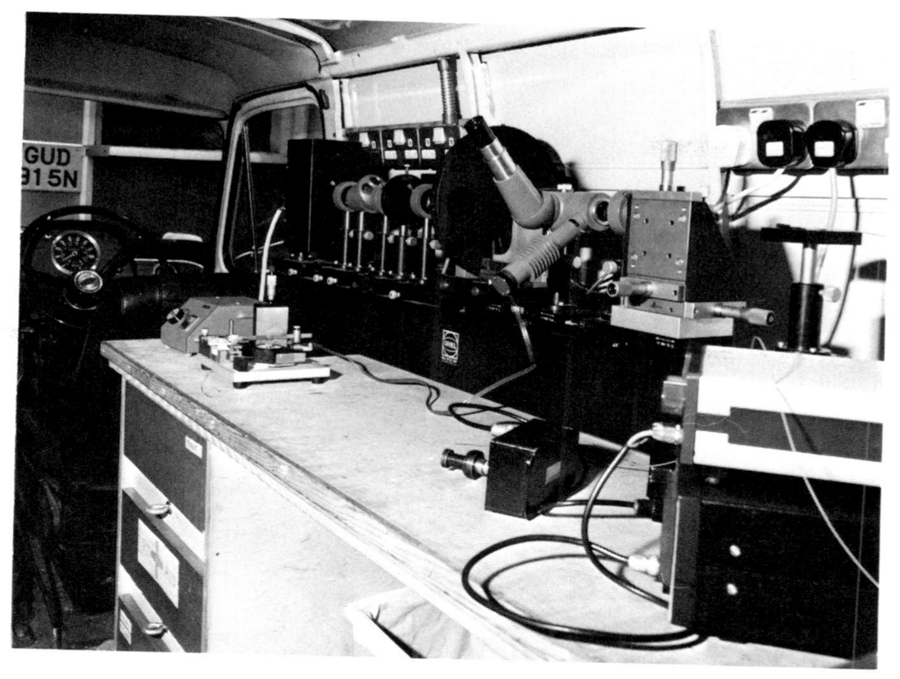

Figure 108. Launch equipment set up in mobile test facility

sufficient signal to allow accurate measurements of the half-widths of the pulses to be made. For an output pulse width of 3 ns, the dynamic range of the dispersion equipment is 30 dB with the amplifier included and a gain of 100 in the avalanche photodiode.

The launch dispersion and attenuation equipment, set up in one of the vans, is shown in Figure 108. The sequence of measurements with this method is to complete the dispersion first, aligning the laser spot on the fibre input end using an infra-red viewer and maximizing the receive pulse. Then the laser source is removed from the optical axis, the receiver equipment is interchanged, the fibre input end is realigned for maximum received mean power, and the attenuation is measured.

The 'Overlay' Equipment

The dispersion pulser described above was later modified to give attenuation data as well. In particular, a second drive circuit was added to the laser source to provide a 300-Hz square wave output suitable for attenuation measurements. This new overlay method thus enables attenuation and dispersion measurements to be made using a common source of the system type by simply switching from one drive circuit to the other.

In the attenuation mode the mean power level is of the order 100 μW which gives a dynamic range of measurement of 60 dB. The mean power

Figure 109. Overlay dispersion attenuation source in box

stability is better than 2 per cent in 30 min which is the order of time of a cut-back measurement.

Using the overlay laser and driver shown in Figure 109, there is a considerable reduction in the amount of equipment required. In addition, since there is no need to realign the fibre input end between dispersion and attenuation measurements there is a significant saving in setting up time and it is a more realistic measurement with respect to system performance.

In the 140 Mbit/s field demonstration, after joining the separate cables to produce 3-km sections, each section end was terminated in a jewelled ferrule located in the female half of a demountable connector. Each connector had a facility for fibre adjustment in the X–Y plane. A cut-back attenuation cannot be employed once both ends have been terminated. Instead, the overlay system is used to make a nondestructive measurement. The male connector half on the end of the overlay tail is plugged into the female connector half on the cable section. The long length signal is optimized and detected via a 200-μm short fibre tail and the short length signal is obtained by directly presenting the overlay male connector half to a photodetector at the launch end. The loss of the section can be calculated allowing for the insertion loss of the connector which is of the order 1 dB. The dispersion is measured as before. This equipment is regularly used as a means of checking the terminated cable performance of the 140 Mbit/s system.

Figure 110. Optical fibre tester

AN INSTRUMENT FOR TESTING TERMINATED OPTICAL FIBRES

For general optical communication systems work an equivalent of the classical electrical 'multimeter' is required. An early version of such an instrument[146] for the nondestructive measurement of attenuation in terminated optical fibres is shown in Figures 110 and 111. The instrument has its own GaAs laser source and, in addition to fibre attenuation measurements, it is able to check system lasers and detectors on a go/no-go basis.

Nondestructive attenuation measurements can be made on fibres with a range of core sizes, between 0 and 50 dB with an accuracy of ± 1 dB. The measurement method compares the signal from the test fibre with that from a separate short reference length of a similar fibre. All test and reference fibres must be terminated in the standard ferrules for which the instrument is calibrated and care must be taken to avoid the transmission of cladding

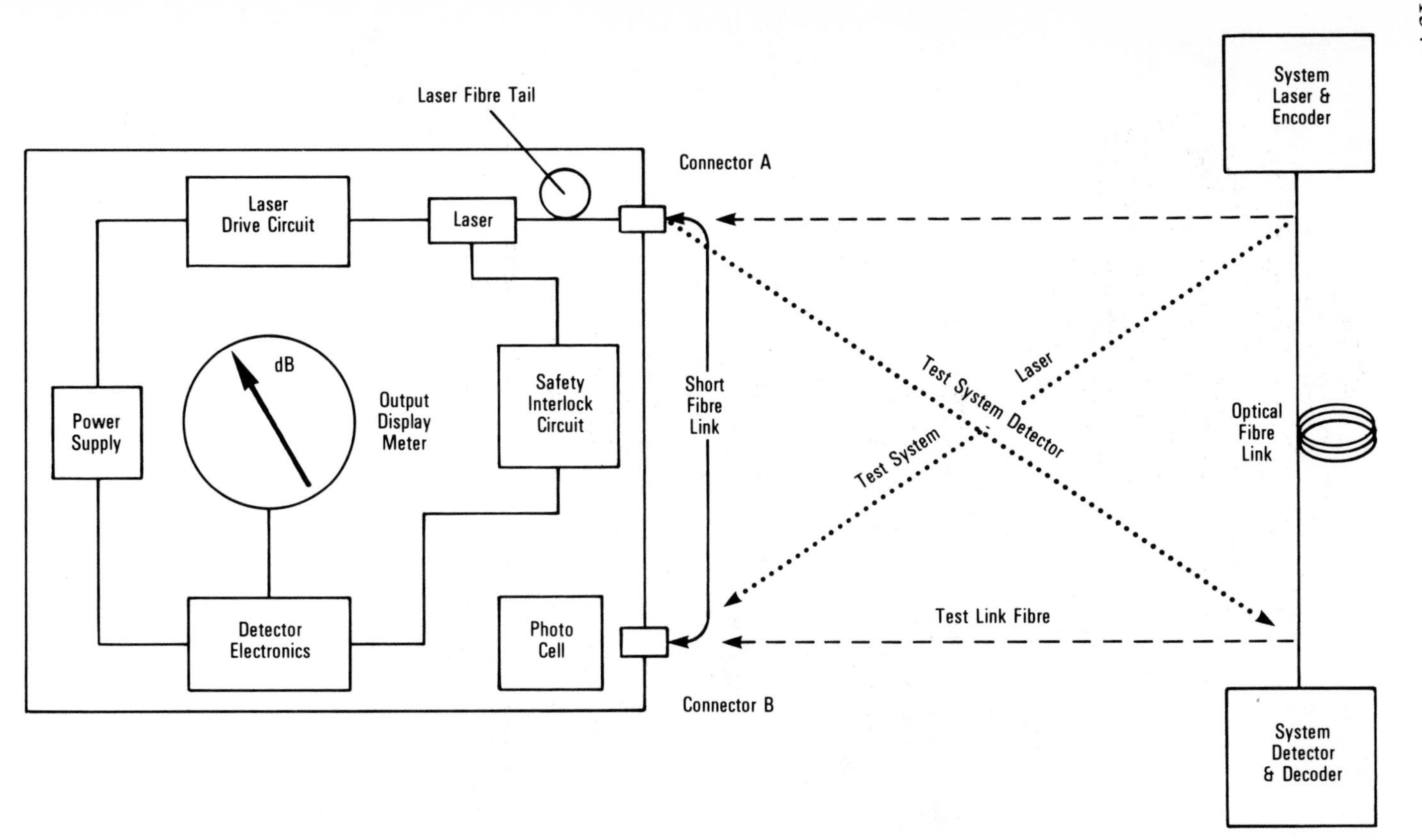

Figure 111. Block schematic of an instrument for testing terminated optical fibres

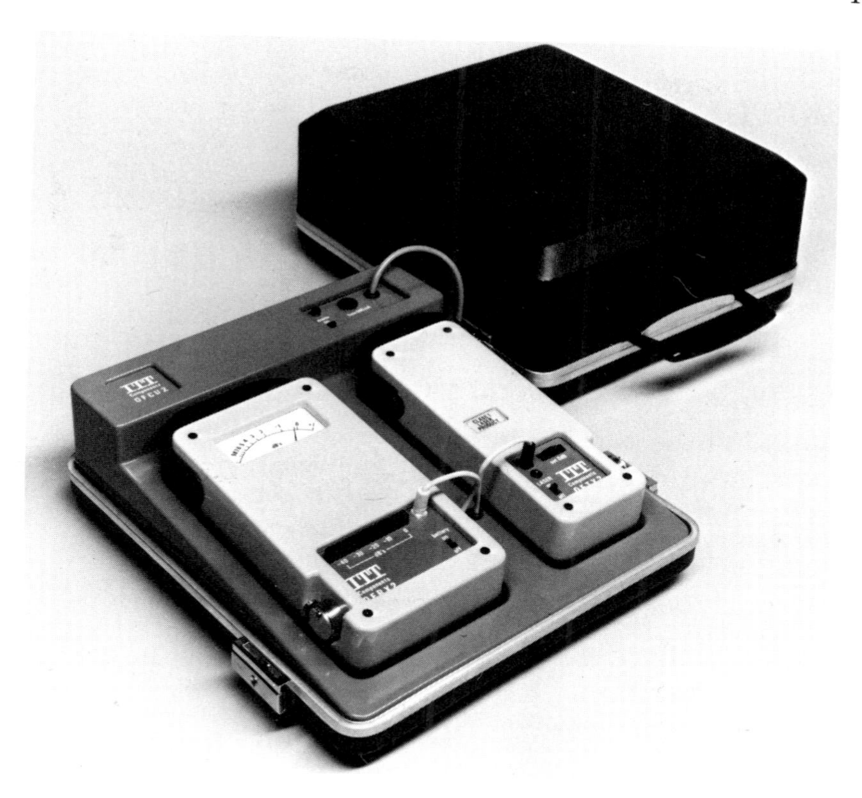

Figure 112. Optical fibre test set

modes through the reference fibre which would lead to excessively pessimistic attenuation values.

A second version of this instrument is shown in Figure 112 in which the transmitter and receiver sections are separate. After a simple calibration procedure the two units can operate independently and can be connected to the ends of an installed cable which are a considerable distance apart.

CHAPTER VII

Semiconductor Laser Light Sources for Optical Fibre Communications

INTRODUCTION

Semiconductor lasers are rapidly being developed to a stage at which they will make ideal sources for wideband optical communications systems. Despite early concern about reliability, recent investigations of degradation mechanisms have led to considerable improvements in working life, and many laboratories have reported 20,000 hours continuous wave (c.w.) operation. Extrapolations now predict lives well in excess of 100,000 hours.[147–149]

Although many of the proposed optical communication systems can operate satisfactorily with a simple light emitting diode (LED) or high radiance LED, the advantage of a laser's much higher radiance frequently makes it a more attractive proposition even in relatively narrow bandwidth or short distance systems. For example, if a higher power can be launched, more loss can be tolerated in cables and connectors, perhaps leading to an overall cost advantage. However, it is in wideband, long distance systems that the higher power output, higher modulation efficiency, narrower spectral linewidth, and narrower emission patterns of the laser make it the only practical source.

This chapter describes the fabrication and performance characteristics of stripe geometry lasers, suitable for use with small core single mode or graded index fibres in systems requiring modulation at rates up to and in excess of 250 Mbit/s. Also included is a summary of recent developments likely to lead to performance improvements in future commercial lasers.

LASER STRUCTURE

A stripe geometry SiO_2 insulated laser[150] is shown in Figure 113. The junction structure is a conventional double heterostructure with $Ga_xAl_{1-x}As$ passive layers containing about 35 per cent Al and an active region with about 5 per cent Al. This latter quantity minimizes the strain in the active layer caused by thermal expansion mismatch between the GaAs substrate and the $Ga_xAl_{1-x}As$ passive layers. The junction is produced by liquid phase epitaxy using the multiple bin sliding graphite boat shown in Figure 8

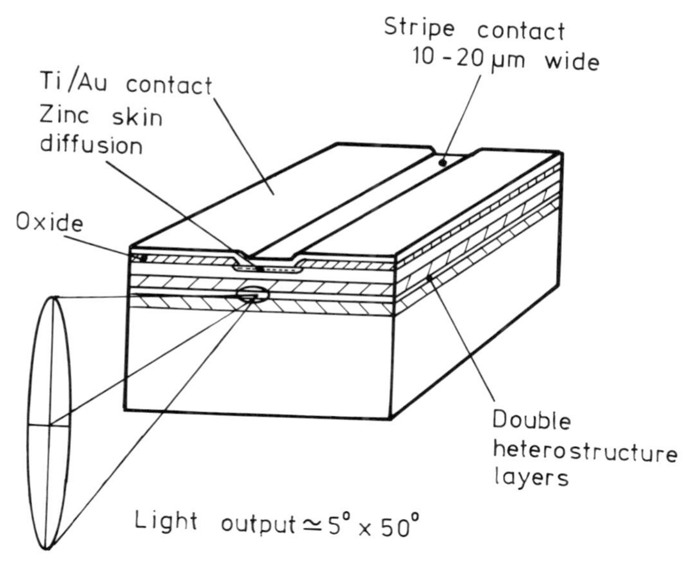

Figure 113. SiO_2 insulated stripe geometry laser

(Chapter I). Typical layer dopings are shown in Table X. One important design consideration is that the p-GaAl/As layer should be doped sufficiently to prevent electron leakage at high temperature[151] but not so highly doped as to cause excessive transverse conduction of current away from the stripe region.

After growth the epitaxial wafers are carefully selected for freedom from crystallographic defects, such as dislocations and stacking faults, because they cause rapid degradation during c.w. operation.[152,153] Suitable wafers are coated with SiO_2 using radio frequency plasma deposition and ~20 μm wide stripes are then opened using conventional photolithography. A shallow Zn diffusion produces a highly doped p^+ surface in the stripe region which assists the formation of a low resistance ohmic contact by subsequent evaporation of Ti–Au layers. After forming the contacts the wafer is cleaved

Table X
Layer compositions of a typical double heterostructure Ga_xAl_{1-x} wafer

Layer		Dopant concentration in melt	Carrier concentration estimate (cm^{-3})
Substrate	0.0		$\sim 10^{18}$
n-passive	0.35	5 at.% Sn	$\sim 10^{17}$
Active	0.05	0.3 at.% Si	$\sim 10^{17}$
p-passive	0.35	0.3 at.% Ge	$\sim 10^{17}$
p-contact	0.0	0.3 at.% Ge	4×10^{17}

into strips to produce the reflecting surfaces, then cleaved again to form dice approximately 400 μm square with the stripe contact lying across the p-side of each die. The dice are bonded to a copper heat sink using indium solder; a soft metal such as indium is used in order to minimize residual strain in the chip.

LIGHT/CURRENT CHARACTERISTICS

The 'theoretical' light/current characteristic of a laser is as sketched in Figure 114. In practice the shape of the characteristic is very dependent on the electron and photon distributions in the plane of the junction. One of the most common problems is that of the appearance of 'kinks' in the curve, as shown in Figure 115. It has been shown experimentally[154,155] that these are usually associated with a sideways displacement of the light filament and both the near-field and far-field patterns change in the vicinity of the kink. This lateral instability is caused by the interaction between the optical and carrier distributions which occur because the refractive index profile is determined to a large extent by the carrier distribution in the active region.[156] A detailed theoretical analysis[157] shows that unless the optical mode is stabilized by a built-in lateral waveguide then an infinitesimally small amount of asymmetry in the gain or refractive index distributions across the stripe region will cause a kink to occur. The strength of the waveguide required for stability depends on the degree of asymmetry present and hence on the uniformity of such parameters as doping level, injection level, Al concentration, and layer thickness across the stripe region.

Several practical methods have been developed for avoiding kinks. These include the use of a very narrow stripe region (e.g. <5 μm)[158,159] in order to limit the possible movement of the filament, the use of transverse junction

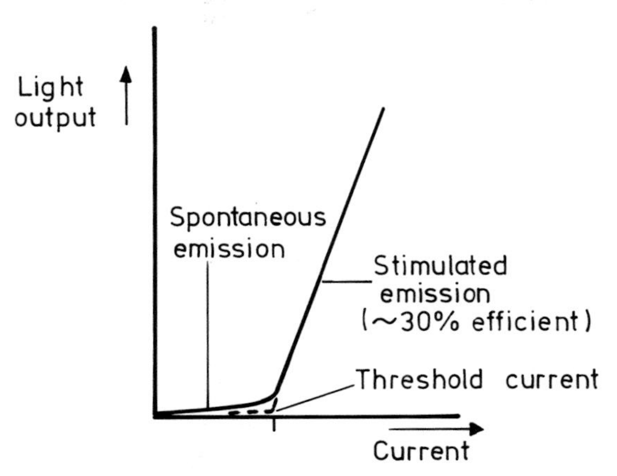

Figure 114. Ideal light–current characteristics of a semiconductor laser

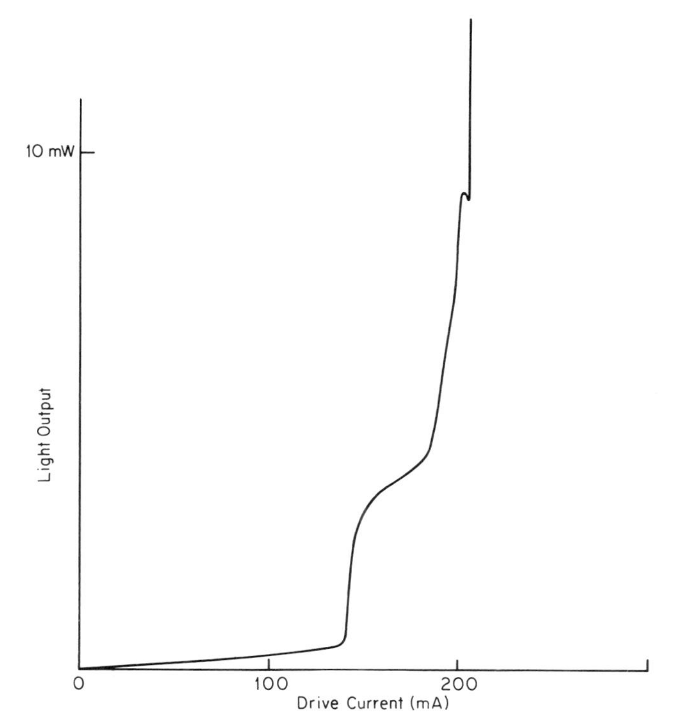

Figure 115. Light–current characteristics of a laser exhibiting a nonlinear region (or 'kink')

lasers,[160] or the use of a built-in lateral waveguide to stabilize the mode. Techniques for imposing a lateral waveguide have varied from the use of $GaAs/Ga_xAl_{1-x}As$ steps in the buried heterostructure, to the use of a deep Zn diffusion, making use of the refractive index difference between different p- and n-doping levels (see section below on New Developments).

In the SiO_2 insulated 20 μm stripe lasers described in this paper the behaviour can best be explained in terms of a significant built-in parabolic refractive index variation across the stripe, probably resulting from bi-refringence induced by the residual stress in the SiO_2[161] layers. This built-in waveguide can be strong enough to stabilize the lasing mode and prevent the type of kinks associated with a lateral movement of the mode.

In addition to the stress-induced waveguide the characteristics of these lasers are also strongly affected by self-focusing.[162] In a laser that has an optical mode width significantly narrower than the pumped stripe width, then at high output intensities the electron density at the stripe centre is 'clamped' to near the threshold level, whereas near the stripe edges the light intensity is lower, and so the electron density is higher. This increased electron density results in a reduction in the refractive index at the edges of the stripe, so adding to the existing waveguide and providing stronger confinement for the lasing mode.

The overall effect of the above mechanisms depends very strongly on the width of the defined stripe region. We describe here two basic devices which illustrate the effects: first, a device with a 20 μm wide stripe where self-focusing plays a major part and, secondly, a device with a 4 μm stripe width where self-focusing is eliminated.

20 μm Stripe Lasers

In a laser where the built-in guiding is too weak or where the nonuniformity is too great, the output characteristic shows kinks with the associated sideways movement of the near and far fields. In an optimized device, however, the mode stability can be excellent and the light/current characteristic is smooth, as shown in Figure 116(a).

Two significant features are common in this type of laser; first, there is almost always significant curvature on the initial part of the characteristic. This is a result of the change in optical guiding that occurs above the threshold as the self-focusing becomes stronger at higher light intensities. The effective gain requirement reduces as the light intensity increases thereby giving a high apparent incremental efficiency just above the threshold and a more normal incremental efficiency at higher powers when the improved waveguide has been formed. Secondly, the waveguide strength usually increases by self-focusing to the point where the first order mode can

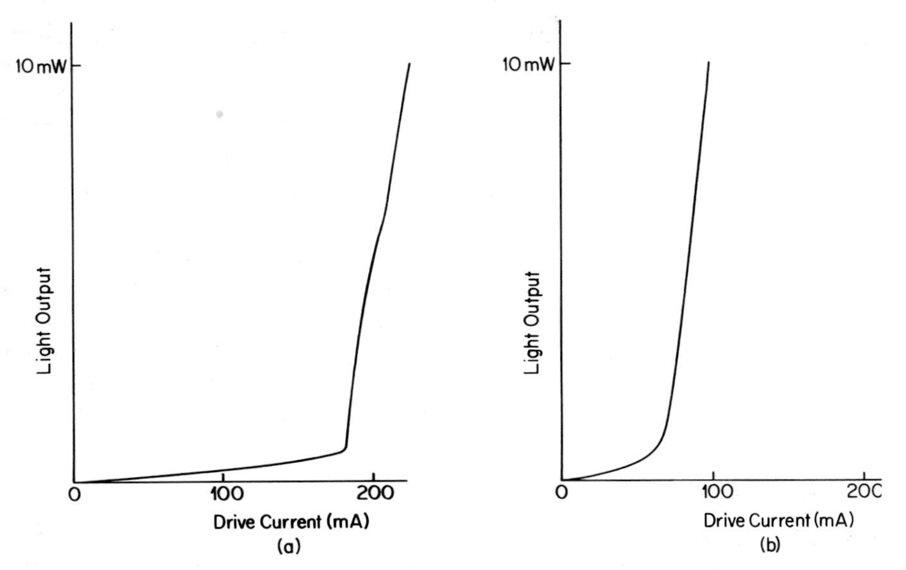

Figure 116. Light–current characteristic of (a) a well behaved 20 μm SiO_2 insulated stripe laser and (b) a 3 μm wide stripe laser

also propagate. This shows as a second steep region in the characteristic as shown in Figure 116(a) and the power at which this occurs is usually between 3 and 10 mW depending very much on device fabrication parameters such as stripe width, layer doping levels, and cavity length.

For reliable operation (see section below on Laser Reliability) it is unwise to operate the lasers at an output higher than 5 mW. It has been quite feasible to obtain a compromise between the various design parameters to make 20 μm wide stripe lasers which have a single mode output free from kinks up to output powers in excess of 5 mW.

Narrow Stripe Lasers

Two main problems occur with the 20 μm laser. First, even in a properly stabilized device without kinks the curvature on the output characteristic can prove inconvenient for analogue modulation and, secondly, if very low threshold currents are sought by shortening the cavity then the first order mode appears at a lower output power.

The above effects are eliminated by using very narrow stripe structures. In order to achieve significant improvements in overall performance it is necessary to use a stripe region less than about 5 μm wide. In SiO_2 insulated lasers as narrow as this there is a strong dip in the real part of the refractive index under the stripe region. This is partly because the injected carriers reduce the refractive index and partly because at these widths the sign of the stress-induced guiding has changed. The result is strong 'anti-waveguiding' and in this case the lasing mode is generally described as being 'gain guided'.[163] In a gain-guided situation only those rays travelling almost parallel with the stripe are amplified, others are diffracted out into the fringe regions and lost. The effect looks from the outside like a guided mode except that the wavefront is cylindrical rather than plane. The 5–15 μm range of stripe widths appears to be rather prone to instabilities, probably because this region covers the change-over from gain-guiding to self-focusing behaviour and small changes in parameters can have a large effect on the laser behaviour.

Figure 116(b) shows the light/current characteristic of a 30 μm wide SiO_2 insulated laser. Comparison with the 20 μm device of Figure 116(a) shows that the threshold region is 'softer' but that the linearity of the output curve is very much better, with no sign of slope change due to higher order modes appearing. Lasers of this type have been found to give excellent results with analogue modulation (see section below on Modulation).

Effect of Temperature

Provided that the junction structure has a low threshold current density and reasonable temperature coefficient and that the thermal resistance of the heat sink bond is low (10–20 K/W), c.w. operation is easily achieved at

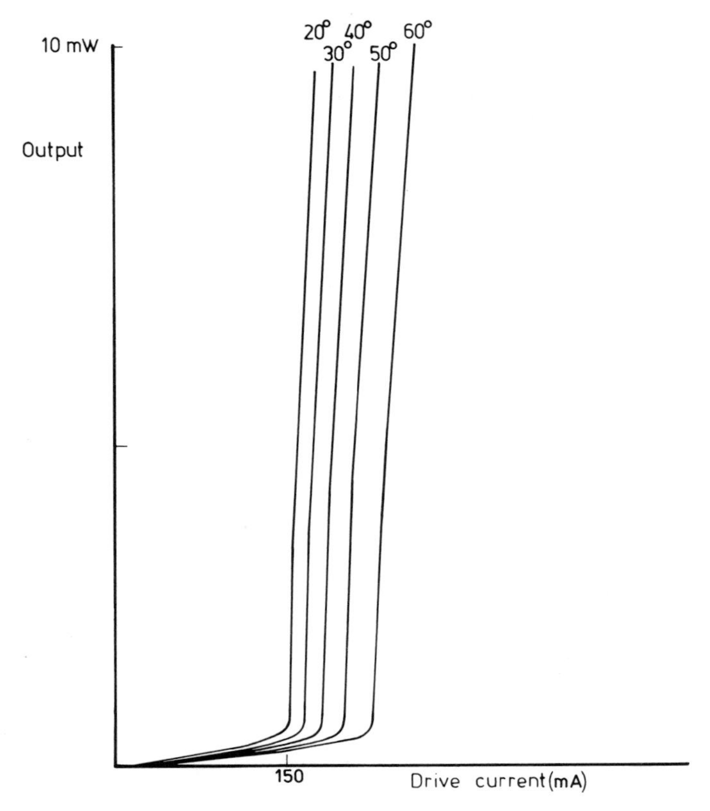

Figure 117. Light–current characteristics of a 20 μm stripe laser over a range of temperatures

temperatures up to 60°C. Figure 117 shows the measured light current characteristics for a range of temperatures for a typical 20 μm wide laser.

LASING MODE STRUCTURE

In most practical systems it is important that the laser has a stable mode structure. Changes in mode structure with current or time can, for example, introduce excess noise into fibre optic systems.

Measurements of the optical distribution across the laser facet (near-field) or as a function of angle in the output beam (far-field) are frequently used as diagnostic techniques to investigate mode behaviour in lasers. We restrict ourselves here to a description of the mode behaviour as it affects the laser applications.

20 μm Stripe

Figure 118 shows the measured intensity across the facet of a 20 μm stripe laser at various currents corresponding to output powers between 0.5 and

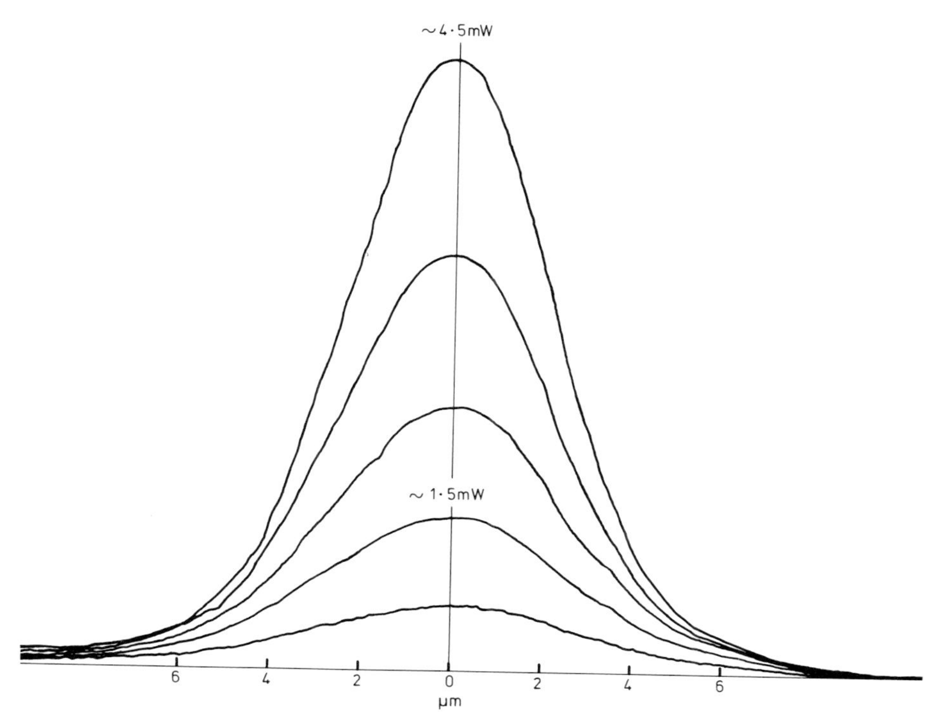

Figure 118. Near-field intensity distribution of a 20-μm laser over a range of power output levels

5 mW. The principal characteristics are a significant amount of narrowing from about 8 μm wide at threshold down to about 5 μm wide at the full output, and a very stable position—the peak moving by less than 0.2 μm over this power range. It should be said that not all lasers have such good mode stability but the yield of stable lasers can be made satisfactorily high by attention to material uniformity and care in processing.

When self-focusing produces a waveguide strength great enough to guide the first order mode, the output consists of a mixture of both modes. The two modes operate at different wavelengths and can be measured separately by spectrally resolving the near field. Figure 119 shows the results of such a measurement and also shows the distribution of spontaneous emission measured by selecting a wavelength about 300 Å less than the lasing wavelength. The fact that the spontaneous emission width is very much wider than the lasing width is easily seen, and also apparent is the dip in the centre of the spontaneous distribution caused by the much higher stimulated recombination rate in that region.

The far-field beam pattern of a 20 μm stripe laser is shown in Figure 120. The beam divergence parallel with the junction increases slightly with increasing current and is normally in the range 3–4°. The divergence in the

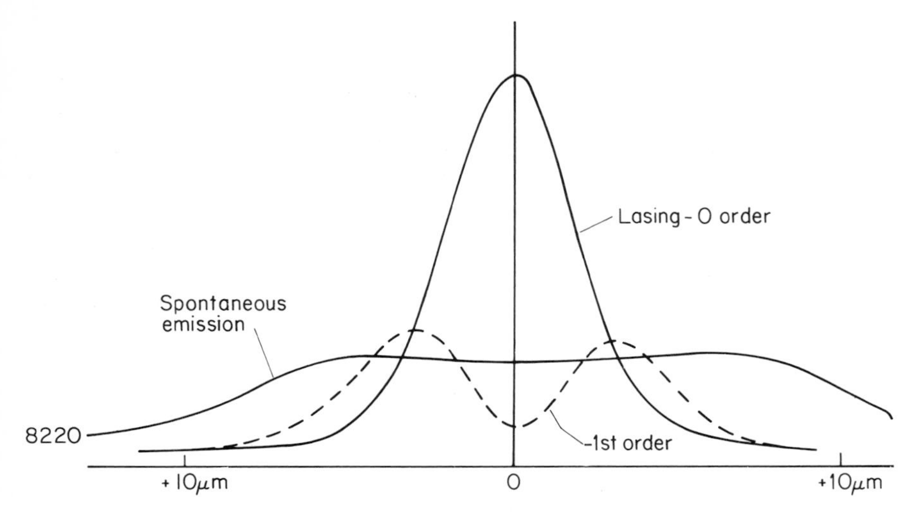

Figure 119. Near-field intensity distributions of the zero and first order modes of a 20 μm laser. Also shown is the spontaneous emission intensity (measured on a more sensitive scale).

plane perpendicular to the junction is usually around 50° and is independent of drive current, being controlled by the large refractive index steps between the GaAs and (GaAl)As layers. The beam shape is consistent with there being an approximately parabolic variation of the dielectric constant across the stripe region and the beam waist is near the laser facet. In fact the dip in gain in the stripe centre produces a slightly concave wavefront[162] which causes the beam waist to be about 10 μm in front of the facet, but the difference in width between the facet and the waist is small.

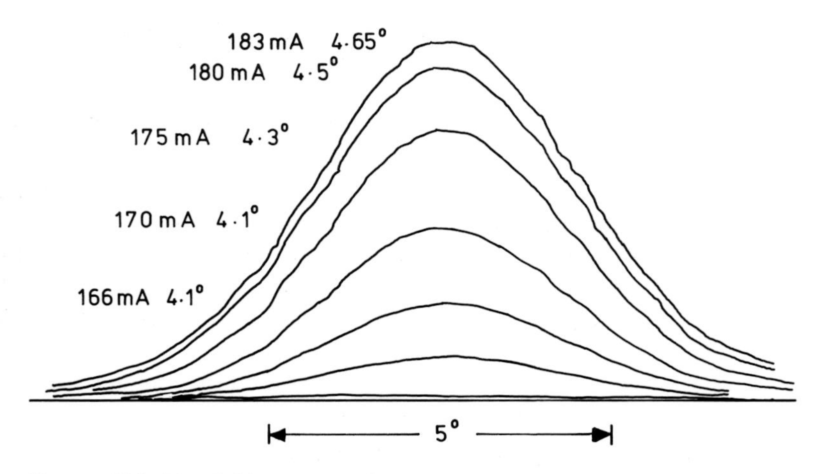

Figure 120. Far-field pattern of a 20 μm stripe laser at various drive levels

3 μm Stripe

The near-field pattern of a 3 μm wide stripe laser is shown in Figure 121. It is perhaps surprising that the width (8 μm) is somewhat greater than for a 20 μm wide stripe. The reason for this is the strong defocusing resulting from the dip in the refractive index at the centre. The convex cylindrical wavefront produces a very narrow (1–2 μm wide) virtual waist to the beam, apparently some 10–30 μm behind the facet of the laser. Since the beam waist in the perpendicular plane is actually at the laser facet this difference may need to be allowed for if a lens is used to refocus the emission onto a fibre end for example.

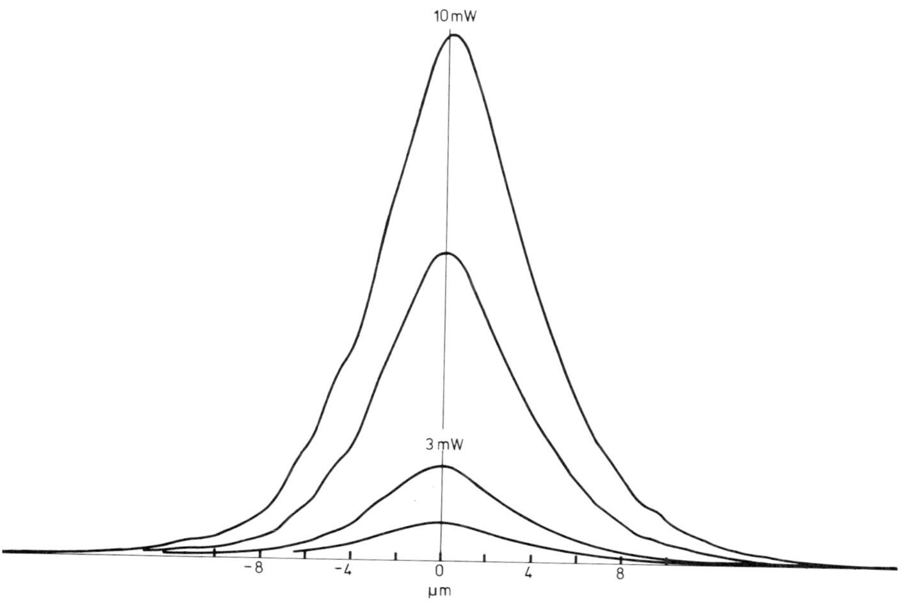

Figure 121. Near-field patterns of a 3 μm stripe laser at various output powers

The far-field pattern of a very narrow stripe laser is usually 20–25° broad. Figure 122 shows a typical beam pattern. The double lobed distribution is common in this type of device and is not a higher order mode but simply a result of the lossy gain-guided wave in an antiwaveguiding structure. Optimization of layer doping levels and stripe width is likely to produce a single lobed far field.

LASING SPECTRA

Figure 123 shows the spectrum of a good 20 μm wide SiO_2 insulated stripe laser operating c.w. at various output powers. Although a few minor peaks occur, the main power is concentrated in a single longitudinal mode when

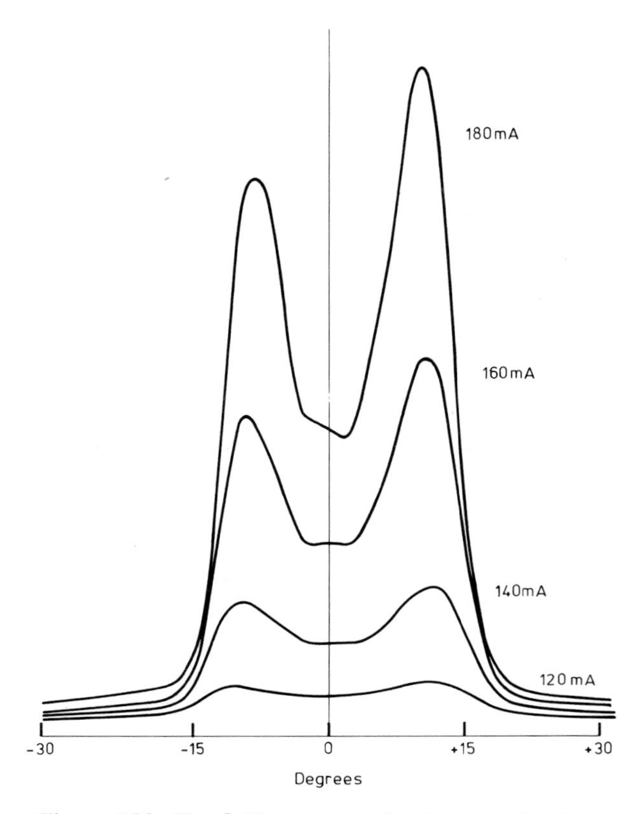

Figure 122. Far-field patterns of a 3 μm stripe laser at various drive levels

the output power is a few milliwatts. This behaviour is typical of good uniform devices although frequently two or three modes may be present in some current ranges. The drift to longer wavelengths at higher powers is partly due to heating and partly due to the reducing gain requirement as self-focusing becomes stronger. Figure 124 shows the spectrum of a 3 μm wide laser and here one can see the considerable difference between this device and the wider lasers. In all very narrow lasers the spectrum is much broader than in 20 μm lasers operated under c.w. conditions; for this reason these devices have been found to be very useful in cases where excess system noise (modal noise) is generated by sources which have only one or two spectral lines (see Chapter IX).

INTENSITY NOISE

In semiconductor lasers there is always a significant noise amplitude superimposed on the steady light level. This noise is due to amplification of

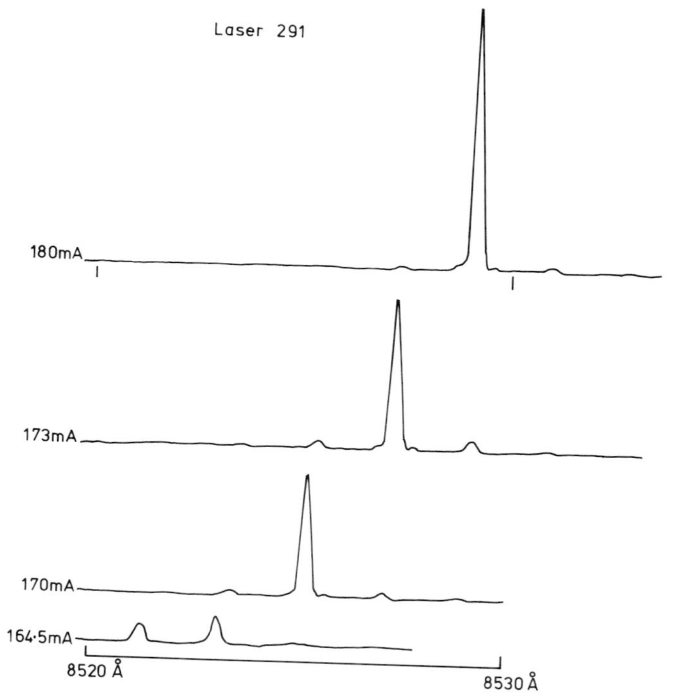

Figure 123. c.w. lasing spectra of a 20-μm stripe laser operated at various output power levels

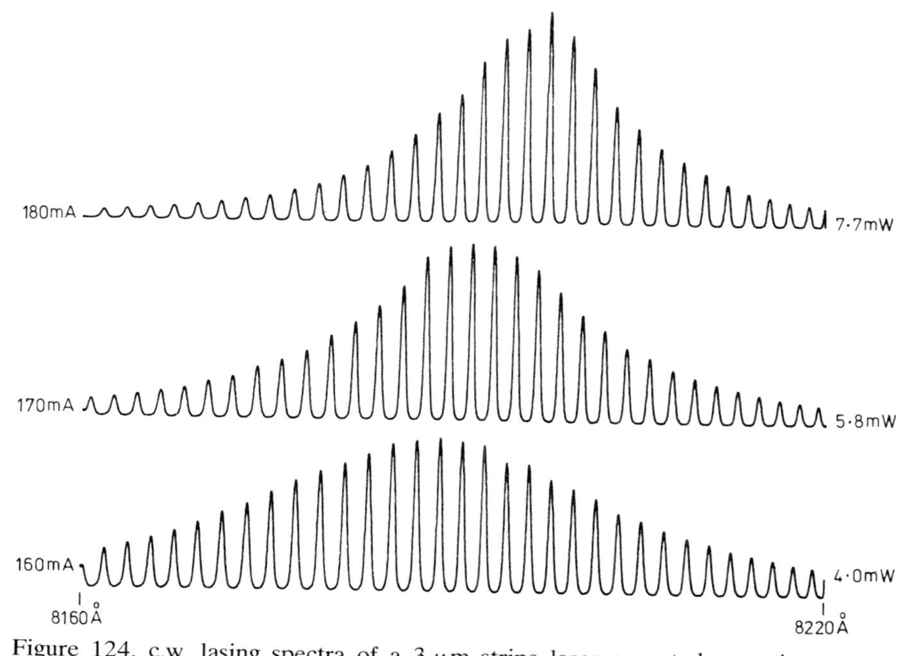

Figure 124. c.w. lasing spectra of a 3 μm stripe laser operated at various output power levels

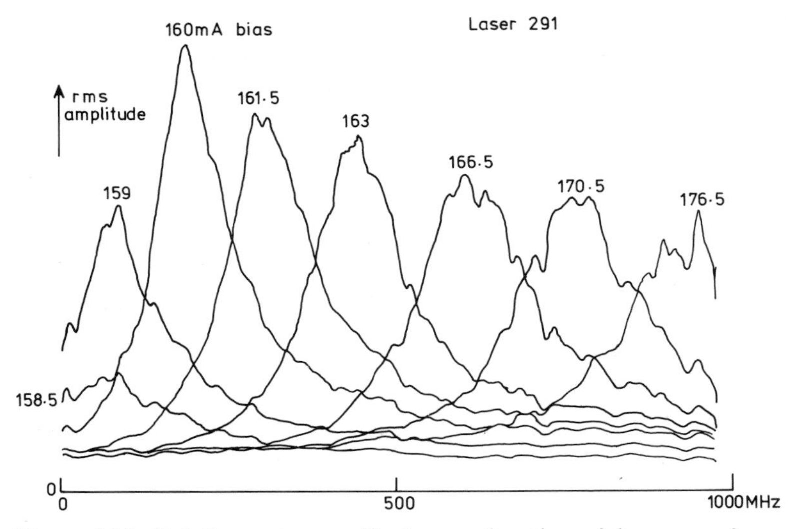

Figure 125. Relative noise amplitude as a function of frequency for a 20 μm stripe laser operated at various drive currents

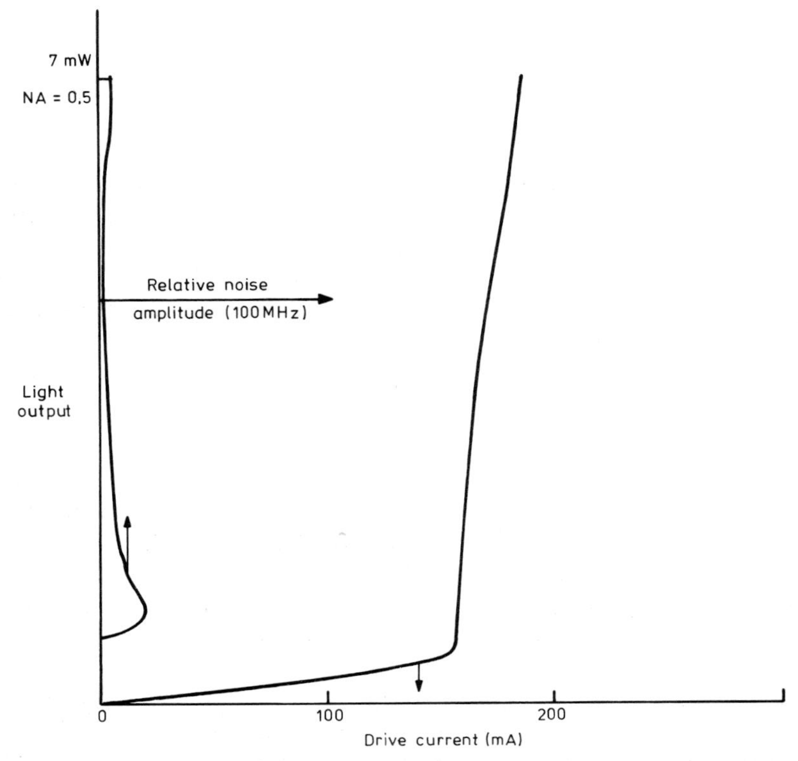

Figure 126. Noise power in 100 MHz bandwidth plotted as a function of d.c. light output

quantum fluctuations and shows a resonance-like frequency variation related to the small-signal response (see next section) where a peak response occurs at a frequency lying between about 100 MHz just above threshold to around 1000 MHz at an output of 5–10 mW.

The relative noise amplitude as a function of frequency is shown in Figure 125 for a 20 μm laser at various drive currents. As a result of this behaviour the noise power in the bandwidths appropriate to a practical 140 Mbit system peaks close to the threshold. This value is plotted in Figure 126 on the normal light current curve of a typical 20 μm laser; the peak value is usually less than 5 μW r.m.s. Some unstable devices show very much higher noise levels and at some currents up to 100 per cent modulation of the output can occur. This is usually referred to as self-pulsing and is commonly seen in lasers with 'kinks' and unstable mode patterns. Selected well-behaved SiO_2 insulated lasers described here do not usually show excess noise or self-pulsing at output powers less than 5 mW, and so these excess noise effects need not be considered in a practical system.

The very narrow stripe lasers usually show lower overall noise levels and the peak in the noise spectrum is less pronounced than in 20 μm lasers. This is probably related to the greater damping observed in these lasers when subjected to modulation (see next section).

MODULATION

Analogue Modulation

When a laser is biased continuously above threshold and a small modulation signal is applied to the current drive, the response can be calculated from the rate equations. The response is expected to show a resonance-like behaviour as a result of the interaction between the photon population and injected carrier concentration. The resonance frequency f_R at a drive current I is approximately given by

$$f_R = \frac{1}{2\pi}\left(\frac{1}{\tau_n \tau_s}\,\frac{(I - I_t)}{I_t}\right)^{1/2} \tag{41}$$

where τ_n = electron lifetime, τ_s = photon lifetime, and I_t = threshold current. The exact shape of the response curve depends very much on the damping in the laser. The sources of damping are the lateral diffusion of carriers and the presence of spontaneous emission. The detailed calculation of the response curve is complicated.[164,165] Qualitatively the results indicate that more damping will be seen in lasers of very small cross-section This is seen in practice.[166]

An experimentally determined modulation response of a 20-μm laser is shown in Figure 127. The laser, which had a threshold current of 190 mA, was driven with a constant current of 205 mA and superimposed on this was a 12 mA peak-to-peak sinusoidal modulation, swept in frequency from 10

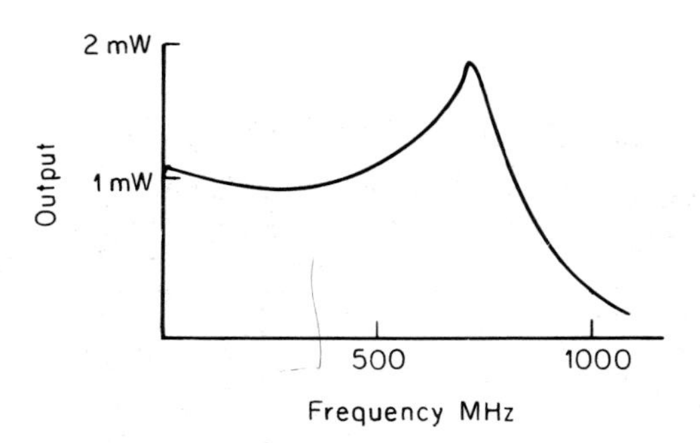

Figure 127. Frequency response of small signal modulation of a 20 μm stripe laser

to 1000 MHz. The output was detected using a photodiode with a response time of less than 200 ps and displayed on a spectrum analyser. The shape of the response curve is as expected, with a flat frequency response up to 500 MHz and a resonance at 700 MHz. The output level is of the order of 1 mW peak-to-peak at lower frequencies.

The frequency of the resonance varied from 100 MHz at 1–2 mA above threshold (measured, of course, with a much smaller input signal) to 900 MHz at 215 mA (25 mA above threshold). The value closely followed the theoretically expected variation given in equation (41).

Owing to the nonlinearity of the output characteristic of the 20 μm stripe laser, significant distortion was observed. However, by carefully choosing the operating point the third order intermodulation distortion could be limited to −35 dB for a 1 mW peak-to-peak signal.

Very narrow stripe lasers show a rather different response to modulation with a very much less pronounced peak in the curve at the resonant frequency. Also, owing to the more linear output characteristic, the modulation linearity is significantly better than for a 20 μm laser. With 1 mW peak-to-peak putput less than −60 dB third order inter-modulation distortion has been achieved (see Chapter IX).

Pulse Modulation

Although many modulation techniques have been proposed, the one of most practical interest is that of directly driving the laser with a pulse code modulation (p.c.m) current.

When a laser is pulsed from below threshold the most significant feature is the appearance of damped oscillations (spiking or ringing) in the light output. This has been treated theoretically and experimentally[167,168] and the main results are as follows:

(1) The oscillation frequency depends on the final degree of overdrive and is related to the small signal resonance at that overdrive level.

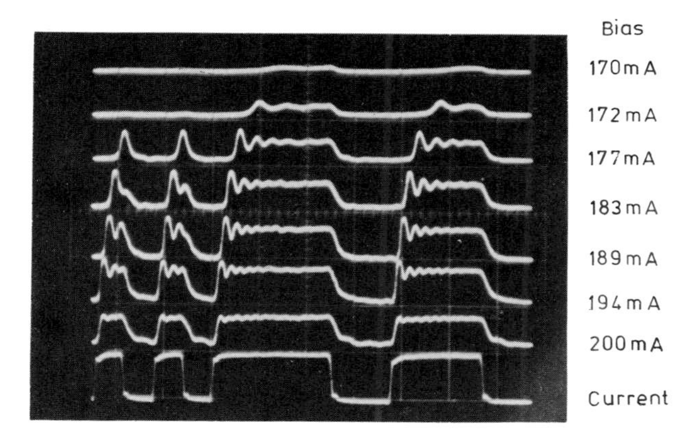

Figure 128. Response of a stripe laser to pulse modulation at various d.c. bias levels. Modulation current is 24 mA (shown on bottom trace). Horizontal scale is 5 ns per division and the vertical scale is 5 mW per division

(2) For currents near threshold, the decay time of the oscillation is approximately the electron recombination time (~3–5 ns).

(3) The amplitude of the output spikes is determined mainly by the change in current imposed by the modulating signal.

(4) The lasing delay between the application of a current step and the appearance of the first spike is decreased by increasing the bias current.

The principal features can be seen in the experimental results shown in Figure 128 for an SiO_2 insulated stripe laser driven with a variable d.c. bias upon which was superimposed a 250 Mbit/s non-return-to-zero p.c.m. signal. The c.w. threshold was 190 mA and the drive pulse amplitude 24 mA. The top trace, with a bias of 170 mA, shows that lasing only occurs in the longer pulses; the first spike appears with a delay of 5 ns. At 177 mA, spikes are just appearing in the shorter pulses and the delay is reduced to 2 ns; at 183 mA, two spikes can be seen in the shorter pulses, and at 200 mA, above threshold, the modulated light output is an excellent replica of the input signal.

Modulation at higher frequencies requires care since spiking will occur at frequencies comparable to the modulation frequency.

Effect of Modulation of Lasing Spectrum

It has been reported[169] that lasers show a very broad spectrum during the first few nanoseconds of the output pulse when modulated with a step increase in current from below threshold. This is due to the considerable upward swing in electron density in the time intervals between the output spikes which causes a larger number of longitudinal modes to have above threshold gain than would occur during steady lasing.

Figure 129(b) shows the lasing spectrum measured during an output pulse obtained by applying a 2 ns, 40 mA pulse superimposed on a d.c. bias just below threshold. At least 15 modes are excited giving a spectral band width of 3 nm. Figure 129(a) shows the spectrum obtained with the same input pulse superimposed on a d.c. bias current slightly above threshold. The spectrum in Figure 129(a) is almost identical to the normal c.w. spectrum, except that the mode intensity is increased by the increase in drive current. With a real pulse code modulation drive, transients and short-term temperature drifts cause some broadening and variation in the peak wavelength. Experiments show that with present lasers the broadening and drift are contained within an envelope about 1 nm wide, the maximum variation being observed between a long sequence of ones and a sequence of zeros.

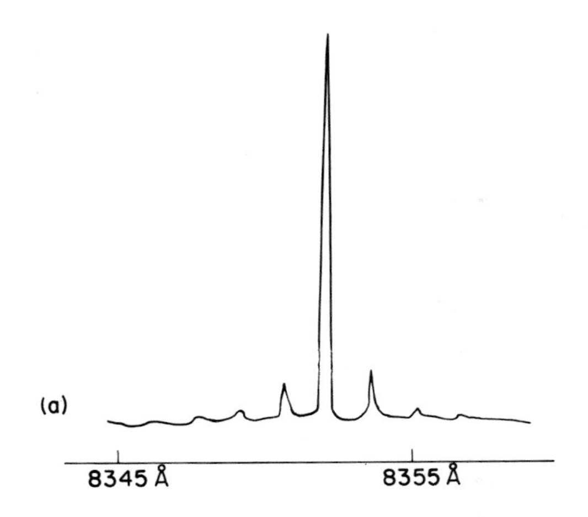

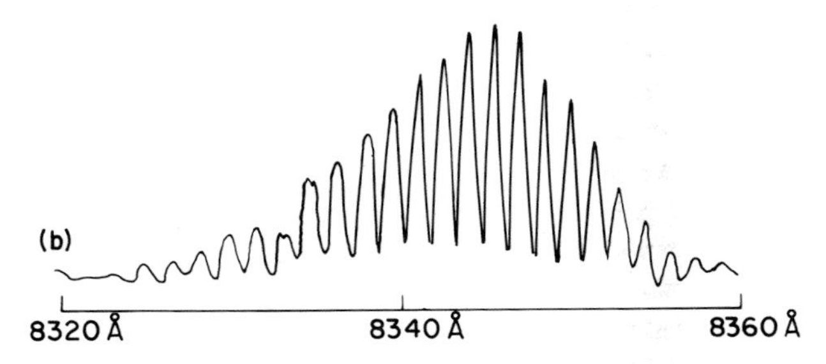

Figure 129. Lasing spectrum of a 20-μm stripe laser measured during a 2 ns pulse: 40 mA pulse amplitude. (a) d.c. bias just above threshold. (b) d.c. bias just below threshold

At first sight the above results would suggest that a bias above threshold would be optimum since it gives a narrower spectrum and hence less material dispersion. In practice it has been found (see Chapter IX) that excess system noise can be introduced by a laser with few spectral modes and so biasing below threshold and using RTZ pulses is usually more appropriate.

Effect of Modulation on Mode Structure

If a stripe laser can support more than one transverse mode the modulation response of the two modes may be different and this could have adverse effects on system performance. This has been treated experimentally and theoretically by Buus *et al.*[170] In the 20 μm stripe lasers reported here the zero and first order modes behave almost independently in the way expected from their relative amounts of overdrive above their respective thresholds. In pulse response, for example, if the laser is pulsed from below threshold to a current above the threshold for the first order mode then the zero order mode appears after a short time, as seen in Figure 128, and after a few nanoseconds the first order mode appears, ringing with an amplitude and frequency appropriate to its lower level of overdrive. In analogue response if the steady d.c. bias is above the threshold for the first order mode then the peak response for the zero order mode occurs at a high frequency, say 700 MHz, while that of first order mode occurs at a lower frequency, say 200 MHz.

As a result of the complicated response of the two modes it is often unwise to operate in a region where more than one mode occurs.

Other much smaller effects can occur in lasers supporting only a single mode. For example, in a laser where self-focusing is important the amount of transverse optical confinement can vary with light output. This may show as a variation of the mode near- and far-field widths during the period of ringing in the early part of an output pulse.[171] This is only likely to be important in very high bandwidth systems.

LASER RELIABILITY

Over the past few years the degradation mechanisms which seriously limited the life of early c.w. lasers have been widely studied. Much has been written about degradation effects and it is now clear that the main cause of the short lives of earlier devices was the formation and growth of dark line defects, or related effects. These are areas which appear dark when the device is operated as a spontaneous emitter and viewed with an infrared microscope.

Present evidence indicates that dark line defects are initiated at crystal defects in the active region and that their rate of growth depends, among other things, on residual mechanical strain in the device. In the devices life tested at STL, every effort has been made to eliminate crystal defects and to use processing techniques that minimized any residual mechanical strain. The main factors in avoiding crystal defects are:

(a) choice of low dislocation substrates;

(b) avoidance of scratching or other damage during processing and epitaxial growth; and

(c) avoidance of oxygen contamination during epitaxy.

The techniques used to minimize residual strain are:

(a) use of 5 per cent Al in the active layer;

(b) minimizing the strain induced by the SiO_2;

(c) avoidance of alloyed metal contacts near the junction; and

(d) use of a soft metal (indium) for heat sink bonding.

After fabrication of the laser chips visual inspection can be used to select only those which are free from crystal defects. This step screens out those lasers which would otherwise die rapidly due to dark line defect propagation[172] and so the reproducibility of subsequent life tests is improved. Figure 130 shows photomicrographs of two lasers: (a) has a dislocation reaching the surface near the stripe region and (b) has no surface features at all. Devices like (b) *all* show good life test results.

Mounted lasers have been life-tested for periods now exceeding 20,000 hours. In these tests the operating current was continuously adjusted to maintain a constant light output of 2–4 mW and the change in threshold current was monitored during operation. Figure 131 shows the results for a typical batch of lasers operated at 20°C. Two phases of degradation are apparent in these devices: a first stage, in which the threshold current increased by about 10 per cent over a few thousand hours and a much slower stage yielding a degradation rate of about 0.5–1.0 per cent per thousand hours over the 10,000–20,000 hours range.

The first, more rapid stage of degradation has been found to be closely related to residual mechanical stress in the lasers. For example, reducing the SiO_2 induced stress was found to give a factor of two reduction in the initial degradation rate.

The second stage has been shown to be partly a result of cumulative facet damage during operation. The light intensity at the facet is very high and even in an inert nitrogen atmosphere the GaAs surface becomes stained, possibly as a result of a photoinduced chemical reaction. The use of Al_2O_3 facet coatings has been reported[173,174] to overcome this problem and initial tests on the lasers described here confirm this result.

At least part of the long-term residual degradation may be due to some internal effect since tests with the lasers operating as LEDs just below threshold show a very low but still significant degradation. That this is only a small effect at room temperature is verified by the fact that lasers constructed with low strain and with Al_2O_3 facet coatings have recently operated for several thousands of hours with no measurable change in threshold.

Since the final room temperature degradation rate is very low we have the problem of predicting the ultimate working life of these lasers. Accelerated

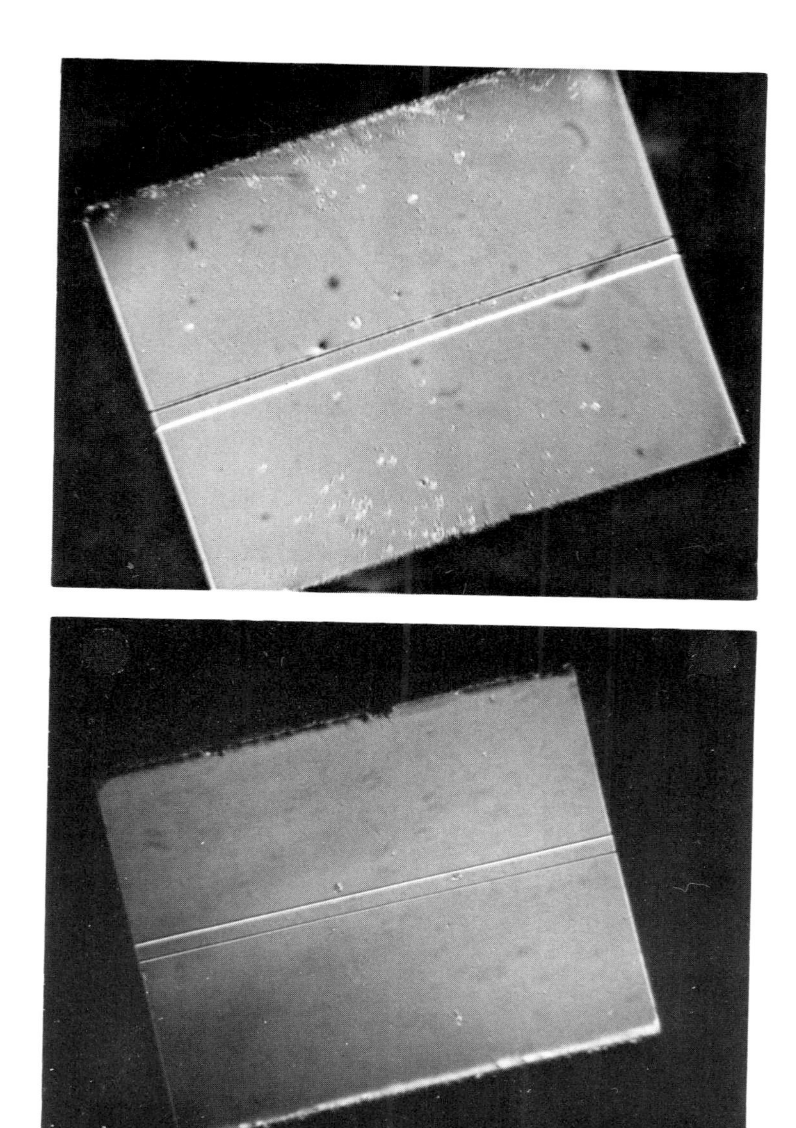

Figure 130. Photomicrographs of two stripe geometry lasers: (a) showing a dislocation near the stripe region and (b) with no surface features

life tests at elevated temperatures have been carried out by Hartman and Dixon[147] and used to predict that their lasers will have a median life at 20°C of 300,000 hours. Similar tests on the present SiO_2 insulated lasers[175] have shown that different degradation mechanisms may dominate at higher temperatures and so this type of acceleration technique is unsuitable until all the mechanisms have been identified.

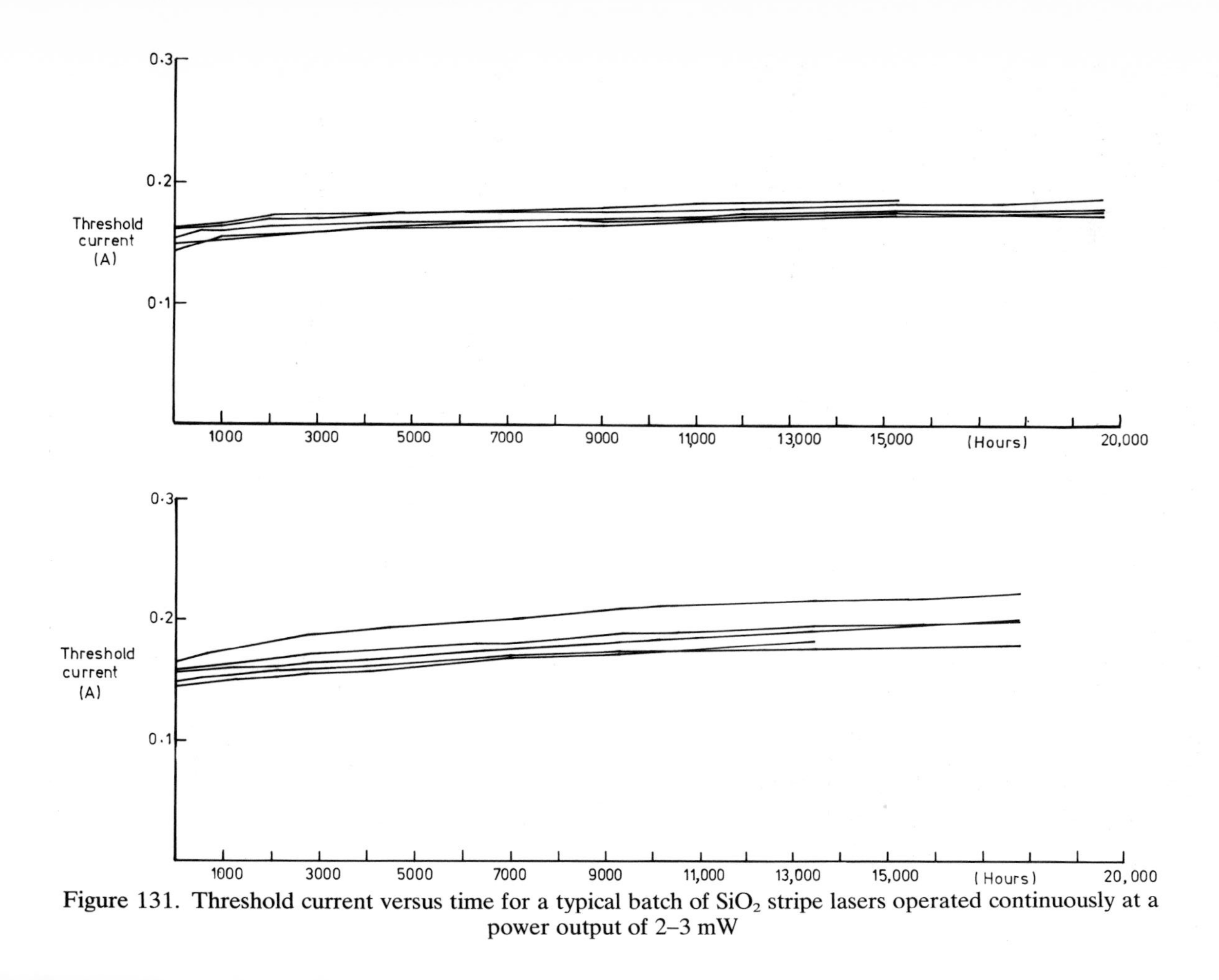

Figure 131. Threshold current versus time for a typical batch of SiO_2 stripe lasers operated continuously at a power output of 2–3 mW

Extrapolations can be made from the 20°C life tests by assuming that the degradation rate will remain constant and that the laser will remain operational until the threshold current has risen to a point where heating becomes excessive. All lasers will easily tolerate a 50 per cent increase in threshold current and some will still operate c.w. after a factor of three increase in threshold. Using these two extremes, the estimated operating lives of lasers from an early fabrication batch are shown on a log-normal plot in Figure 132. It can be seen that a conservative estimate of the median life is 74,000 hours with many lasers probably able to operate for over 200,000 hours. The standard deviation of predicted lives is small, probably as a result of the careful fabrication and inspection of dice before mounting.

Preliminary tests at elevated temperatures show that the life will be shorter by about a factor of two at about 40°C. However, the above results were obtained on lasers made early in the development programme and improvements in technology are expected to lead to significant life improvements in more recent devices. There is little doubt that even at the present state of development these lasers are suitable for many years of unattended operation.

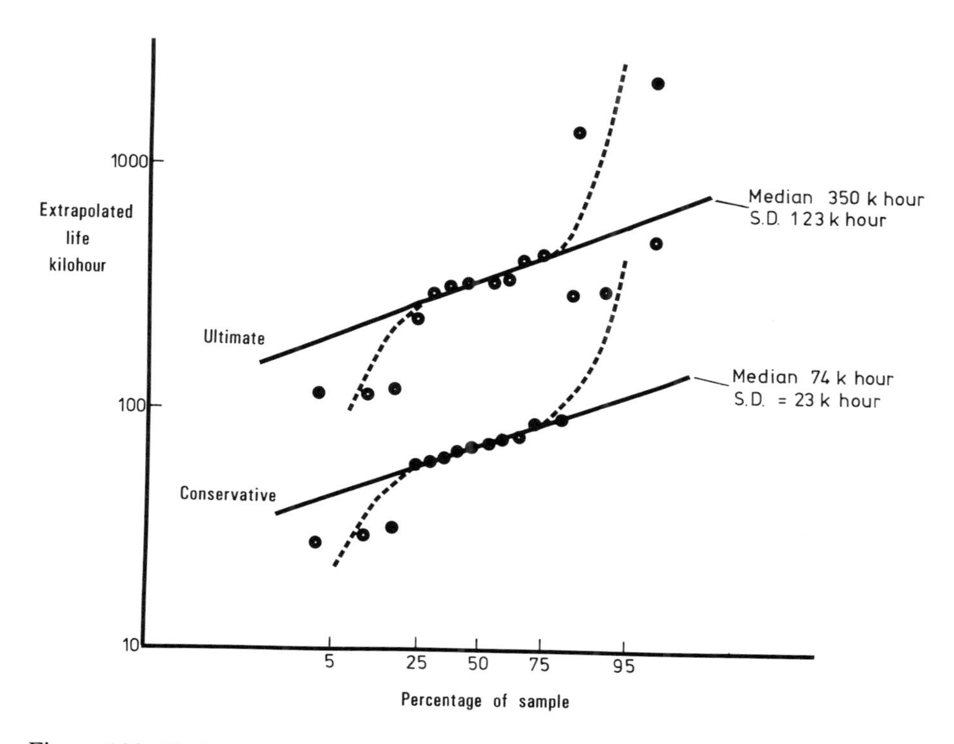

Figure 132. Estimated operating life of a batch of stripe lasers deduced from 22°C life tests

DEVICE PACKAGING

Most of the measurements described in previous sections were made with lasers mounted in experimental packages, as shown in Figure 133. Device (a) is not encapsulated and life-tests were performed using this configuration mounted in an enclosure containing flowing dry nitrogen. An alternative model (b) has a glass window type of encapsulation and can be used in normal environments.

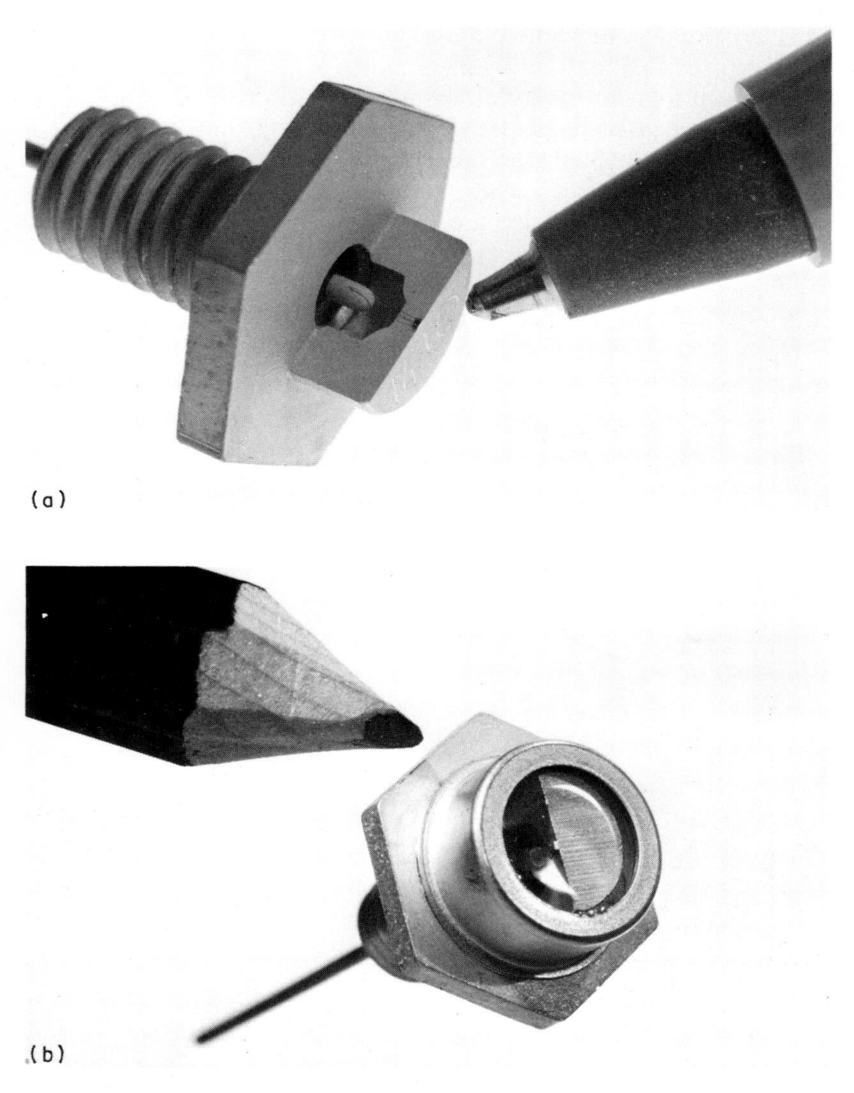

Figure 133. Stud laser packages: (a) unencapsulated and (b) encapsulated

Neither of the above methods is really suitable for use in a fibre-optical system since no provision is made for attachment to the fibre. Figure 134(a) shows a prototype mounting in which the copper heat sink (D-shaped) is attached to a substantial steel mounting onto which the fibre-housing can be connected. An important feature of the package is the window in the rear of the mounting through which the emission from the back surface of the laser can be monitored for controlling the drive current through a feed-back circuit (see Chapter IX and XI). Another feature is the low inductance strip-line current input to allow ease of driving at high bit rates. This package is suitable for either butt-connection to a fibre or for use with lens launching.

Possibly more convenient to use than either of the above packages is that shown in Figure 134(b). In this the laser chip is on a heat sink which is housed in a standard ITT optical cable connector. A short length of fibre brings the output of the laser to a position where it couples to a standard fibre termination. This type of housing has been built with an integral photodetector for monitoring the laser output or with a rear window for external mounting.

The design of laser packages for use with fibre-optics presents a considerable challenge since the requirements for current input, good heat sinking, optical monitoring, and hermeticity are difficult to reconcile with the fundamental requirement of good rigid coupling to the optical fibre. No entirely

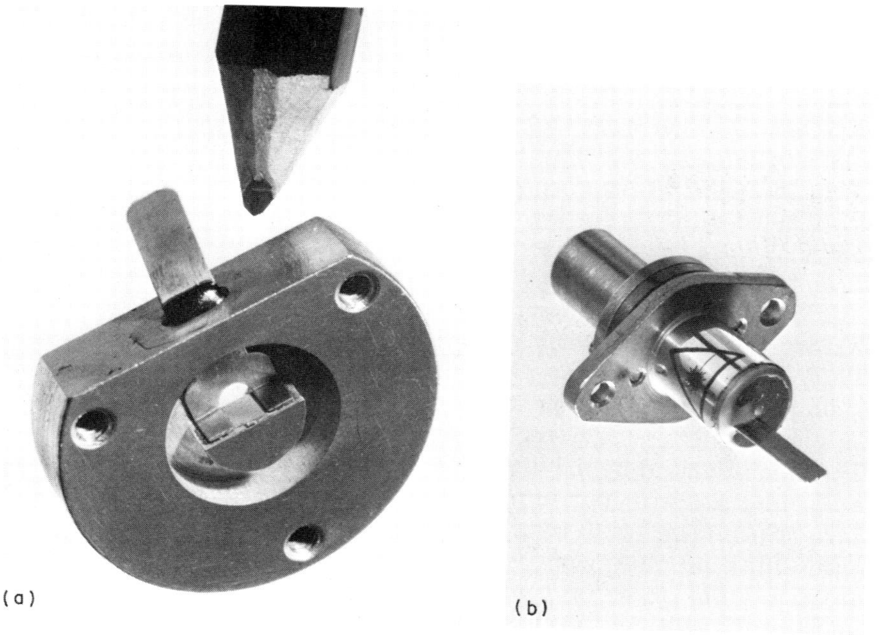

Figure 134. Laser packages suitable for fibre-optic coupling: (a) a laboratory model and (b) a type designed around a standard optical fibre connector

satisfactory solution has been found and more development work is required before the ultimate aim of standardization can be attempted.

NEW DEVELOPMENTS

New Structures

The devices described in the previous sections are typical of those currently entering production phases and so will be important for the initial development of optical systems. However, considerable research is being carried out around the world on new device structures and is likely to have a significant impact on the type of performance parameters available during the next decade.

Two features of the present lasers require improvements. First, it would be an advantage to have lower operating currents; this is particularly important for long-haul optical routes where remote repeaters must have power supplied from conductors in the cable. Secondly, in many devices the transverse mode stability is inadequate, as discussed in the previous section on light–current characteristics.

The threshold current of a conventional stripe geometry laser can be reduced by making the stripe width narrower and the cavity length shorter. This has been successfully carried out by Steventon *et al.*[176] who achieved a threshold current of 28 mA with a stripe width of 5 μm and a cavity length of about 60 μm. This would appear to be near the ultimately achievable figure for this type of device since lateral diffusion of carriers prevents any advantage from narrower stripes and cavity lengths much below 100 μm make handling and mounting inconvenient.

If much lower threshold currents are required the injected current must be restricted to a region less than 5 μm wide and a transverse optical waveguide must be used to retain the optical distribution within the region of gain. The lowest threshold currents yet achieved have been with the 'buried heterostructure' laser reported by Tsukada *et al.*[177] who obtained c.w. thresholds of less than 10 mA with the structure shown in Figure 135(a). Here the (GaAl)As and GaAs epitaxial layers outside the stripe region were etched away and replaced by n-(GaAl)As. Injected carriers are now unable to diffuse sideways out of the active region and the strong lateral waveguide produced by the low refractive index (GaAl)As layers ensures a narrow optical distribution in the transverse plane. A possible disadvantage of this structure is that to achieve a single mode in the transverse plane the stripe width must be only about 1 μm wide. This means that the total power output is lower than for conventional lasers, typically about 1 mW. A significant advantage is extremely good mode stability owing to the fact that the refractive index steps are very large compared with the small changes caused by nonuniformities, and carrier injection effects.

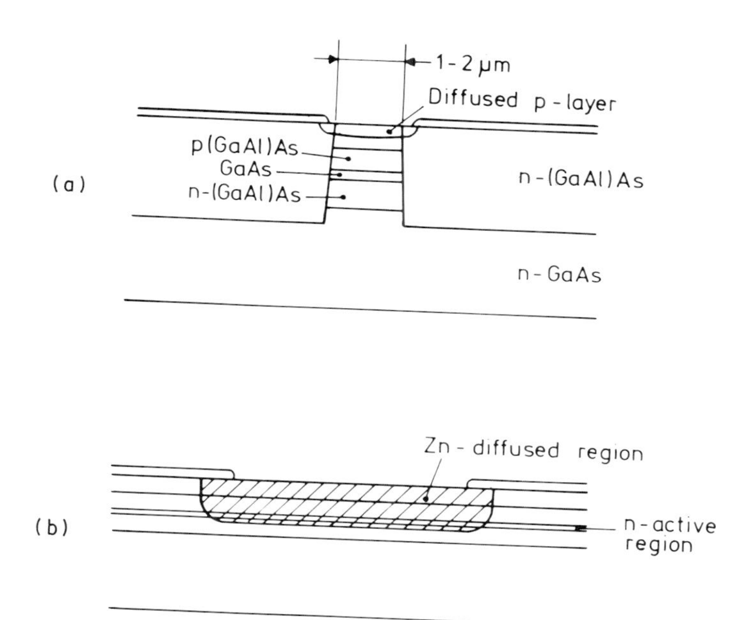

Figure 135. New laser structures. (a) The buried heterostructure. (b) The deep-diffused laser

Many other novel structures have been successfully made. These include the transverse junction laser[160] which achieved a threshold of 21 mA, the channel-substrate laser[178,179] and many others. One particularly convenient method of both confining the current and building in an optical waveguide is to use a deep-diffused laser structure[154,180] Figure 135(b). This makes use of the fact that there is a difference in refractive index between moderately doped p-type GaAs and more highly doped n-type GaAs. Thus, by diffusing zinc (a p-dopant) into an n-doped active region it is possible to introduce a step-index waveguide into the structure. Very careful control of doping concentration and depth is required, but Kobayashi *et al.*[154] have shown that very good mode stability can be achieved using this technique. Recently, Thompson *et al.*[180] reported threshold currents as low as 25 mA with a 4 μm wide, deep diffused laser with a cavity length of 120 μm. These devices had a very stable single transverse mode with a half-intensity width of 3 μm and a far-field width of 10°, making the device very suitable for single mode optical transmission systems.

At the present time it is not possible to predict which of the new laser structures will finally emerge as the best devices for optical communications. Such factors as ease of manufacture, yield and cost will be as important as the ultimate performance specification. It is certain, however, that current research will lead to significant improvements and low threshold, mode-stable lasers will eventually be available commercially.

New Materials for Longer Wavelengths

As a result of the lower fibre losses and lower material dispersion at wavelengths between 1.0 and 1.5 μm, there is a need for lasers (and detectors) operating in this wavelength region. The GaAs–(GaAl)As system which has been used for virtually all semiconductor lasers in the past is only of use in the 0.8–0.9 μm region and a new material system must be used.

The success of the GaAs–(GaAl)As system has been a result of the fact that changing the Al concentration changes the band-gap and refractive index while retaining a crystal lattice spacing very close to that of GaAs. When searching for a new material to use for making semiconductor lasers operating in the required wavelength region it is necessary to look for a system in which the active region can have an energy gap around 1.0 eV and can be bounded by p- and n-regions with an energy gap of at least 1.3 eV and the same lattice spacing. A further practical requirement is that good quality substrates should be readily available for the epitaxial growth and, lastly, but importantly, the technology should ideally be similar to that previously used for (GaAl)As lasers.

Figure 136 shows the lattice parameters and energy gaps of various III–V compounds. The lines joining the binary compounds represent the parameters of solid solutions which are known to exist. It is clear that no suitable compounds exist for use with GaAs in the 1.0 eV region. The system which has shown the greatest promise in the early stages is the InP–(GaIn)(AsP)

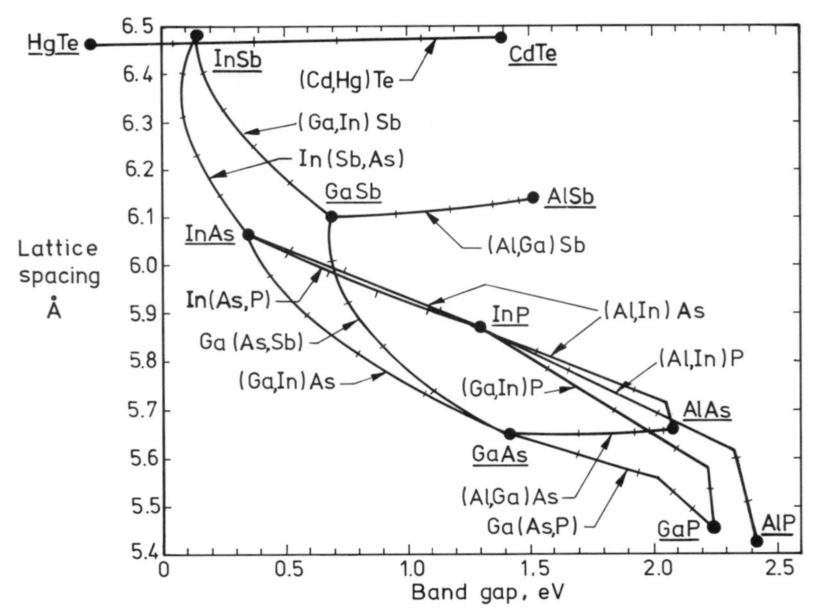

Figure 136. Lattice parameter versus energy gap for various III–V compounds and solid solutions.

series of solid solutions. By appropriate choice of composition it is possible to grow layers of material with energy gap between 0.73 eV (GaIn)As and 1.3 eV (InP).

Double heterostructure lasers have been made by Hseih[181] and others using an InP substrate, p- and n-InP passive layers, and a (GaIn)(AsP) active region. These were successfully operated c.w. with performance very similar to that of conventional (GaAl)As devices. Preliminary life tests were encouraging with 6000 hours being obtained from early devices.

Many laboratories are now working on lasers for use at longer wavelengths and within the next few years we may expect to see these becoming commercially available.

CONCLUSIONS

We have described the performance characteristics of currently available SiO_2 insulated stripe geometry lasers. These have already been developed to the stage where an acceptable yield can be obtained from devices which are very well suited to the requirements of optical communications systems spanning a very wide range of bandwidths. One of the first applications of these devices has been in a 140 Mbit/s field demonstration system (see Chapter XI) and the performance of the devices has been entirely satisfactory.

Research and development is continuing and we have outlined recent work which is leading to improved performance parameters such as reduced operating currents, better mode stability, and longer operating lives. Also of importance is the new work on longer wavelength devices which will ultimately lead to systems with longer section lengths without repeaters. During the next few years we can look forward to these improvements being incorporated in commercial devices so that the system designer of the 1980s will have a range of devices available to him which will be ideally suited to his requirements.

CHAPTER VIII

Detectors for Fibre-optic Communications Systems

INTRODUCTION

The fibre-optic communications system has become established with semiconductor diode light sources based on GaAs (e.g. GaAlAs, GaAsP) and silicon diode detectors. The GaAs diode light source, either LED or laser, was virtually the only source suitable for use in fibre-optic systems. The silicon photodiode was by no means the only photodetector possible for use with fibres, but it was already well developed for use in other photodetection applications. It has become the pre-eminent photodetector and is likely to keep this position for some years.

The semiconductor photodiode can be of small size, well compatible with optical fibres and, if made of a suitable material, will have excellent sensitivity and response time characteristics. It should also be a robust device, and have a very long life under normal operating conditions.

It is a fortunate coincidence that silicon photodiodes are most sensitive to light of around the wavelength produced by GaAs diodes, although the peak is not sharp. Such detectors[182] can also benefit from the vast wealth of silicon technology built up in other fields, with the result that excellent silicon photodiodes are available at costs down to a few pounds. The tendency therefore is to stretch the capability of silicon detectors to wavelengths where another material should theoretically give better performance.

The basic operation of a semiconductor photodiode is very simple: a current is generated within the device which is closely proportional to the level of illumination. Having said this, the detailed operating characteristics of the device depend on numerous factors including several complex processes, and some knowledge of these is desirable in order to be able to make best use of the available devices. This chapter discusses the essential characteristics of photodiodes which are relevant to fibre-optic communications.

SEMICONDUCTOR PHOTODIODE FUNDAMENTALS

Light travels in the form of discrete packets of energy or 'photons'; the energy per photon is given by the following expression:

$$\varepsilon = \frac{hc}{\lambda}$$

where

h = Planck's constant,
c = velocity of light } in a particular medium, usually
λ = wavelength of light } assumed to be free space.

When a photon is absorbed, the energy of which is equal to that of the semiconductor band gap, it can excite an electron so that it moves to a higher energy state, in this case from the valence band to the conduction band (see Figure 137).

Exciting an electron out of the valence band leaves an empty 'hole' behind, so photons normally generate free electrons and holes in pairs. In p-type material holes are the majority carriers, so the photoelectron of a hole–electron pair is a minority carrier, and in n-type material the polarities would be reversed.[183]

Now in a reverse biased p–n junction the electric field developed at the junction acts so as to stop majority carriers crossing from one side to the other when the device is reverse biased. However, the field accelerates minority carriers from both sides to the other side of the barrier, and it is this transport of minority carriers that comprises much of the reverse leakage current of a junction diode. The electric field developed in a reverse or zero biased diode sweeps mobile carriers to their respective majority sides, so that a 'depletion region' is created on either side of the metallurgical junction. Photo electron–hole pairs generated in the depletion region rapidly separate and move to their respective sides, and one electron flows in the external circuit for every pair generated. The minority carrier of the pairs generated in undepleted material close to the junction can also

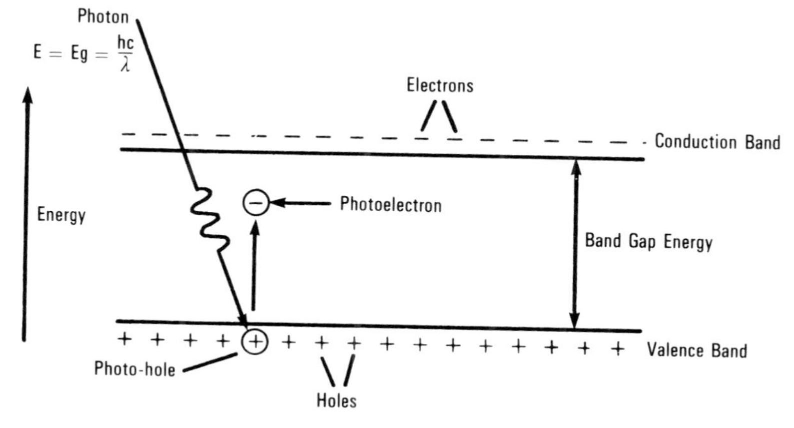

Figure 137. Energy band diagram for photodiode

contribute to the device photocurrent, but it must reach the junction by diffusion so this only applies to pairs generated within about one diffusion length of each side of the depletion region. Thus, photons of suitable wavelength absorbed at or near the semiconductor junction may each cause one electron to flow in the external biasing circuit; in practical photodiodes 100 absorbed photons may cause between 30 and 95 electrons to flow round the circuit, and the 'quantum efficiency' of the detector is said to be between 30 and 95 per cent. The response time of the device is strongly affected by whether absorption takes place in the depletion region or outside since carrier diffusion is a slow process over the distances involved: this is discussed in a later section on response times.

For a photon to excite an electron from the valence to the conduction band, the photon energy needs to be equal to the band gap energy but not much greater, or other effects will cause the detection efficiency to drop again, as discussed later. This means that silicon detectors have a peak response at about 800 nm, germanium ones at about 1400 nm and so on. As the material band gap is narrowed to cater for longer wavelengths, the reverse current flowing across the junction, even in the dark, increases rapidly, particularly at high temperatures. This 'dark current' is usually no problem with silicon photodiodes, but it becomes a severe problem in longer wavelength detectors.

SENSITIVITY OF SEMICONDUCTOR PHOTODIODES

Suitable wavelength photons incident on the active area of a photodiode, such as that of Figure 138, will be absorbed in the depletion depth and contribute to the device photocurrent.

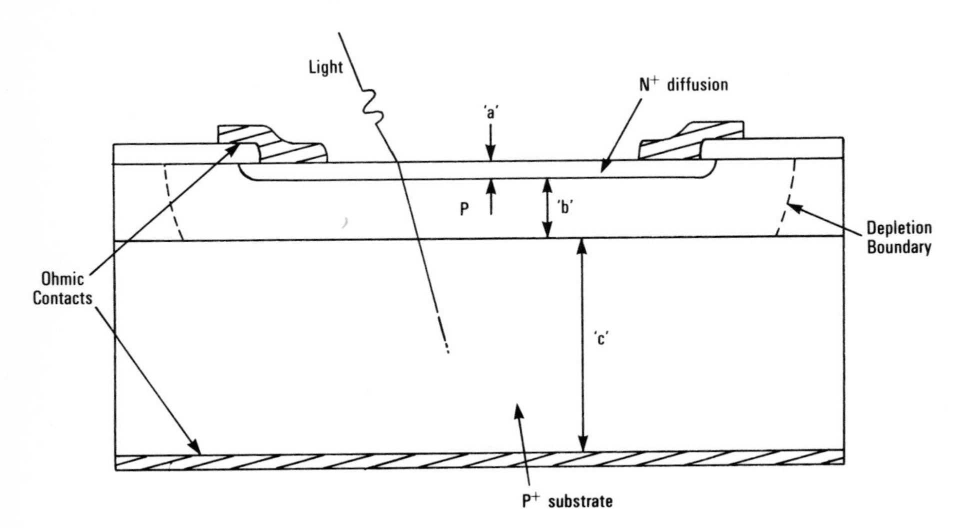

Figure 138. Structure of the planar photodiode

If the photons are absorbed outside the depleted volume, they will create surplus minority carriers which diffuse slowly to the depletion region. We will first assume for simplicity that all absorption takes place in the depletion region, and consider the effects of absorption elsewhere later.

Hole–electron pairs are continually being generated throughout the semiconductor by thermal means. These thermally generated carriers give rise to an unwanted dark current when the device is reverse biased. Dark current has several different contributions which we will consider later.

Avalanche photodiodes (APDs) are designed so that the highest electric field when the device is biased to its operating region occurs in a thin region below the whole of the sensitive area of the device. When photocarriers pass through the thin high field region they receive sufficient energy from the electric field to generate secondary free carriers when they hit the fixed atoms in the semiconductor lattice. So, if the mean multiplication factor is M, then we expect $n \times M$ electrons to flow in an external circuit for every n primary photocarriers. M is only a mean value averaged over many multiplied carriers: the multiplication is a discrete statistical process, and any one primary carrier may generate many more or many less carriers than the mean value. This statistical variation is a source of noise in the multiplied signal current, and complex analyses have been made by McIntyre,[184] Personick,[185] and others to characterize avalanche photodetection. For simplicity we will assume that the multiplication process causes Gaussian noise and the amplitude follows a power law, i.e.

$$\text{multiplication noise power} \propto M^2 F$$

where F is the excess noise factor and varies from under $M^{0.3}$ for good silicon devices[186] to about $M^{1.0}$ for germanium devices. These assumptions hold well at moderate multiplication factors under most operating conditions.

The noise which limits receiver sensitivity is of two main types, as follows.

(a) Thermal, Johnson (or Nyquist) noise.[187] A resistance R at absolute temperature T can supply a noise power kTB into an external circuit. If we represent the noise sources as a noise current generator in parallel with a noiseless resistor, having the equivalent circuit shown in Figure 139 we get:

$$\langle i_{\mathrm{th}}^2 \rangle = \frac{4kTB}{R}$$

where B = bandwidth in Hz, k = Boltzmann's constant, and $\langle x \rangle$ denotes mean value of x.

(b) Shot noise.[187] This is due to current flowing as discrete electrons rather than a continuous electric fluid. The shot noise current due to a d.c. current I flowing over a potential barrier such as a semiconductor junction is given by

$$\langle i_{\mathrm{sn}}^2 \rangle = 2eIB$$

where e is the electronic charge, 1.6×10^{-19} C.

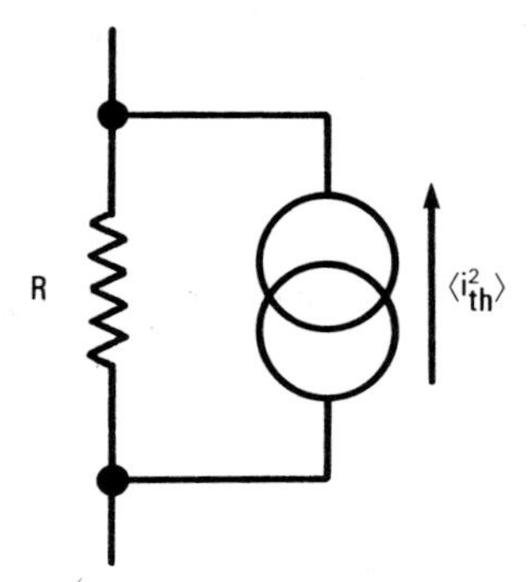

Figure 139. Equivalent circuit of noise source

We have already mentioned avalanche multiplication noise, which is related to shot noise. Generation/recombination noise may be important with photoconductors, and flicker (or $1/f$) noise may have an effect in low frequency systems.

The sensitivity of an optical receiver depends on the modulation method and equalization as well as the photodetector. For simplicity here we will work with instantaneous received optical power and noise power which is independent of modulation. This also gives the same ratio of signal-to-noise as a steady unmodulated optical carrier.

We can now write an equivalent circuit for the photodetector (see Figure 140) where MI_s is the multiplied signal current

$\langle i_n(1)^2 \rangle = 2eFM^2 I_s B$ = multiplied shot noise in signal,

$\langle i_n(2)^2 \rangle = 2eFM^2 I_{DB} B$ = multiplied shot noise due to dark leakage current, and also unwanted background radiation,

$\langle i_n(3)^2 \rangle = 2eBI_{DS}$ = noise due to nonsignal current in the detector which is not multiplied,

$\langle i_n(4)^2 \rangle = \dfrac{4kT_{eq}B}{R_{eq}}$ = thermal noise,

R_{eq} = total equivalent resistance of preamplifier input,

T_{eq} = equivalent noise temperature of preamplifier input circuit.

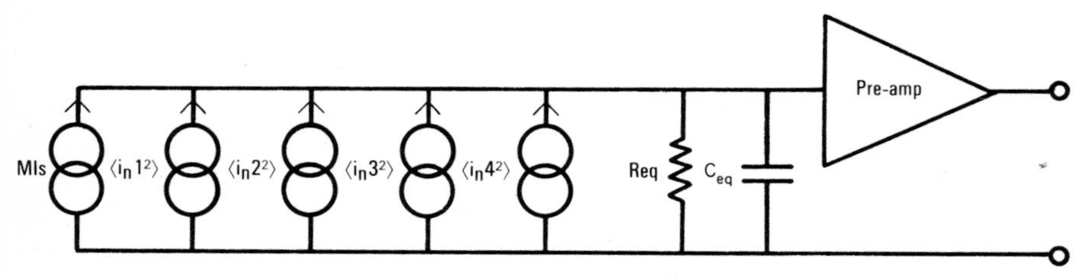

Figure 140. Equivalent circuit of photodiode and preamplifier

The ratio, C/N, of instantaneous signal power to noise power in the preamplifier input circuit is then

$$\frac{C}{N} = \frac{(I_s M)^2}{2eBFM^2(I_s + I_{DB}) + 2eBI_{DS} + 4kT_{eq}B/R_{eq}} \tag{42}$$

This needs to be transferred into terms of optical power input by

$$I_S = RP_L \quad \text{where} \quad R = \frac{e\eta}{hc/\lambda} = \text{responsivity}$$

and

P_L = optical power,

h = Planck's constant,

c = velocity of light,

λ = wavelength of light,

$\frac{P^L}{hc/\lambda}$ = photon arrival rate,

η = quantum efficiency.

As a 'rule of thumb' R is about 0.5 A/W for silicon detectors at 900 nm.

We see that for maximum C/N ratio at a gain input power and bandwidth, I_{DB} and I_{DS} are device parameters, but R_{eq} is variable by the receiver designer. Increasing R_{eq} decreases the RC bandwidth of the detector, but if a perfect equalizer were available, best sensitivity would always be obtained with $R_{eq} \to \infty$. (This is the principle of the so-called 'integrating' optical preamplifier.)[185] For simplicity we assume that the system bandwidth is the RC limit given by

$$B = \frac{1}{2\pi R_{eq} C_{eq}}$$

although we note that an improvement in sensitivity by a few decibels is possible by increasing R_{eq} and equalizing.

The equivalent temperature of the preamplifier input resistance includes the effect of the preamplifier noise factor.[187] In a single path amplifier the equivalent noise parameters of the amplifier increase T_{eq} over $T_{ambient}$, typically by a factor of 2, but a transimpedance type of preamplifier can theoretically have a T_{eq} less than $T_{ambient}$, though in practice it is usually greater.

Referring back to equation (42), we see that the deleterious effect of thermal noise and the dark current would be minimized by $M \to \infty$ if the excess avalanche noise factor were unity. Since F always increases with M there is *always* an optimum avalanche gain, M_{opt}. It is possible under some conditions for M_{opt} to drop almost to unity, but given a proper choice of M_{opt} an avalanche detector should never have worse sensitivity than a nonmultiplying one, assuming other device parameters are equal.

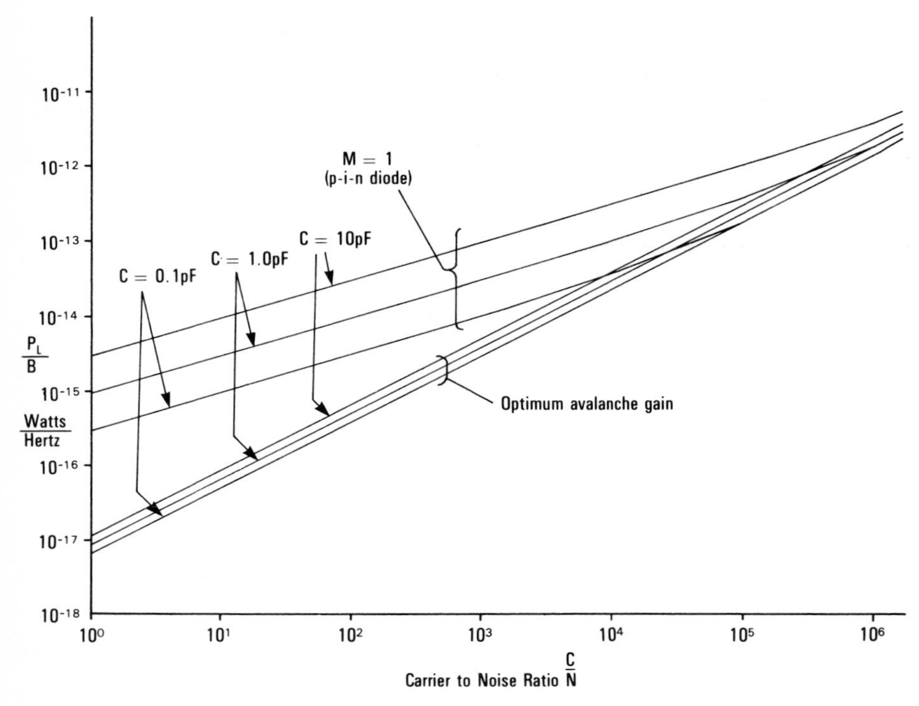

Figure 141. Optical detector power requirements

If we take the restricted case of silicon photodiodes detecting (GaAl)As laser or LED light, the dark current can usually be neglected for system bandwidths above the audio range. In this case equation (42) can be written

$$\frac{P_L}{B} = \frac{\sigma C/Nhc}{\eta\lambda}\left[F + \left(F^2 + \frac{8\pi k T_{eq} C_{eq}}{e^2 M^2 C/N}\right)^{1/2}\right] \tag{43}$$

so we can plot P_L/B explicitly as a function of C/N, with the other terms as parameters. This has been done in Figure 141 assuming $\eta = 70$ per cent, $\lambda = 850$ nm, $T_{eq} = 300$ K, and $F = M^{0.3}$. We see that avalanche gain only helps at the lower end of C/N ratios owing to the effect of shot noise in the signal. We also see that avalanche detectors are less sensitive to device capacitance than p–i–n ones, assuming M is adjusted to

$$M_{opt} = \left[\frac{4kT_{eq}C_{eq}}{e^2 C/N}\right]^{1/2.6}$$

as plotted in Figure 142.

The effect of the dark current on avalanche and p–i–n detectors can be illustrated by plotting the value of the dark current necessary to cause a 3 dB increase in received optical power over the value given in equation (43). The dark current plotted is an equivalent value which is assumed to be all multiplied by the avalanche gain factor. This equivalent dark current may

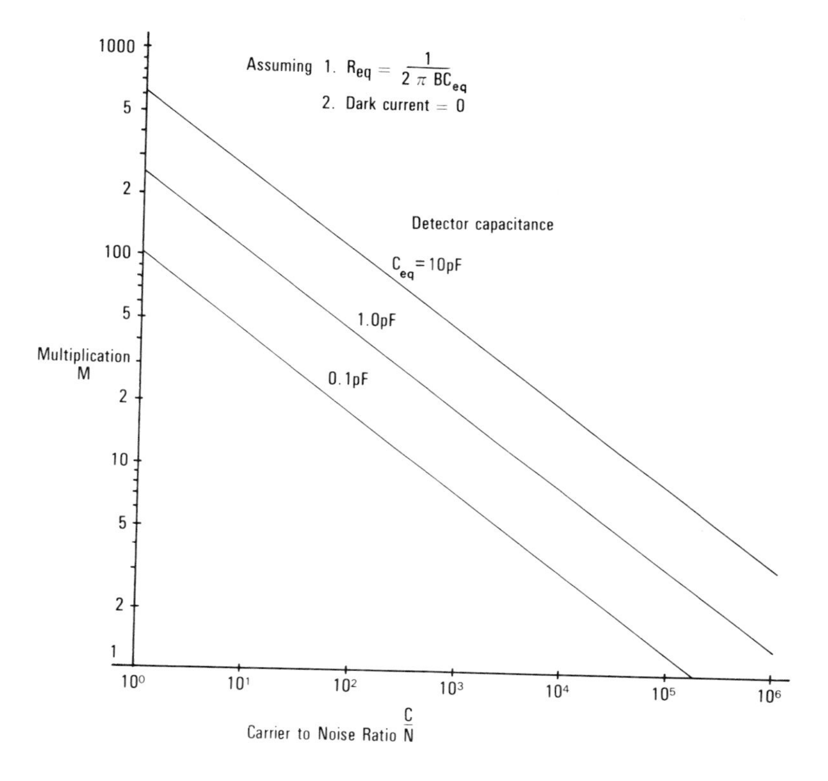

Figure 142. Optimum avalanche gains

be from one to three orders of magnitude lower than the dark current actually measured for an avalanche device that includes surface leakage components which are not multiplied. This partly compensates for the larger effect of the dark current on avalanche detectors than on p–i–n ones, as shown in Figure 143.

In low bandwidth systems the effect of the dark current can be dominant over thermal noise (as in APDs at high gain or photomultipliers for example). Sensitivity curves assuming $\lambda = 850$ nm, $\eta = 70$ per cent, and dark current dominance, are plotted in Figure 144.

The expressions we have derived for sensitivity have been derived in terms of the dark current, multiplication factors, and capacitance. Values frequently used elsewhere are expressed as

noise equivalent power (NEP) in $W/H^{1/2}$,
detectivity $(D) = 1/NEP$, and
area weighted detectivity (D^*).

NEP only scales with $\sqrt{\text{frequency}}$ if the detection system is dark current limited, or if the amplifier input impedance is fixed, and so is little use in

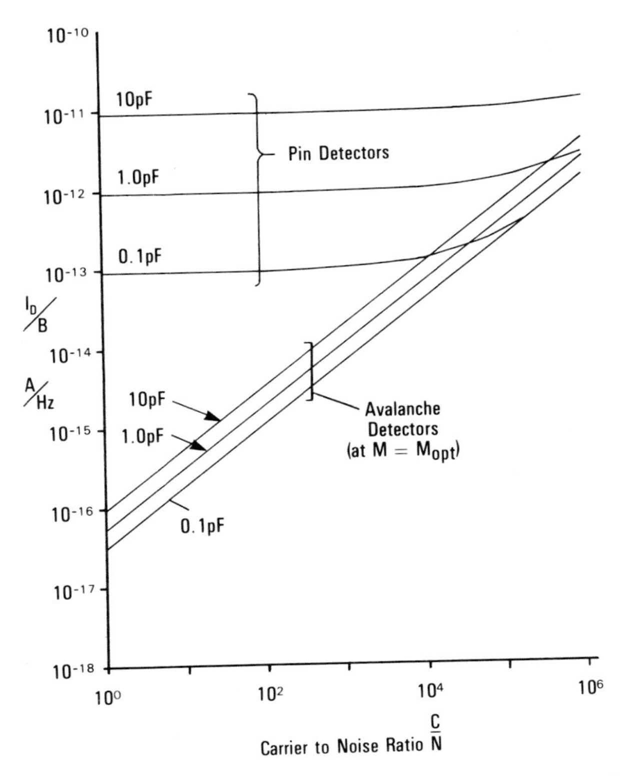

Figure 143. Dark current to cause a 3-dB increase in required optical power level (per unit bandwidth)

calculations referring to photodiodes used in fibre optic systems. Detectivity (D) is simply the reciprocal of NEP and is not very useful either. D^* is a figure of merit to compare detectors in optical systems where the detector is much smaller than the received photon beam. In this case (as for example in star detectors)[199] the greater the area of the detector, the greater the signal. However, in fibre-optic systems the detector can usually collect most of the light from the end of the fibre, so D^* is irrelevant and may be misleading.

RESPONSE TIMES

We have assumed previously that the response time of the detector was determined only by the RC time constant of the photodiode and its preamplifier. In most cases this should be true, but in some cases other mechanisms will dominate, particularly above 100 MHz.

In some photodiodes, e.g. very fast ones where the depletion region is very thin, carriers generated in undepleted material cause a significant part of the device's response to light. In this case the response will have distinct slow and fast components, as shown in Figure 145. The fast component of

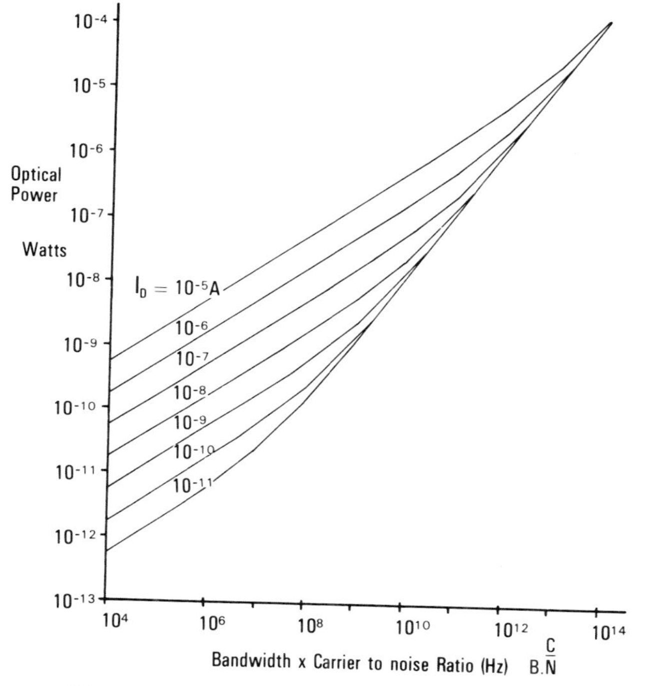

Figure 144. Dark current limited receiver sensitivity

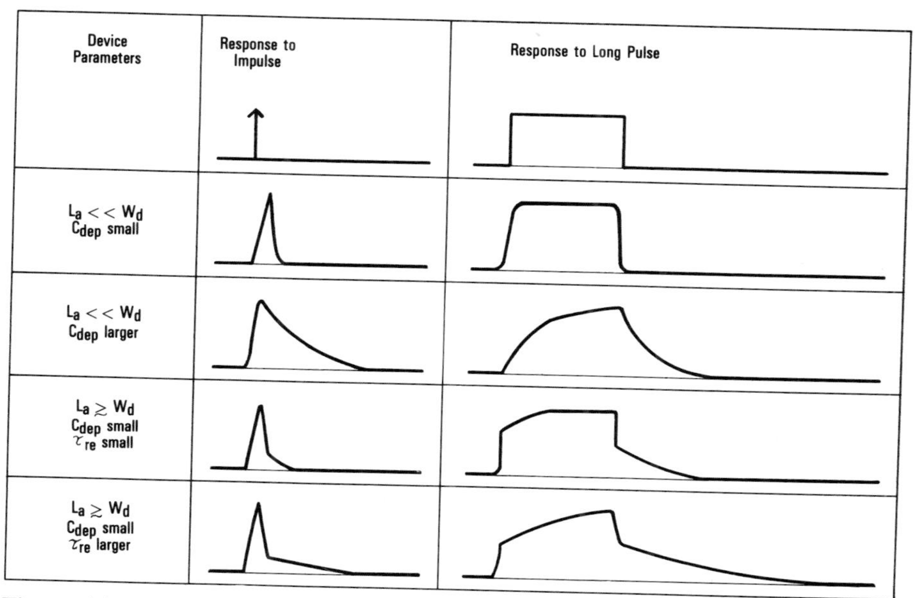

Figure 145. Photodiode pulse responses. L_a = absorption length; W_d = depletion width; C_{dep} = depletion layer capacitance; τ_{re} = semiconductor recombination time

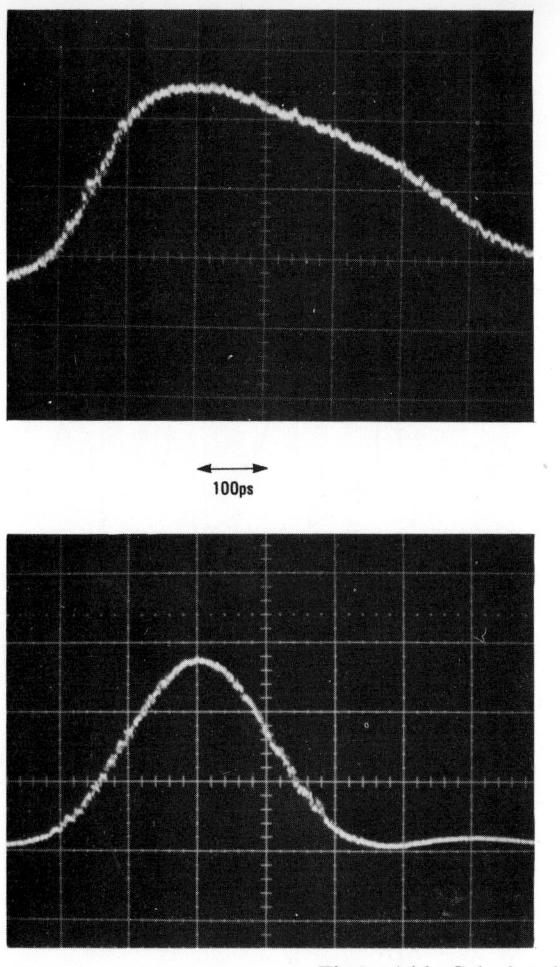

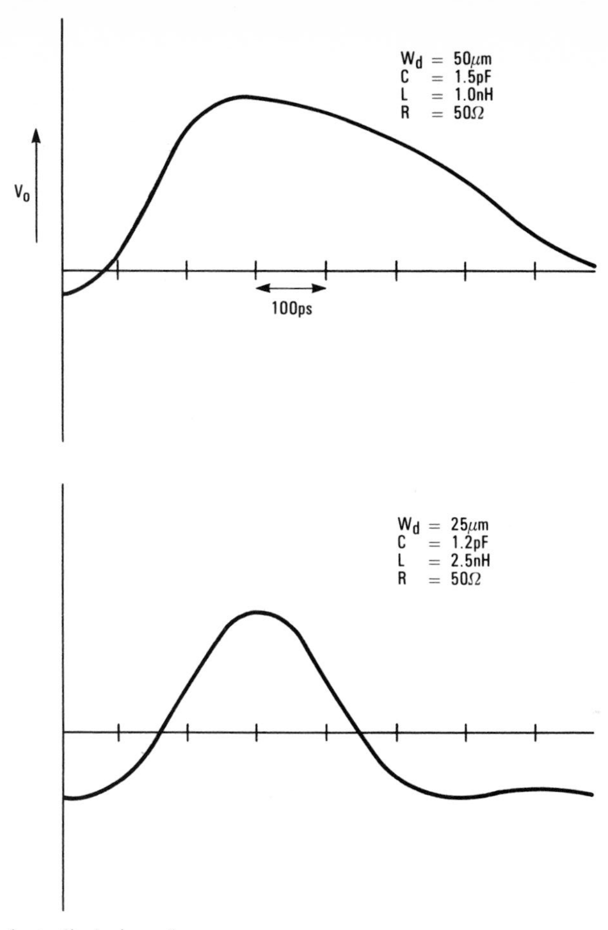

Figure 146. Calculated and measured photodiode impulse responses

the response is due to absorption in the depletion region, whereas the 'tails' are caused by photocarriers generated in undepleted material slowly diffusing to the junction. The decay time of the tails is normally equal to the minority carrier lifetime of the undepleted absorbing regions, so we get more prominent tails from devices diffused in long-lifetime bulk material than in short-lifetime epitaxial material. If a detector with a two-part response is used at or below the cutoff frequency of the slow component then the tails have no adverse effect. However, if the detector is used at a frequency where only the depletion region component is fast enough to be seen, the slow component effectively only causes excess dark current which, as seen previously, reduces the system signal-to-noise ratio. The problem of a two-part response is avoided by 'reach-through avalanche photodiodes' (RAPDs) which have no substrate. For applications where a faster response than that obtainable from RAPDs is required, laboratory examples of substrateless photodiodes down to 6-μm thick have been fabricated.[188]

The depletion layer response time is fast, but not always negligible. Both holes and electrons are subject to a maximum 'scattering limited velocity' which causes a 'depletion layer transit time' of about 10 ps/μm in silicon. Therefore a typical p^+n avalanche photodiode with a 10-μm depletion width would have a minimum rise time of about 100 ps. More significant is the 100-μm thick drift region of reach-through avalanche photodiodes, giving a rather slow rise time of about 1 ns.

A slight caution occurs here when a photodiode is used to detect pulses shorter than its rise time; in this case the rising part of the pulse response as observed on a fast oscilloscope only lasts for the total length of the input light pulse, which may be much less than the true rise time of the device. Under these conditions the rise and fall times of the photodiode are apparently different, and the true rise time is approximately equal to the sum of the observed rise and fall. Even at microwave frequencies semiconductor photodiodes are well behaved, predictable devices. The measured impulse responses in Figure 146(a) compare well[188] with the calculated values using the parameters of depletion width, depletion capacitance, absorption length, load impedance, and package equivalent series inductance for these devices in the equivalent circuit in Figure 147.

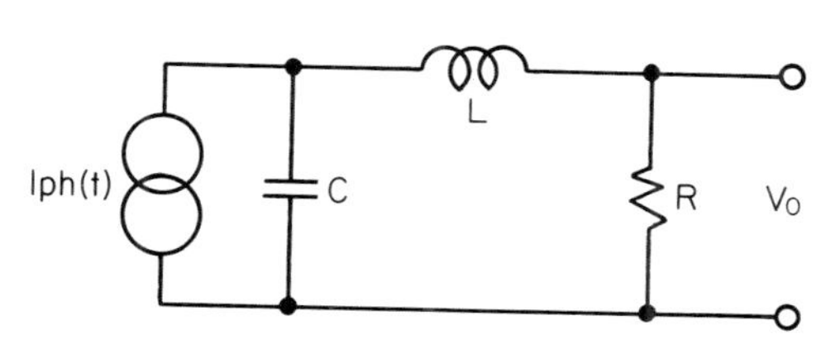

Figure 147. Equivalent circuit for impulse response calculation

PHOTODIODE STRUCTURES, MATERIALS AND FABRICATION

The vast majority of photodiodes are simple devices diffused into silicon. We describe this simple type and then consider the others as variations of this. A heavy surface diffusion is made into a lighter doped material of the opposite type through a window in an SiO_2 masking layer (see Figure 148). A contact ring is made to the outer edge of the diffused area and the central portion then becomes the light sensitive area. If the diode is diffused into epitaxial material, the lower contact can be made directly to the heavily doped substrate, otherwise a heavy contact diffusion may be required.

If the diode is diffused into lightly doped epitaxial material the depletion boundary will 'punch through' to the heavily doped substrate at a low reverse bias and then stop. This is the p–i–n photodiode structure which has a high, uniform electrical field over the whole of the region where light is absorbed. In most cases there is little to choose between the p–i–n and $p^{\pm}n$ diodes.

Photons absorbed in the heavily doped material above the metallurgical junction cause photocarriers which are likely to be captured by surface recombination and do not contribute to the device photocurrent. However, it is difficult to make very high quality shallow junctions, so a compromise junction depth is arranged, usually between 1.0 and 0.1 μm, with the shallow junction devices being said to be 'optimized for blue detection' when made in silicon.

For very short wavelength detection the absorption length would be very small, and so a very shallow junction would be required. An alternative approach here is to use a semitransparent Schottky barrier instead of the diffused junction layer, e.g. Pt on Si with the intermetallic Pt_2Si the actual

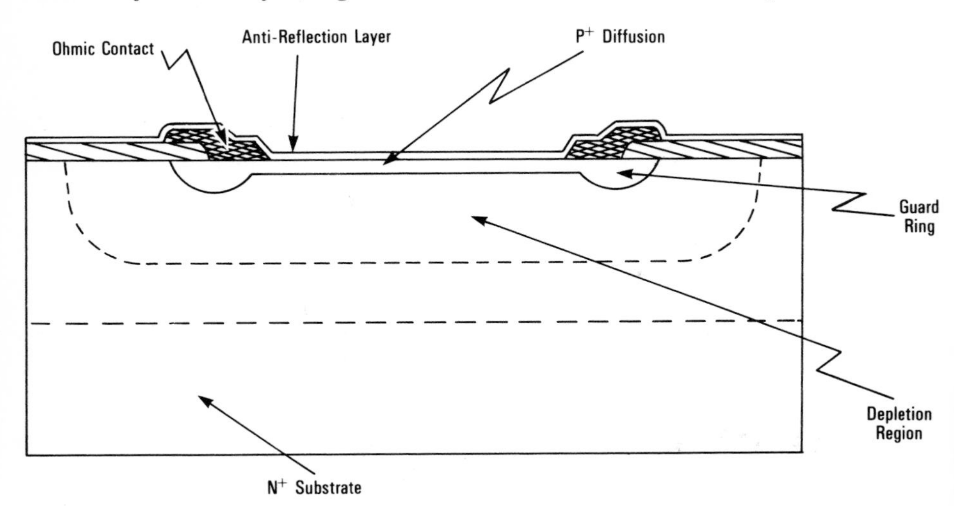

Figure 148. Diffused photodiode with anti-reflection coating and guard ring

barrier. However, the Schottky barrier is liable to have a larger leakage current than the diffused diode, even when a guard ring is diffused around the periphery of the barrier in order to reduce the high peak fields there.

Schottky photodiodes can be operated in the avalanche mode if a guard ring is diffused to cause the peak field to occur away from the edges of the device—the classic $p^{\pm}n$ diffused avalanche photodiode also benefits from such a guard ring.

The leakage current in a photodiode[183] is caused by:

(a) Shockley reverse leakage past the potential barrier as a result of minority carrier generation;

(b) generation/recombination in the depleted volume of the device; and

(c) surface and other extraneous leakage.

The surface leakage component, (c), is usually dominant, and so is approximately equal to the dark leakage current measured at the device terminals. All the components depend on the device processing and packaging as well as the structure and material, and all normally increase quasi-exponentially with temperature, but surface leakage often departs significantly from this behaviour.

For the best sensitivity we have shown that the detector capacitance must be minimized, even for low frequency detection. This can be done either by reducing the device area or by increasing the i-region width p–i–n diodes. The active area of the detector may not be decreased indefinitely or alignment and manufacturing problems become severe, and also the small resulting areas for ohmic contacts can result in excessive contact resistances. The i-region width should not be increased too much either because this increases the depleted volume, which in turn increases the depletion layer generation leakage current. Also, large depletion widths can cause excessively slow transit times and hence detector rise times in some fast detection applications.

For any application it is possible to optimize the device area, doping densities and junction widths, although a silicon p–i–n device with a diameter of 200–500 μm and 100-μm i-region width will satisfy very many fibre-optic requirements.

Avalanche Photodiode Structures

The prime requirement in an avalanche detector is that photogenerated carriers should pass through a region of very high electric field which, for reasons of response time and noise performance, should also be narrow. There are two common structures which do this:

(a) the $p^{\pm}n$ (or $n^{\pm}p$) diode, and

(b) the $p^{\pm}n$–i–n^{+} (or $n^{\pm}p$–i–p^{+}) diode.

As shown in Figure 149 the sharply peaked field profiles ensure that ionizing

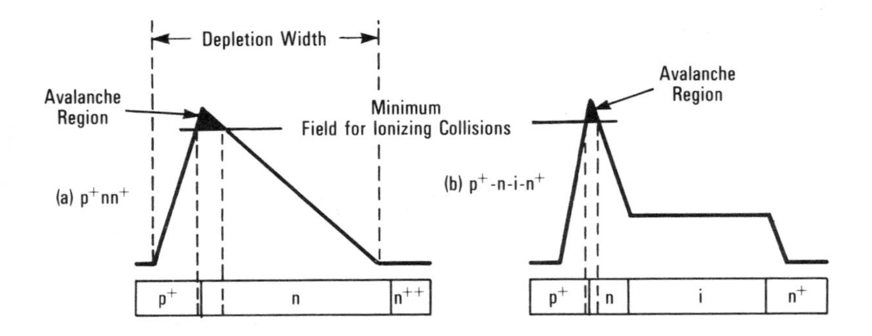

Figure 149. Avalanche photodiode field profiles

collisions only take place in a narrow region. Structure (a) is very much easier to fabricate, but (b) allows wide depletion regions suitable for long wavelength detection without the need for huge operating voltages.[191]

We can assume to a first order that the multiplication factor M is a function only of the peak junction field $\hat{E}$ over the breakdown field of the material E_B, i.e.

$$M = f_1\left(\frac{\hat{E}}{E_B}\right)$$

and in turn the breakdown field is a function of temperature. Similarly, it can be shown that a change of multiplication ratio, δM, depends to a first order on

(i) a change in device temperature, and

(ii) change in the ratio of the bias voltage to the depletion width, i.e. δM is a function of $(\delta T, \delta V_b/W_{dep})$.

This relationship implies that the temperature coefficient of the bias voltage for constant M is approximately proportional to the depletion width in any given material. So wide-depletion APDs have high temperature coefficients; however, extending the same argument will show that an APD with a wide depletion region will change in multiplication with bias voltage more slowly than would a narrow one.

These bias voltage variations are important when the APD has to operate over a wide temperature range. Temperature-tracking bias voltage regulators cause an undesirable increase in the complexity of an optical receiver subsystem. However, in a fibre-optic system it is often convenient to keep the mean received optical power level constant, and in this case an AGC loop can be used to control the bias voltage so that the mean APD current keeps to its expected value. In the event of an interruption in optical power, the bias voltage will increase until the APD goes into hard breakdown, but this is unimportant if the current is limited to some low value.

A neat way of applying bias with AGC to the APD is to feed it from a constant current source decoupled at signal frequencies (Figure 150). If the

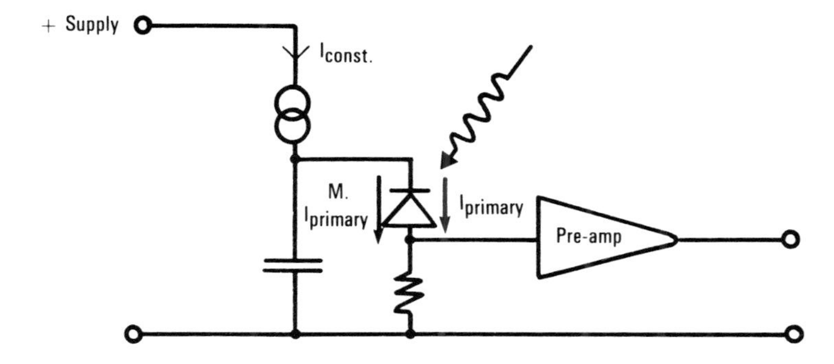

Figure 150. 'Constant current' biasing circuit

photodiode dark current is negligible then the bias voltage is forced to the value which causes the APD multiplication to reach its desired value, i.e. the equation

$$I_{\text{const}} = (M+1)I_{\text{primary}}$$

must hold at all times.

In low bandwidth optical links the APD dark current may not be negligible with respect to the primary (unmultiplied) photoelectron current. In this case some other bias control system must be used, as is also the case when the mean received optical signal level is not expected to be constant. The method used by EMI to overcome this problem is illustrated in Figure 151. Two APD chips are mounted on the same header, but only the one mounted centrally is exposed to the optical signal. The other chip is biased to breakdown and is used as a temperature compensated voltage reference for the first (active) device.

Silicon avalanche photodiodes commonly have total dark currents at room temperature in the region 10–100 nA. However, the majority of the measured terminal dark current will normally be contributed by surface leakage, and the bulk leakage current originating from the active volume of the device may be orders of magnitude lower. Since it is the noise due to the multiplied *bulk* dark current that is most significant in APD sensitivity calculations, such detectors may still give good performance when the total dark current is relatively high.

It should be appreciated that full specification of APD performance is difficult, and manufacturers' data in general only give enough information to roughly predict the sensitivity of their devices in a real system. It can easily be shown that nonuniformity of gain has an adverse effect on noise performance, e.g. by considering the detector as two separate devices operating at different gains and then adding r.m.s. noise powers. Small areas of high gain, or 'spikes' in the gain profile often exist, particularly at high gain, and these have a disastrous effect on noise performance. Small areas of low gain have much less effect. Figure 152 shows the result of scanning one

Figure 151. EMI double chip avalanche photodiode

APD with a light spot in a raster pattern, and displaying its response as a pseudo three-dimensional image. We see here that this device has a line of reduced gain running across an otherwise uniform active area. While this looks bad, its noise performance would be little worse than that of a similar diode with a perfectly flat spatial gain profile, assuming that the device is uniformly illuminated.

Multiplication Noise Statistics

If we consider a $p^{\pm}n$ APD with an operating voltage of 100, since the ionization rate is a very sharp function of field strength the voltage dropped across the region where ionizations are likely may be only 10–20 V. And since a primary carrier must receive at least the band gap energy before it can cause an ionization we see that a maximum of about 10–20 ionizing collisions would be possible from a single primary photocarrier. This emphasizes the discrete statistical nature of avalanche multiplication.

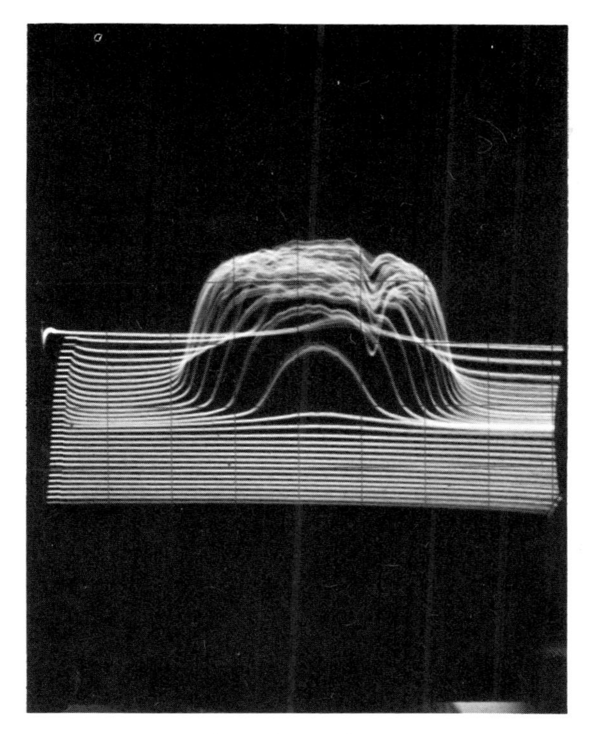

Figure 152. Spatial gain profile of an avalanche photodiode

If we take a material where the electron ionization coefficient, α, equals the hole ionization coefficient, β, e.g. germanium, then the secondary photocarriers (see Figure 153) always transverse the same total width between them, and the probability of tertiary and further photocarriers is independent of the position of the primary ionizing collision. At breakdown there will clearly be one further carrier generated for every traverse of the

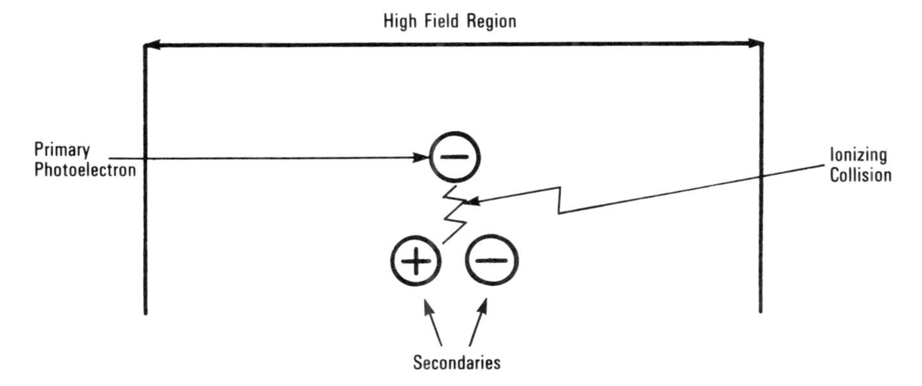

Figure 153. Avalanche electron–hole pair generation

high field region, so

$$NP(N)+(N-1)P(N-1)+(N-2)P(N-2)+\cdots+2P(2)+1P(1)=1$$

where N is the maximum number of ionizations possible per traverse and $P(M)$ is the probability of M secondaries.

We can therefore work out $P(1)$, the probability of ionization at a single collision, and hence $P(0)$ the probability of a primary photocarrier traversing the high field region without collision. In this case, where hole and electron ionization coefficients are equal, we find that almost half of all generated photocarriers will cause no secondaries at all.

In a material where only electrons cause ionizations (i.e. $\alpha/\beta \to \infty$), the multiplication must clearly take place in a single pass, and if N ionizing collisions are possible, the maximum gain is 2^N. Also note that the avalanche response time is only one avalanche width transit time, and that classical junction breakdown never takes place.

Now we can work out the probability of ionization of a single collision for a given mean gain, and again work out the probability of a primary photoelectron passing right through the ionization region without generating secondaries. The probability of no ionization is much less when $\alpha/\beta \to \infty$ than when $\alpha = \beta$, so it is not difficult to see that for best avalanche noise performance α/β should be as large as possible if electrons are the predominant primary photocarriers.

If light is absorbed on the side of the junction such that holes are the injected photocarriers in the avalanche region, where $\alpha \gg \beta$ the noise performance will be correspondingly poor.

Materials

Silicon and germanium have been used much more than other materials to make junction photodiodes, and silicon has been used far more than germanium, mainly because the leakage currents of silicon diodes can be as low as a few picoamps at room temperature, whereas germanium diodes are quite good if they have leakages less than 1 μA. Also, α/β in silicon under normal APD operating conditions is about 20, whereas $\alpha \simeq \beta$ in germanium, so the avalanche performance of silicon is much better.

However, silicon detectors are becoming inefficient when working at 1.06 μm, and optical communications systems using even longer wavelengths are being proposed. It is therefore necessary to develop detectors from a material (other than Ge) with a smaller bandgap than Si. We find that there are no useful single or binary semiconductors with bandgaps in this range,[192] so tertiary and quaternary materials have to be considered.

Encouraging results have been reported from $Ga_xIn_{1-x}As$[193] and $GaAs_{1-x}Sb_x$[194] devices where use has been made of the ability to vary the bandgap across the thickness of the device, so that absorption takes place where it is wanted. In the long term it would appear best to use a quaternary material in which the bandgap and lattice constant can be varied independently. The possibility then exists of 'tailoring' basically the same device to work best at one of a range of wavelengths (in the range 0.5–3 μm for GaInAsP). As mentioned before, it is an advantage for avalanche devices if the hole and electron ionization coefficients are widely different. However, too much weight should not be put on achieving $\alpha \neq \beta$ because

(a) useful gain is still available if $\alpha = \beta$, and

(b) referring back to the sensitivity curves, we see that if a p–i–n detector amplifier could be made with a total equivalent capacitance of about 0.1 pF, the difference between p–i–n and avalanche sensitivities at a typical minimum system C/N ratio of 18–20 dB would only be about 10 dB. This difference between p–i–n and APD is further reduced to about 4 dB if advanced preamplifier and equalization techniques are used.

We conclude that it is more important to have a good low leakage detector with high quantum efficiency at the right wavelength than one with optimum avalanche noise performance.

Detectors Other Than Semiconductor Photodiodes

While the semiconductor photodiode is likely to remain the most popular photodetector is fibre-optic systems, there are circumstances in which other detectors or detection methods may be more appropriate.

The following paragraphs give brief descriptions of other photodetectors so that they can be compared broadly with semiconductor photodiodes; more detailed accounts will be found in the references.

Vacuum Photodiodes and Photomultipliers

Whereas in a semiconductor photodiode the excited photocarriers move from one energy band to another, in a vacuum photodiode the photoelectrons are physically emitted into the space between the photocathode and the anode, and are attracted through a vacuum to the latter. Vacuum photodiodes are usually very much larger than their semiconductor counterparts, and they can combine large active areas without excessive anode/cathode capacitance since this latter depends on shape and not size.[195] Since electrons *in vacua* are not limited in velocity by collisions with a crystal lattice, response times of vacuum photodiodes are no larger than those of their much smaller semiconductor counterparts.

A major drawback to the vacuum detectors in optical communications is that most photocathodes are only sensitive to visible or near-visible light. Only recently have photocathodes with quantum efficiencies greater than about 1 per cent for GaAs laser light become available. These new devices, incorporating negative electron affinity photocathodes (as with other vacuum photodiodes), are much larger, more fragile, and more expensive than semiconductor photodiodes, and normally require much higher operating bias voltages. However, their large sensitive areas make them very useful in some applications.

The avalanche photodiode also has its vacuum equivalent in the photomultiplier. In these devices photoelectrons are accelerated and then multiplied by collisions with special surfaces known as 'dynodes' before collection by the anode. Gains are commonly between 10^4 and 10^{10}, with very low noise and bandwidths in the gigahertz range for 'crossed field' and 'channel plate' types.[196] Therefore, despite their inconvenient operating voltages, bulkiness, and expense, photomultipliers can be very high performance devices, and will continue to have numerous specialist applications.

Photoconductors

Some semiconductors have a high resistivity at room temperature so that if contacts are made to them and a voltage applied, only a small current will flow, even without the barrier of a p–n junction.[197] As in a junction photodiode, incident photons cause band-to-band movements of carriers and extra current can flow in an external circuit when light is incident.

There are two main areas where photoconductors are preferred to photodiodes:

(a) in the mid and far infrared where junctions are difficult to make successfully, and

(b) in slow response applications with visible and near-visible light. This is because the photoconductor gives an effective gain, G, which is roughly equal to the ratio of the mobile carrier transit time τ_t to the effective carrier lifetime T, i.e.

$$G = \frac{T}{\tau_t}$$

G may be 10^5 or more, but the response time is T or greater where this is usually milliseconds or more. Therefore such photoconductors are normally limited to low audio bandwidths.

It is possible to 'pump' photoconductors with an electrical signal greater in frequency than the optical signal to be detected, in which case the useful response can be extended into the microwave region. However, the noise performance is worse than a good junction photodiode, so this mode of operation is only sensible when good junction devices are not available.

Bolometric Detectors

If optical energy is allowed to fall on the blackened junction of a thermocouple, the absorbed light will cause a temperature rise which in turn causes a photocurrent to flow. This type of detector has the advantage that it is not selective of the wavelength of detected light. However, the frequency response is likely to be very low. Better performance is often gained from pyroelectric detectors, where the temperature difference between two sides of a special crystal causes an external current to flow. These devices are usually (but not always) limited in frequency response[198] and the sensitivity is low when compared with junction photodiodes, so they are unlikely to find many applications in fibre-optic systems.

Other Detection Methods

Various other devices and techniques for optical detection exist, e.g. bipolar and field effect phototransistors have been made, and detection by photomixing with coherent light has been proposed.[199] These methods of detection can be useful in specialized applications, but the fibre-optic system designer is better off using discrete photodiodes and amplifier transistors, and photomixing can be ignored for the present. Since avalanche photodiodes are close to fundamental quantum detection limits anyway in the near infrared, no great improvements are possible in detector sensitivity by any means. Similarly, if visible light is used in fibre-optic systems, vacuum photomultipliers cannot be much improved upon by other detection methods.

CHAPTER IX

System Design

INTRODUCTION

In any line communication system, there is a maximum distance that can be accommodated between transmitter and receiver. Where terminals are more widely spaced than this, as in many trunk and junction systems, it is necessary to insert repeaters at intervals, as shown in Figure 154. Each repeater consists of a line receiver, an amplifier, and a line transmitter. In addition, it may perform regeneration and retiming for pulse code modulation (PCM) signals, and also provide supervision, alarm, and engineering order wire facilities.

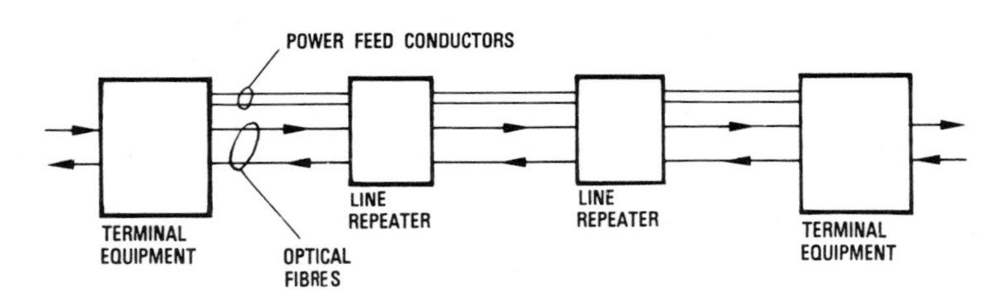

Figure 154. Use of repeaters in a line communication system

Widespread research and development on fibre-optic communications systems for PTT applications is a result of the projected economic advantage over alternative systems, such as PCM on coaxial cable. Economic comparison is complex but the advantage of optical fibre systems is largely due to their increased repeater spacing compared to coaxial systems (see Chapter X). Thus, in long haul systems the total number of repeaters is correspondingly reduced. In junction systems and future long haul systems the economic effect may be even greater because distant terminals can be directly connected by low loss optical cables without the need for a line repeater. The length of an optical link without the need for intermediate amplification depends on both the fibre characteristics and the performance of the optical transmitters and receivers.

This chapter discusses the design of optical fibre systems with special emphasis on digital systems.

OPTICAL TRANSMITTERS

Optical Sources

Semiconductor optical sources are now generally accepted as the correct choice for fibre optical systems. These are semiconductor diodes that produce light when a current is passed. Direct modulation of the intensity can be achieved by modulating the drive current; the response is fast enough for very high information rates.

Two types of semiconductor diode source are widely used in optical communications systems: injection lasers and light emitting diodes (LEDs). Both devices work by emitting photons as electrons fall from the conduction to the valence bands, the essential difference being that the laser structure gives rise to stimulated emission, whereas LEDs work solely by spontaneous emission. As a result, the laser is much brighter, enabling more optical power to be launched into a given fibre; it also has a faster reponse. The line width (or range of operating wavelengths) of a laser is narrower than for a LED, thus increasing the maximum bandwidth available from fibres—an important factor in the highest capacity long haul systems. LEDs have generally been considered to be more suitable for linear systems than lasers since they exhibit a smooth nearly linear power/current law whereas the laser response can exhibit nonlinearities. However, as these are related to multimode effects, the development of single transverse mode lasers is likely to provide excellent linearity over a useful working range.

The life and reliability of semiconductor optical sources have received intensive attention during the evolution of optical communications, since at one time it seemed possible that the poor life of high brightness sources would seriously restrict the applications. Improvements have been so great that lasers from several laboratories have been working continuously for 10,000–20,000; extrapolation has led to predictions of 10^5 h working life, as discussed in Chapter VII on optical sources.

These increases will largely result from the rapidly developing understanding of failure mechanisms. For LEDs, extrapolation from results at 190°C has led to a prediction of over 10^7 h operation at room temperature.[230]

The wavelength of operation depends primarily on the bandgap and hence on the device material. Gallium aluminium arsenide (GaAlAs) sources, presently the most advanced, generally operate at around 800–900 nm, depending on the Al concentration. There is a growing interest in materials operating above 1100 nm so as to exploit the improved attenuation and bandwidth properties of fibres at these wavelengths.

Optical Transmitter Circuit

In direct modulation of semiconductor sources the optical power output is simply controlled by the applied current. The starting point for the optical

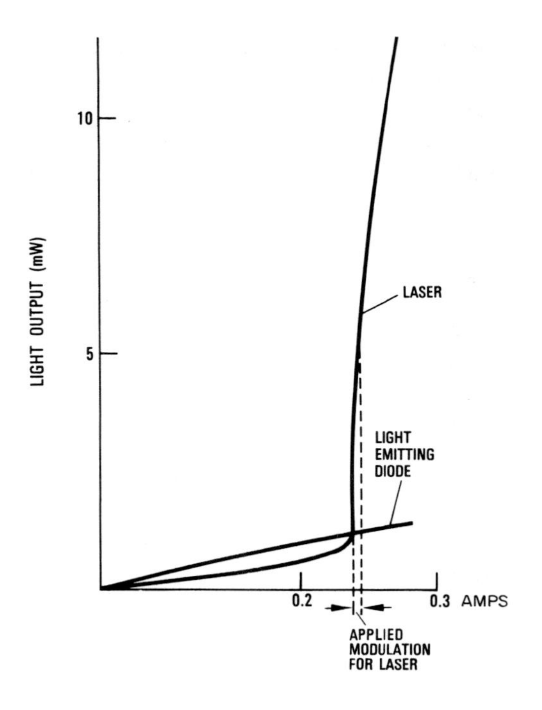

Figure 155. Typical output/drive characteristics for a continuous wave laser and a LED

transmitter design is the power/current characteristic, shown for an LED in Figure 155. Also in the figure is the corresponding curve for a laser which shows the existence of a threshold current above which stimulated emission occurs.

The transmitter circuit switches the current that modulates the device. In the case of a LED operating under a binary (two-level) PCM, this may simply involve switching between zero and some preset maximum. In the case of the laser, some spontaneous emission occurs at low currents where the device behaves just like a LED. As the current is increased a threshold is reached, I_{th}, beyond which lasing occurs and the slope efficiency steepens considerably. For the device to be used as a laser the current must exceed this threshold.

There are various limitations which must be taken into account when considering the means of modulation. The first limitation is thermal. There is a practical limit to the power which can be dissipated by the laser chip via its heat sink and, moreover, the threshold current increases with junction temperature, thus requiring more and more drive current which increases the junction temperature, and so on. If the threshold current is greater than the current at which the junction heating becomes significant, as is the case for most broad contact lasers, then the laser will not be capable of continuous operation (without cooling). Such lasers are, therefore, operated in the pulsed mode only. Lasers capable of continuous operation have been developed in more recent years, the necessary improvement being achieved by making the lasing region a stripe narrower than the chip. This reduces the threshold current whilst retaining the low thermal resistance to the heat sink. These c.w. devices are in general superior to the broad contact devices in all respects, except perhaps for peak power and manufacturing cost.

The second limitation is that of peak optical power output. The laser output may become 'noisy' at high power levels and these regions should be avoided if system performance is not to be impaired. If the power is excessive then the cleaved partially reflecting end faces can 'burn off', thereby permanently damaging the laser.

The third limitation is that of turn-on delay. If a pulse step of current is applied to a laser to take it above threshold then there may be a delay of a few nanoseconds before laser emission appears. There is no turn-off delay however, so the output pulses will be shortened. The further over threshold the laser is driven, the smaller the delay. This pulse shortening, together with the thermal limitations on operating duty cycle, significantly limit the maximum operating PRF of pulse lasers, as shown in Figure 156. It will also be seen that there is a lower frequency limit determined by the self-heating time constant of the junction.

The c.w. laser overcomes these maximum operating frequency limitations in two ways. First, the maximum PRF is proportional to the duty cycle and, secondly, it is possible to eliminate turn-on delay completely by prebiasing the laser to the threshold current. This, of course, is not possible with pulse

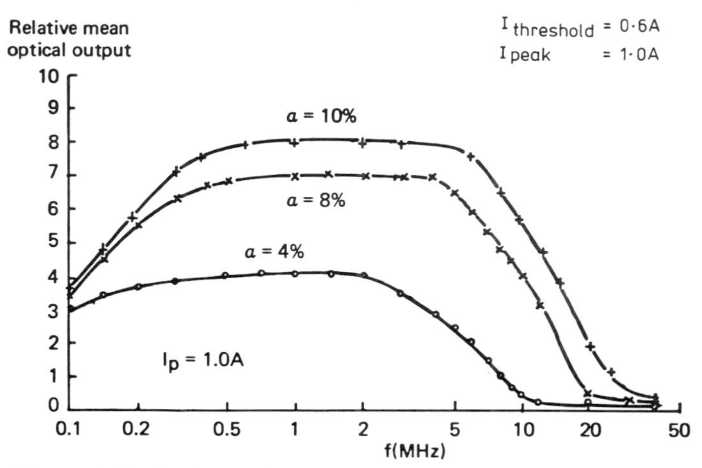

Figure 156. Early broad contact laser response to pulse trains at various frequencies

lasers owing to thermal limitations. The maximum modulation frequency of c.w. lasers when biased in this way is typically several gigahertz, quite sufficient for most applications.

Overall optical feedback is usually used with c.w. lasers but seldom used with the broad contact lasers. The threshold current of injection lasers increases with temperature and time and differs from device to device. This causes the operating point to vary if the drive current is held constant. Whether or not this variation is tolerable is determined by the slope efficiency above threshold and by the tolerable range of variation of the operating point. For example, high slope efficiency, high temperature coefficient of threshold current, large operating temperature range, and the requirement of long unattended operating life are the conditions necessitating optical feedback. It is possible to overcome the temperature coefficient problems by constant temperature operation but this will not compensate for degradation and is impractical for most systems.

The most common form of optical system is the 'mean power' control system in which the average power output is automatically adjusted to be constant (Figure 157). The photodetector mounted next to the rear face of the laser package can be seen in Figure 158 and the whole mounted on the transmitter board of a repeater is shown in Figure 159. This can be a very simple system using a cheap, slow photodetector for the monitor. The performance at low frequencies can be improved by using a normalized time average of the modulating signal as the reference (Figure 160). The time constant of the filter should be equal to the time constant of the photodetector capacitance and the associated load. This technique provides a response down to d.c. and is interesting in that it provides real time feedback at d.c. and low frequencies and mean power feedback at high frequencies. Both techniques are suitable for analogue and digital systems.

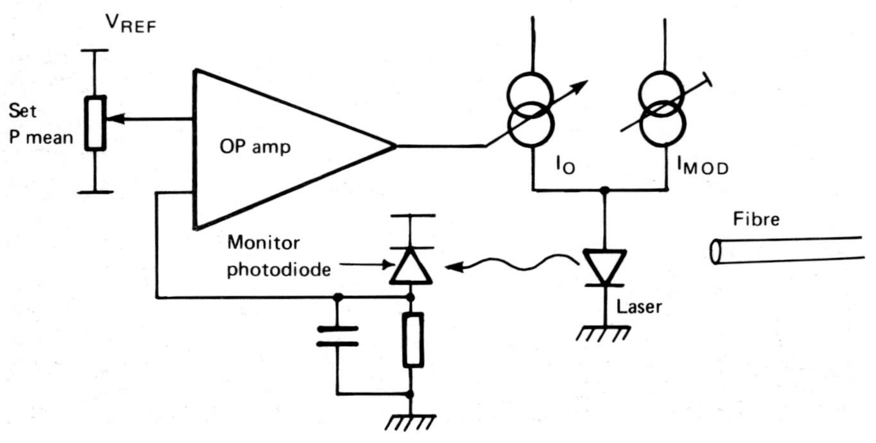

Figure 157. Mean power feedback

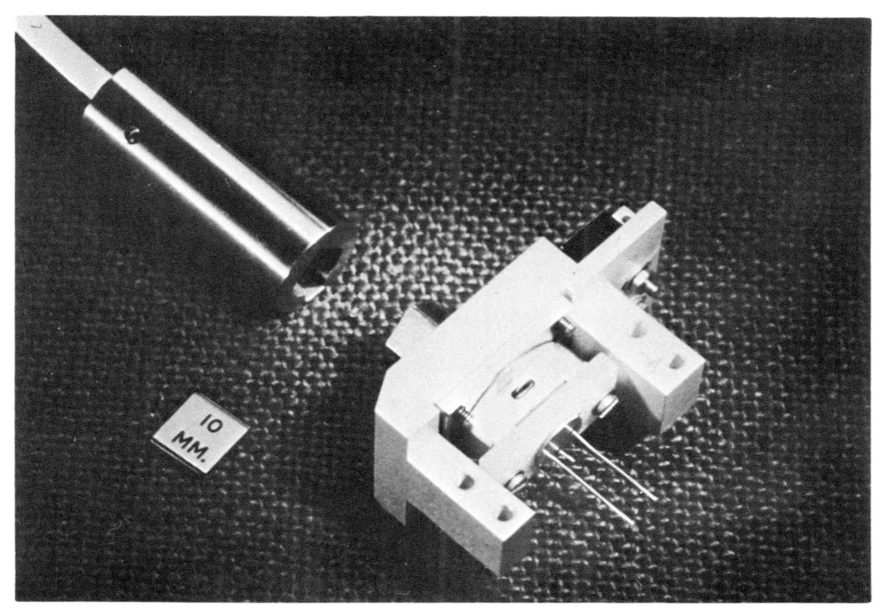

Figure 158. Laser package with photodiode for optical feedback control

Figure 159. Transmitter subsystem

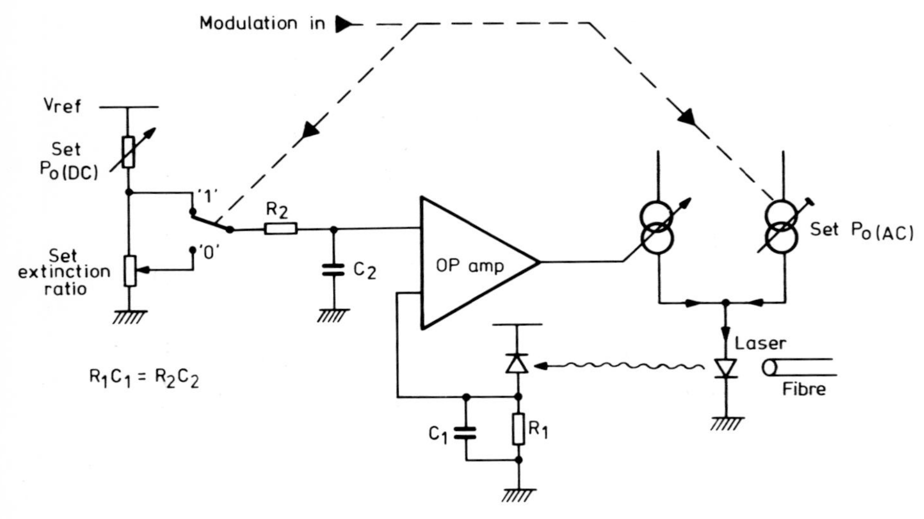

Figure 160. Mean power feedback with d.c. response

The main disadvantage of mean power feedback is that there is no compensation for variations in slope efficiency. The modulation current must be preset for each laser. The slope efficiency varies with temperature for some devices and sometimes changes during life. It also differs considerably from device to device. These variations cause few problems when the laser is driven from zero bias, e.g. low bit rate systems, but for high bit rate systems the need for good control of the zero level may necessitate some form of feedback control of the modulation current. One such approach is the 'sampling feedback control system' (Figure 161) in which the monitored optical waveform is sampled in both the one level and the zero level, compared with the preset demanded values and the bias current and modulation current adjusted accordingly. This method, therefore, accurately defines the optical one and zero levels.

Another method of controlling the zero level is to measure the turn-on delay which occurs when the zero level is set below threshold and feedback is set to a constant fixed delay to control it. Because the delay falls to zero at threshold, the loop will only stabilize for a finite delay, i.e. somewhat below threshold.

The main problem with both these basic methods is the need for a wideband monitor detector and associated circuitry. This leads to increased cost and complexity and puts additional constraints on the laser package design. The basic sampling feedback system needs a bandwidth at least as large as the main transmission system whilst the turn-on delay method requires a bandwidth many times greater to enable the small values of delay to be measured.

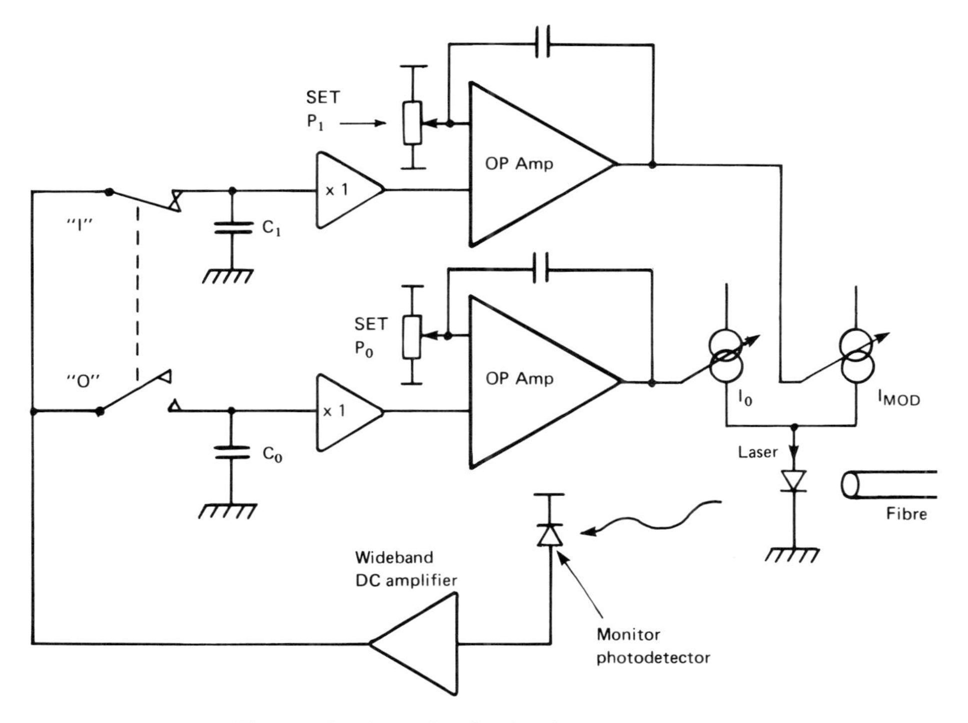

Figure 161. Sampling feedback control system

Fortunately, it is possible to operate a sampling feedback system with a much reduced bandwidth by making use of the fact that sequences of successive zeros or successive ones frequently occur. All that is necessary is to decode the time of these sequences, and to sample the signal only at those instants allowing the slower monitor photodetector circuit time to charge or discharge to the correct level. The control loop bandwidth must be reduced to allow for the lower sampling frequency; however, this presents no difficulties since the control loop only has to follow slow temperature changes and the very slow degradation of the laser threshold.

Turning now to the choice of transmitter duty cycle, low duty cycle operation is very attractive where there is no peak power limitation due to damage or noise. There are three reasons for this. First, the peak electrical power from a receiving photodiode is proportional to the square of the input peak optical power. So, halving the duty cycle for the same mean power would give four times the peak electrical power from the detector whilst the twofold increase in bandwidth required would only double the noise; the result is a doubling of the signal-to-noise ratio at the output of the receiver amplifier. The fibre bandwidth may limit the improvement possible from high peak power, low duty cycle operation. The second advantage is that the ratio of peak current to threshold current becomes large, making the

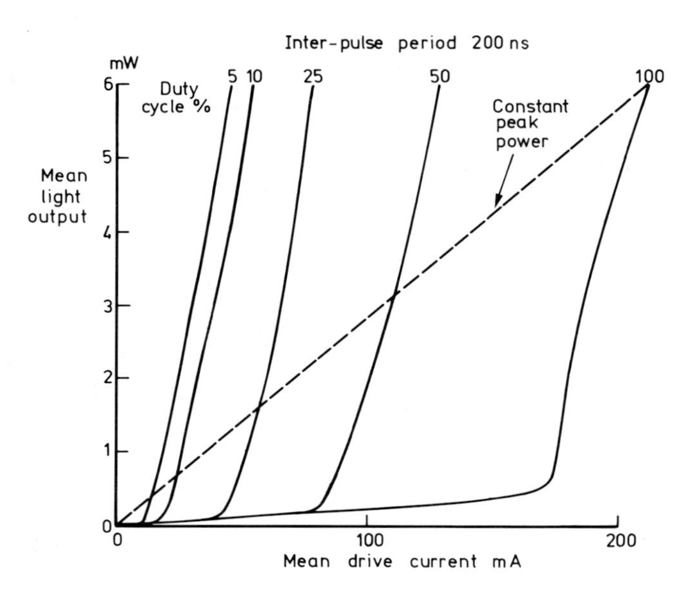

Figure 162. Effective laser characteristics for various duty ratios (20-μm laser)

variations in threshold much less significant. The third advantage is that of overall power efficiency. The efficiency above threshold is many times greater than below threshold, and therefore driving well over the threshold at low duty cycle gives higher overall efficiency. Figure 162 shows the effective laser characteristic when plotted as mean optical power output versus mean current and clearly shows the reduced laser current consumption at low duty cycle for the same mean power output. A further advantage lies in the fact that the laser is a forward biased diode with low series resistance and thus the mean dissipation is reduced at low duty cycle (since the voltage drop does not increase proportionally to the drive current). Advantage can be taken of this by allowing a longer period of degradation before junction heating starts to accelerate the degradation. Alternatively, less heat sinking is required for the same life.

Another consideration for binary pulse modulation of lasers is the extinction ratio. Although the receiver output signal depends only on the swing of optical power between P_0 and P_1, the receiver noise level is affected by the mean optical power, especially for a near ideal receiver. This means that, it is desirable to minimize P_0 particularly for low duty cycle working.

OPTICAL RECEIVER

Photodetection

As discussed in the previous chapter, two types of photodetector are suitable for the detection of high speed signals in the near infrared. These are vacuum tube devices in which a photocathode generates free electrons in response to light, and semiconductor diode devices in which photons excite electrons to the conduction band. In either case a photon can cause an electron to flow in an external circuit (amplifier front end), providing photon current to electron current conversion.

Both types of device are available as photomultipliers in which the electron flow induced by a photon is internally multiplied by a factor M before being applied to the external circuit.

In principle, the electrical output signal from the photodetector can be amplified indefinitely, but if the received optical signal is progressively decreased while the amplifier gain is increased, a limit occurs at which the electrical noise becomes unacceptable. This noise is of fundamental importance since it determines the minimum optical signal required by the receiver; that is to say, it determines its detectivity.

Noise appears from several sources, the principal ones being the following.

(a) Fundamental quantum noise which is simply shot noise in the photon current. In effect, it results from the random time of arrival of the photons. It has the important property that it depends on the signal level.

(b) Multiplication noise, or additional shot noise in multiplication.[231] This results from the statistical nature of the multiplication process in photomultipliers.

(c) Dark current or leakage current, independent of the illumination, that increases the above shot noise effects.

(d) Thermal or Johnson noise in the amplifier—a factor that dominates conventional systems.

(e) Noise in the transmitted signal or generated in the transmission medium. There is no way in which the signal-to-noise ratio of the optical signal reaching the detector can be improved.

For the discussion on optical receivers it will be assumed that this last factor does not apply and that we are dealing with adequately clean signals. (This is generally valid for digital systems, although attention to modal noise, discussed below, may be required to ensure this.)

If nonmultiplying device is used, there is no multiplication noise, and for applications such as PCM where the required signal-to-noise ratio is fairly low, thermal noise generally dominates. On the other hand, if a signal-to-noise ratio above 50–60 dB is required, the fundamental quantum noise becomes dominant because in this case a larger optical signal is required, and quantum noise increases with signal level.

In the low signal-to-noise ratio applications, where thermal noise tends to dominate, the receiver sensitivity can be increased by employing photomultiplying devices since these increase the signal before they are applied to the electronic amplifier. In effect, this reduces thermal noise relative to the signal. The chief penalty is that multiplication noise becomes significant as the multiplication factor M is increased, and so an optimum M is chosen.

This optimum value depends on how rapidly additional shot noise increases with M, and determines the improvement over a nonmultiplying device.

Finally, although the effect of dark current can be negligible for high quality devices, it becomes relatively more significant in low bandwidth system. Furthermore, the rapid temperature dependence of the dark current requires special consideration by the system designer.

For high speed digital systems the most sensitive detector is the semiconductor avalanche photodiode (APD). In this device an electron–hole pair is created by the absorption of a photon. Within a high field region the electron and hole accelerate rapidly in opposite directions thereby creating further pairs: this is the multiplication process. The average number of conduction electrons resulting from each primary pair by this process is the multiplication factor M which is governed by the reverse bias voltage.

The actual number of electrons created will vary from one occasion to the next, since multiplication depends on random ionizing collisions. This is the source of the multiplication noise referred to earlier. It has been shown that this noise is greatest in materials where an electron and a hole have equal probabilities of creating a further pair, and is much less where these probabilities are very different. Thus, for a well designed device, shot noise in multiplication depends mainly on the material. It has long been recognized that silicon is superior to germanium, and this material is almost universally used in commercial APDs for applications within the wavelength range of silicon. For long wavelength applications germanium is the only well established material but there is promising research on quaternary materials such as GaInAsP.

APDs can have an extremely fast response, with a rise time ranging from well below 1 ns to a few nanoseconds. Unfortunately, some devices exhibit a second rise time or tail effect (Figure 163) which can vary with operating conditions. A modified reach-through APD (RAPD) overcomes this problem and is becoming accepted as preferable for most applications.

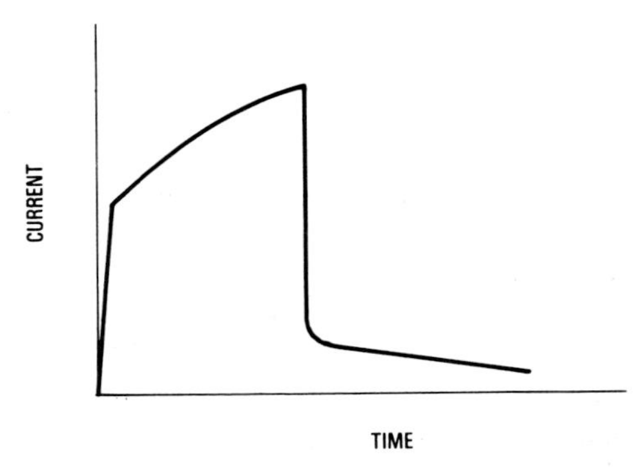

Figure 163. Response of an APD to a square optical pulse showing the unwanted tail effect that occurs in some devices

As pointed out in Chapter VIII, the avalanche photodiode presents a problem of specification for the system designer. The device is extremely complicated to specify, with strong interrelationships between primary quantum efficiency, multiplication M, additional shot noise, surface and bulk leakage current, rise time, tail effect, capacitance variation over the active area, and the dependence of these on operating conditions such as temperature and bias voltage. This complexity also compounds the task of establishing the reliability data required before the device can go into regular service.

Optical Receiver Circuit

It has been seen that to produce a highly sensitive optical receiver it is necessary to minimize the thermal noise, especially when a nonmultiplying photodiode such as the PIN device is used. To the circuit designer, the photodiode is essentially an ideal current source in parallel with a capacitance C (which can be as low as 1 pF). At first sight it might appear that to obtain the required bandwidth, B, a resistance, R, should be connected across the diode, chosen to provide an RC time constant commensurate with the bandwidth, and that the Johnson noise in this resistor is the minimum achievable thermal noise. It turns out, however, that the noise can be cut to several decibels below this figure. The most established, and still generally preferred, circuit for achieving this is the current feedback circuit, shown in Figure 164. Referring to this figure, if the amplifier were ideal, the equivalent input thermal noise would be the Johnson noise in the resistance R_F, whereas the bandwidth corresponds to the much lower input impedance $R_F A^{-1}$. With practical amplifiers, noise figures several decibles lower than predicted can be achieved using this simple calculation. For further improvements, photodiodes and first stage amplifiers can be integrated into a single package, the photodiode active area minimized to reduce capacitance, and transistors tailored for minimum noise when fed by a high impedance current source.

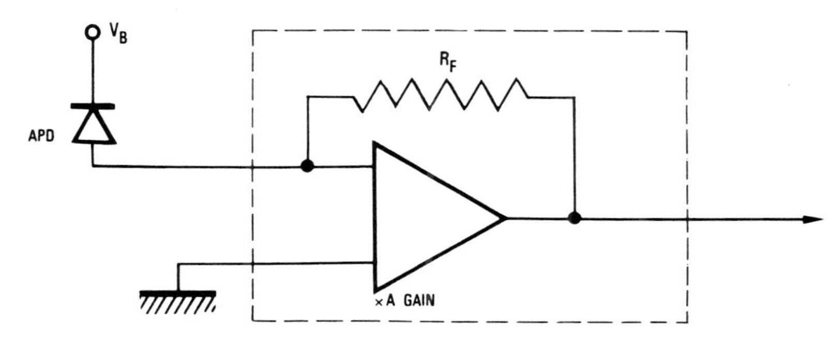

Figure 164. Current feedback amplifier used to minimize noise in an APD

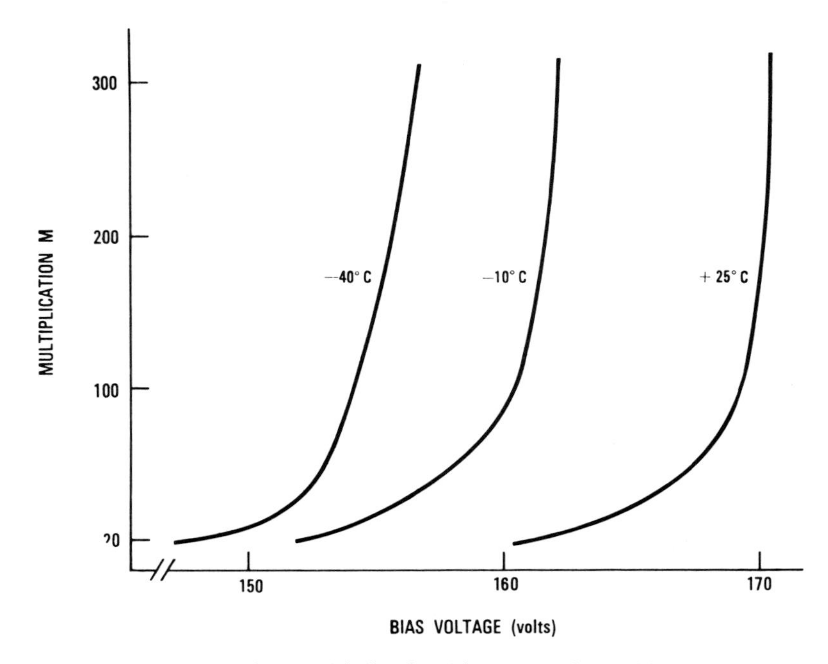

Figure 165. Typical multiplication/bias curves for a silicon APD

The receiver circuit must also provide a steady reverse bias voltage for the photodiode. For PIN photodiode devices there is very little problem, since a noncritical voltage of 5–80 V at extremely low current is all that is required. APDs, however, usually require a bias voltage V_B of 100–400 V, which is somewhat critical since the multiplication factor depends steeply on V_B and on temperature (Figure 165), and also varies between samples of the same device type.

The bias control for the APD must maintain the multiplication factor around an optimum value, which is a trade-off between the thermal noise which dominates at low multiplication and the shot-noise-in-multiplication which dominates at high multiplication.

One approach is to bias the APD with a temperature compensated voltage which is preadjusted, but this is rather limited because the temperature coefficients also vary. A more practical solution to the problem is in some way to force the overall optical receiver gain, defined in amps/watts, to the required value. One such method is to bias the APD with a d.c. constant current source I_B which is capacitively decoupled by C at all signal frequencies (Figure 166). If the mean current is defined by the bias supply and the mean optical power input is known, then the gain in amps/watts is fixed, irrespective of temperature or device. The bias source must be decoupled at signal frequencies to prevent gain modulation. The output from the detector

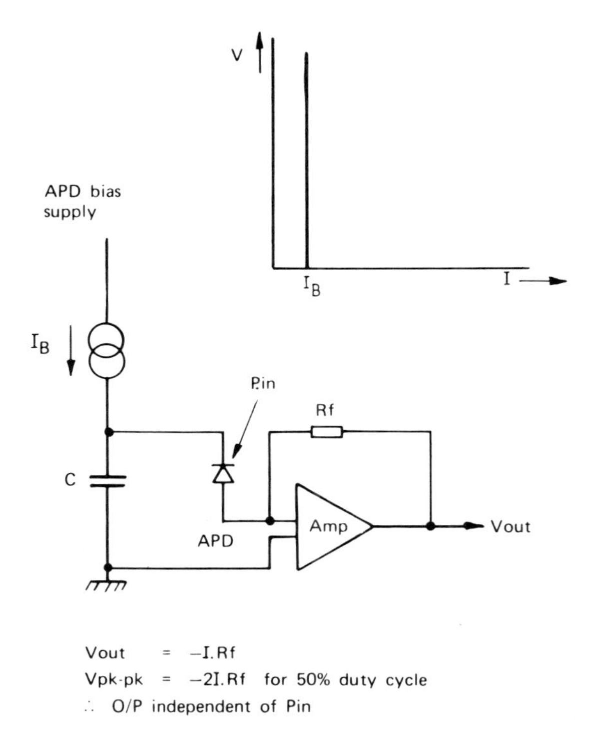

Figure 166. Constant current bias of APD

is defined by the input current which is constant. Another way of considering its operation is as follows. If the bias voltage is below the optimum then the gain will be too low and the product of the gain and the received power will be less than the constant current. Therefore the excess current will charge the decoupling capacitor to a higher voltage and hence higher APD gain until it will stabilize at the required gain.

This method provides 100 per cent AGC since the output current from the photodetector is solely defined by the input current which is held constant. Such AGC is most useful since it considerably reduces the dynamic range of signals applied to the following low noise amplifier thereby increasing the optical dynamic range by 10–20 dB. This technique is most simple, requiring no temperature compensation or preadjustment, but is unsuitable at low bit rates or bandwidths where the dark current becomes a significant fraction of the detector current. What is required is a method that keeps the signal current constant whilst the dark current varies. This can be achieved by peak detecting the a.c. coupled signal after the low noise amplifier, comparing the level with a preset reference, and feeding back to adjust the high voltage bias supply to keep the level constant, thus synthesizing a constant current source with the dark current subtracted.

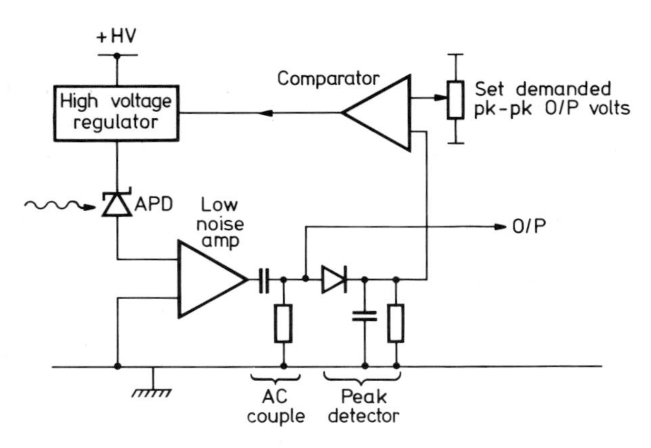

Figure 167. APD bias by feedback AGC

A suitable circuit is shown in Figure 167 and Figure 168 is a photograph of a receiver board. The bias circuit must also be designed to protect the photodiode against possible excess power dissipation at very high input optical power levels, or excess bias voltage for zero applied optical input. The current required by the APD is so low that simple DC-to-DC convertors can supply the bias from low voltage input.

Figure 168. Avalanche photodiode mounted on receiver board of PCM repeater

Optical Receiver Performance

The literature contains reports of performance achieved in practice for digital optical receivers over a wide range of bit rates (in the range 1 Mbit/s to 0.5 Gbit/s). Some of these[200–206] are plotted in Figure 169 in terms of mean received optical power required for an acceptable bit error rate (usually taken to be around 10^{-9}), as a function of the bit rate. Those indicated by (1), (2), and (3) in the figure are derived from References 204, 205, and 206, respectively, and (4) is derived from the results in Chapter XI. The plot also allows one to read off the optical power in an interesting alternative way—the mean number of photons arriving in a bit period. Since the energy of each photon is fixed for a given wavelength (here taken as 850 nm which is a good approximation for all the results plotted), the mean optical power is proportional to the average number of photons per bit period multiplied by the bit rate. The best of the published results lie between 100 and 1000 photons per bit, and it will be clear from the following discussion that these approach the fundamental quantum limit which is due to the granularity of the light.

The simplified 'large-number statistics' implicit in standard noise theory has to be replaced by a more careful statistical approach[231–235] in order to optimize system design with the best available devices. This statement has been supported experimentally with commercially available devices, and will

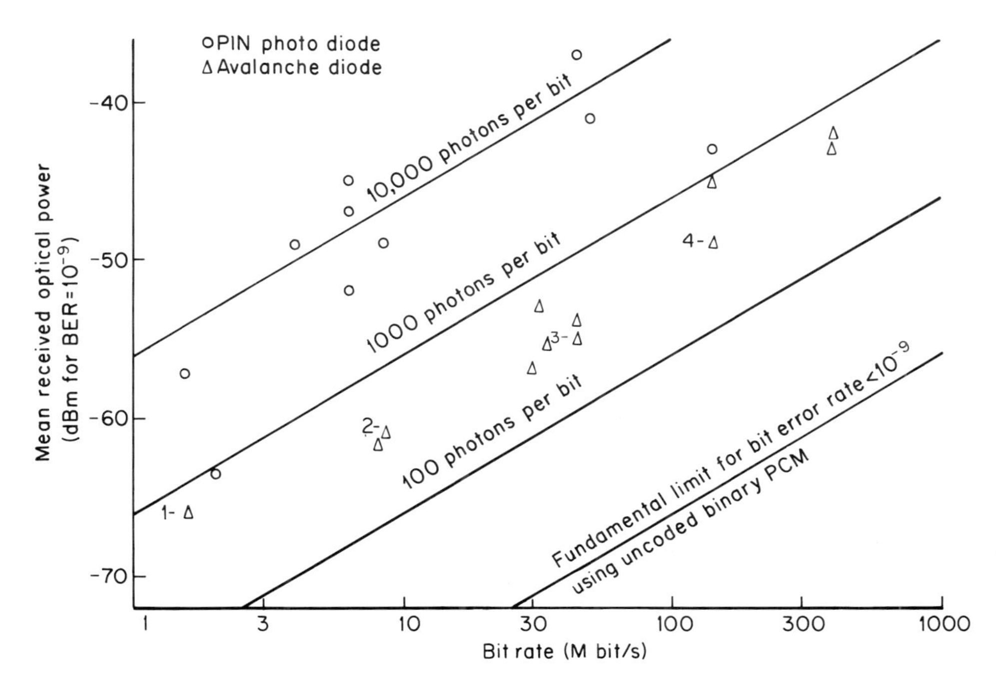

Figure 169. Published performances of optical receivers

be even more significant if the potential for further improvement is to be fully explored.

Figure 169 contains a line corresponding to 10 bits per photon, marked 'fundamental limit, for ber $<10^{-9}$, using uncoded binary PCM'. The explanation of this is as follows: with an equal number of 'ones' and 'zeros', there would be 20 photons, on average, for each 'one', and always none for a 'zero'. If the receiver is thought of as an ideal photon counter, the only source of error is the fact that it is fundamentally impossible to guarantee 20 photons for a 'one', but only to say that the average number will be 20, with a certain probability distribution. Under some realistic assumptions this will be the Poisson distribution. The optimum receiver will be set to record a 'one' if one or more photons are received, and a 'zero' if none are received. The only error which can occur is if a 'one' is transmitted, but no photons are actually received, and the probability of this is given by the Poisson formula as

$$p_k = \frac{m^k}{k!} \exp(-m)$$

where p_k is the probability that exactly k photons will be received. With $m = 20$ and $k = 0$, $p_k = 2 \times 10^{-9}$. Since there are no errors for the 'zeros', we obtain the result that the bit error rate will be 1 in 10^9 for an average received power corresponding to 10 photons per bit.

Bearing in mind the idealistic assumptions made, it is quite remarkable that practical results should approach an order of magnitude of this limit. It should, however, be remembered that many of the results in the region of two or three hundred photons per bit are obtained under fully optimized laboratory conditions, and so for many applications a more rugged design which sacrifices a few decibels of sensitivity might be preferred. For these applications a case may be argued for the use of PIN photodiodes with low noise amplifiers built in. On the other hand, there are applications where complexity and cost of terminals can be accepted in exchange for avoiding repeaters, while achieving the line length demanded by the geography. These considerations would apply, for example, in the case of submerged cable systems.

For such applications the quest for the last few decibels of performance may be justified, and of course there is further potential if coding is optimized with respect to the statistical receiver viewpoint.

OPTICAL COUPLERS

The techniques such as splices and demountable connectors used for coupling lengths of optical cable are discussed in Chapter V, but are not wholly applicable to the coupling problems presented at the repeaters. It is clear that the repeater will be ruggedly housed,and so the electro-optic components will not be accessible at the surface of the housing. It will therefore be necessary to have internal fibre connections. Device-to-fibre couplers should preferably be demountable for ease of maintenance, although an alternative is the production of factory aligned integral assemblies of device and fibre tail. At least two manufacturers of LEDs supply such assemblies. Satisfactory demountable device-to-fibre connectors have been made, however, even for the relatively difficult case of a stripe geometry laser and a 30-μm core graded index fibre.

It is difficult to combine the requirement for hermetic sealing with good coupling efficiency in a demountable device-to-fibre connector. If a satisfactory transparent potting medium or resin can be shown to be free from long-term stability effects on the devices, this would open up many possibilities. The use of lenses, tapers, or other built-in optical elements are all possible solutions.

DIGITAL TRANSMISSION

REPEATERS FOR BINARY PCM SYSTEMS

Most existing work on high capacity fibre-optic systems is based on binary PCM using intensity modulation of the optical source. At the optical transmitter a binary signal in the form of an electrical pulse stream is converted into an optical pulse stream: essentially, an electron current is directly converted into a photon current. The optical signal propagates along the fibre, undergoing both attenuation and distortion, before the photon current is converted back into an electron current at the optical receiver in a line repeater (Figure 170). The electrical output of the receiver is an attenuated and distorted version of the original signal. This received signal is then processed by an electronic regenerator which amplifies, retimes, and reshapes the pulses. In principle, the regenerator is similar to that for a conventional coaxial PCM system, but with some difference in design details. The regenerated and retimed signal is then passed to an optical transmitter for further transmission.

In addition to line repeaters at intervals along the optical fibre, terminal repeaters are used at either end of the path. Figure 171 shows that these share the important features of line repeaters. Most of the discussion in this chapter is common to both types. For some low capacity routes there will be no need for line repeaters.

Also indicated in Figure 171 are the power supply and, as part of the regenerator, the equalization circuit used to correct for distortions of the transmission medium. Additional factors discussed are line coding, supervision, and the engineering order wire.

PCM REGENERATOR

The PCM regenerator reshapes and retimes the binary pulses so that the retransmitted signal is identical to the original waveform. In this way there is

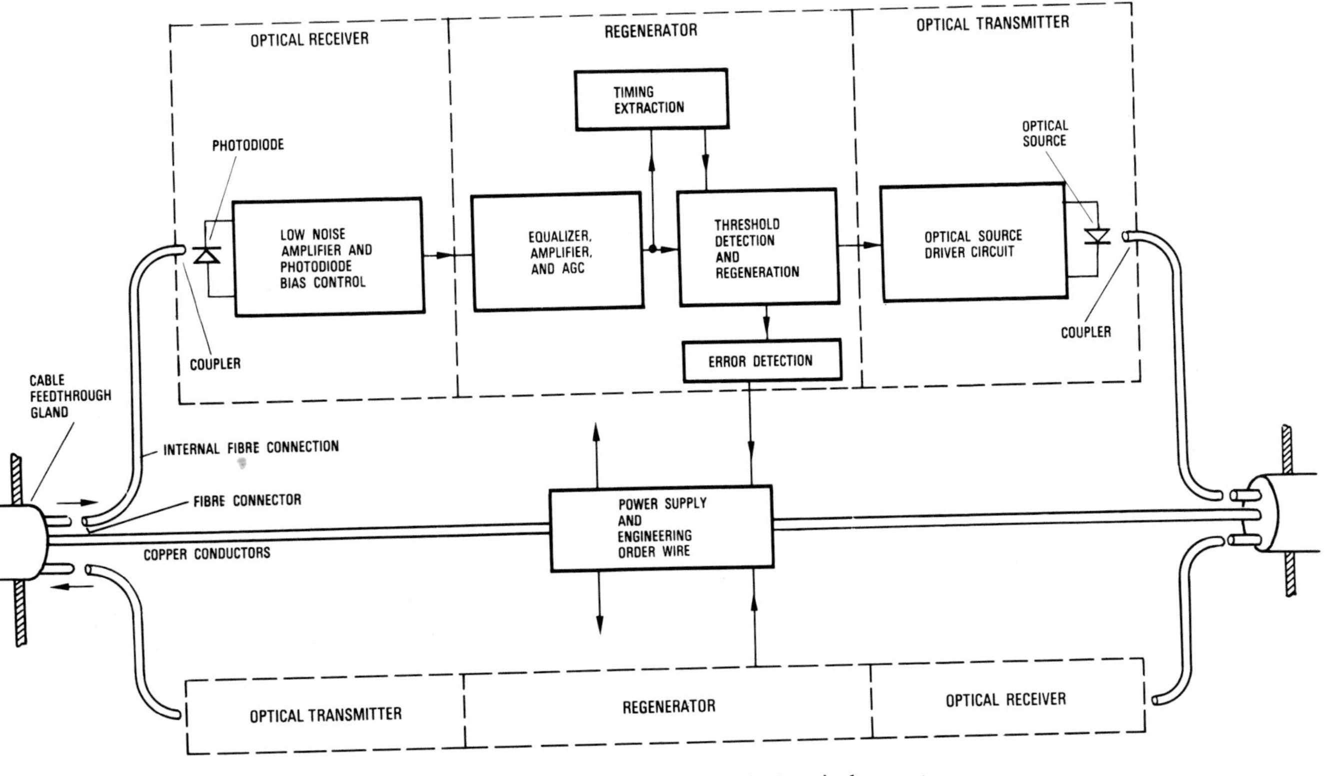

Figure 170. Block schematic of a typical optical repeater

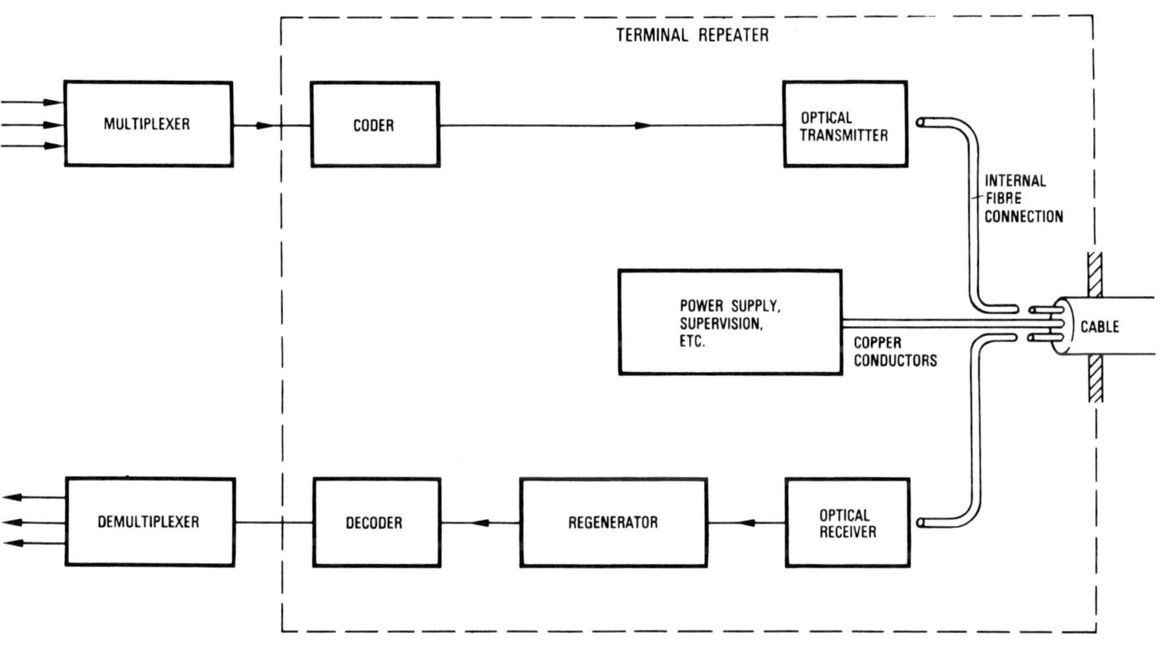

Figure 171. Block schematic of a typical terminal equipment

no cumulative distortion of the signal as it progresses through many repeaters in tandem, except for the accumulation of random binary errors, one for zero, or vice versa. In practice it is possible to reduce the probability of errors, the bit error rate, to negligible proportions. It is common to specify a bit error rate of 10^{-7}–10^{-10} for each repeater.

The function of the PCM regenerator is to sample the train of pulses at a regular frequency equal to the symbol rate, and at each sample to make a decision as to the most probable symbol being transmitted. The same symbol is then retransmitted. In the case of a binary regenerator this is simply a question of choosing one or zero. The most common approach is simply to set a threshold and to retransmit a one if the received signal is above threshold at the decision point, and a zero if it is below (Figure 172).

It is obvious that electrical noise can induce errors unless it is negligible in comparison with the signal amplitude. This will be compounded if inter-symbol interference exists. Consider the input to the regenerator when a one has been transmitted, examples of which are sketched in Figures 173 and 174. To eliminate inter-symbol interference, the response to a transmitted one should pass through zero at all neighbouring decision points.

The effect of inter-symbol interference can most easily be visualized by inspecting an eye diagram. This is obtained experimentally by superimposing on a scope trace the input pulses to the regenerator which have been preceded by a wide variety of pulse sequences. An example is given in Figure 175: the open eye shows the available tolerance in setting the threshold and decision points.

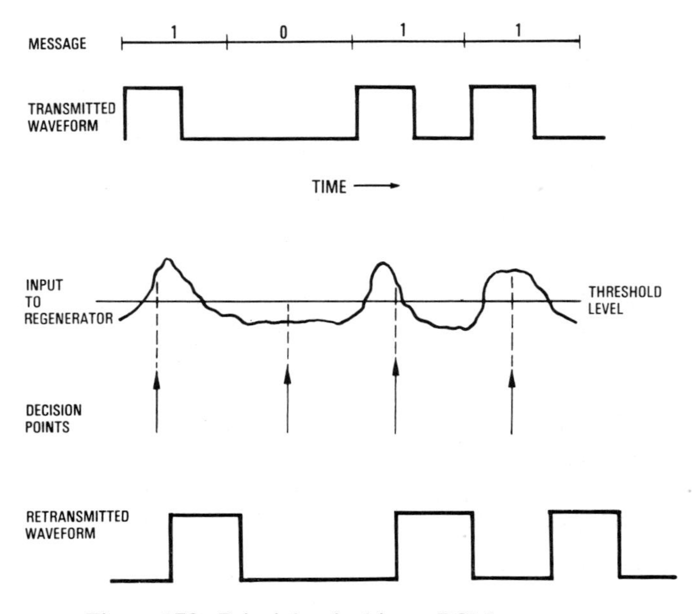

Figure 172. Principle of a binary PCM regenerator

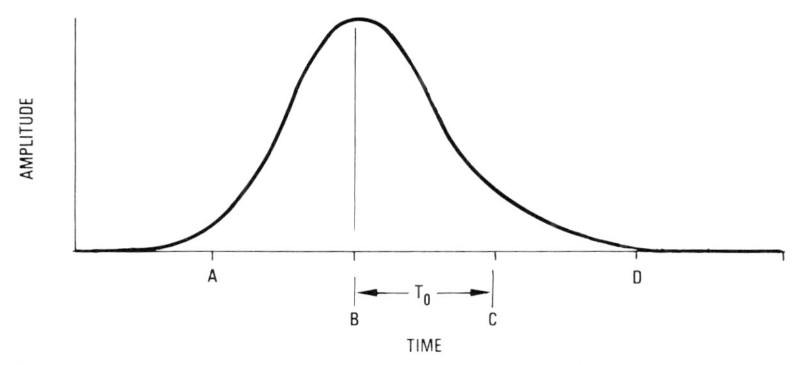

Figure 173. Input pulse to a regenerator corresponding to the transmission of a 'one'

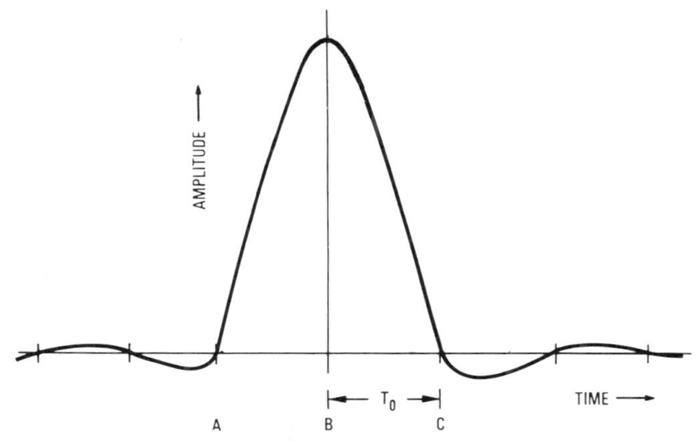

Figure 174. Response to a transmitted 'one' for zero intersymbol interference (zeros at A, C, and D)

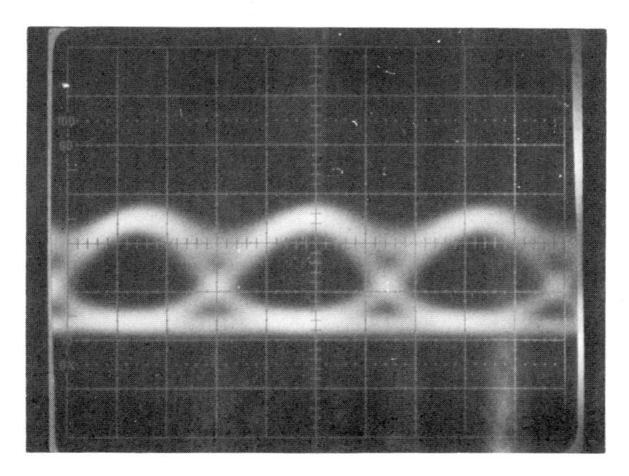

Figure 175. Typical eye diagram. The open eye is a measure of the available tolerance in setting the threshold and the decision point

It is clearly imperative that retiming of the pulses is accurate and uniform: there must be no long-term drift or some information would be lost and the sequence spoiled. The decision points must remain in the correct phase with respect to the received pulse train, i.e. near the widest part of the eye. This is achieved by deriving the timing from the received waveform itself, either using a high Q tank circuit or a phase locked loop. To obtain sufficient timing information it is necessary to ensure that there is an adequate energy content at the frequency T_0^{-1}, where T_0 is the period between decision points, or else to use a nonlinear circuit to exploit other frequency components. It is also necessary to consider the problem of long periods with no timing information such as would occur with a message consisting only of zeros.

Turning again to the bit error rate, the essential performance parameter of a PCM regenerator, it can be seen that this depends strongly on the signal-to-noise ration of the input signal. Assuming Gaussian noise statistics

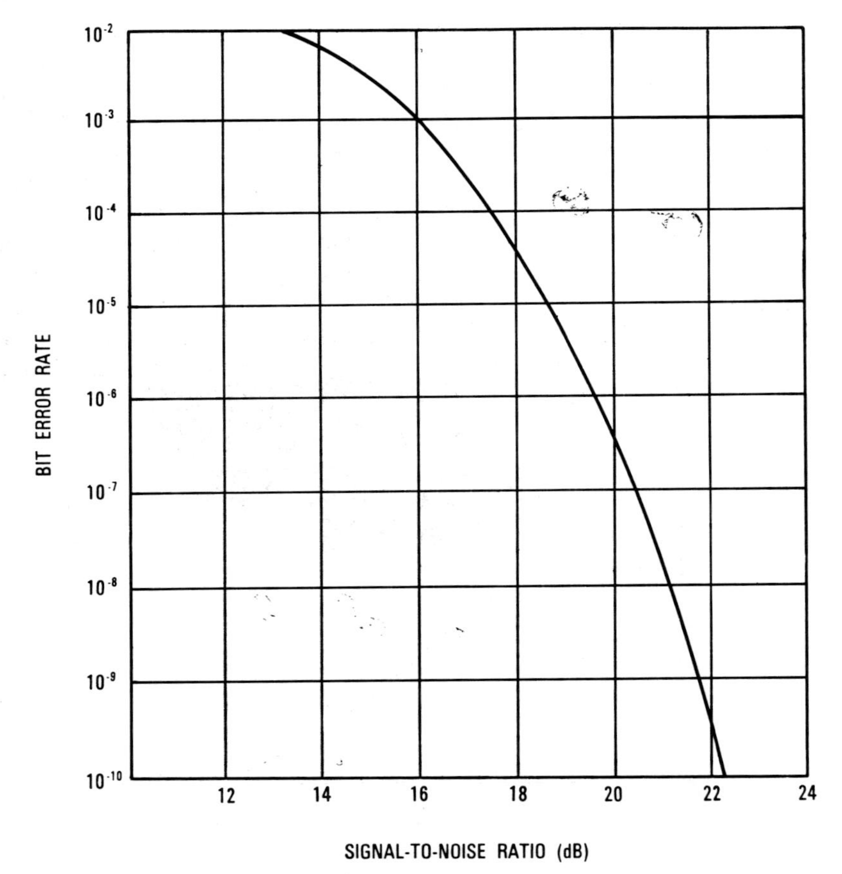

Figure 176. Bit error rate as a function of the signal-to-noise ratio for directly modulated systems. Signal is peak signal and noise is r.m.s. noise

for the receiver, accurate retiming, and optimum threshold setting, the theoretical bit error rate is a well known function of the signal-to-noise ratio (Figure 176). This graph is remarkable for its steepness. To the system designer this means that there is no useful trade-off between bit error rate and signal-to-noise ratio, but rather that there is a required minimum signal-to-noise ratio above which the bit error rate is negligible and below which the system rapidly becomes useless.

EQUALIZER

The equalizer is required to provide an output suitable for the regenerator, with a good open eye diagram. Normally, the equalizer is a linear frequency shaping filter which accepts the frequency response of the combination of transmitter, medium, and receiver, and transforms it into the required raised cosine or other response.

Correction for the receiver requires care where tailing is present in APDs. In very high capacity systems, however, the dominant issue is often the correction for the fibre frequency response which, in the case of multimode fibres, is limited by a multipath effect—the fibre dispersion. It is possible to employ a fibre with a 3-dB bandwidth well below the symbol rate T_0^{-1}, but to transform the resulting receiver output to the required response, such as in Figure 177, it is necessary for the equalizer to boost the higher frequencies with respect to the lower frequency components. This results in more noise at the output. This in turn leads to a trade-off between fibre bandwidth and required optical power at the receiver. The extra power required as a result of equalizing the fibre is generally referred to as a dispersion penalty and is a central factor in system design.[237] Personick[232,233] has derived comprehensive design equations.

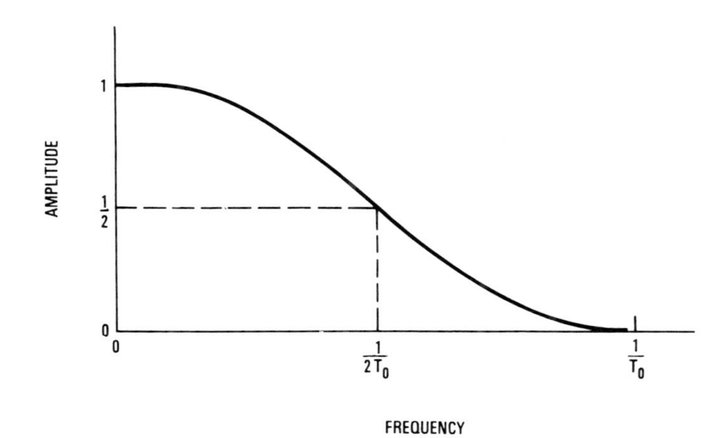

Figure 177. Frequency response corresponding to Figure 174

LINE CODING

The binary signal to be transmitted may be applied directly to the optical link, but this could cause problems of d.c. level restoration (for a.c. coupled circuits) and of timing. To overcome this problem in PCM systems it is usual to employ redundant coding. This may, for example, take the form of encoding every three signal bits into four binary pulses, requiring an increase in the transmission rate of 33 per cent. This redundancy means that each group of four transmitted pulses, of which there $2^4 = 16$ possibilities, can only take one of $2^3 = 8$ allowed sequences. These can be chosen to avoid d.c. coupling and timing problems, and at the same time provide further error detection by checking for nonallowed sequences.

The simplest and most common form of error detection is parity checking. Each block of N pulses can be made to have an even number of ones so that any single error in the block can be identified at the expense of a very small amount of redundancy. Error correction is also possible, but requires considerably greater redundancy and equipment complexity. This is not usually justified by the small reduction in the permitted signal-to-noise ratio (Figure 176) that results from tolerating a higher received bit error rate.

SUPERVISION

The problem of monitoring every section in a line communications system so as to identify failed or marginal units has much in common with coaxial cable PCM systems. In the event of errors occurring in a link, the faulty section can be identified if the bit error rate is monitored at each repeater. This can be done using parity checks; a simple technique has been described by Jessop.[236]

It would be preferable to detect marginal operation before errors occur and this could be done at each repeater by monitoring the eye opening shown in Figure 175. So far, the circuitry involved has made this uneconomical, but cost trends in large scale integration (LSI) could change this situation.

Two supervisory techniques specifically for fibre-optic systems are used at the receiver and transmitter. For a given output current the applied voltage on the APD warns when the input optical power is low, while the laser current (adjusted for a preset power) shows when the laser performance is becoming degraded.

Engineering order wires are normally required in transmission systems for installation and testing. Where copper conductors are supplied in parallel with the fibre for repeater power feeding, they can also be used for the order wire. Alternatively, the order wire can be modulated onto the same fibre using unwanted low frequency components. For greater independence each route could contain a separate fibre pair for the order wire.

MULTILEVEL PCM

Binary modulation has been widely chosen for the first PCM optical transmission systems. Although this complicates the interface with standard electronic PCM equipment, which is normally ternary, it simplifies intensity modulation of the source to on/off as opposed to multilevel intensity modulation.

Stability of the sources and optical feedback techniques are now sufficiently advanced to allow multilevel modulation to be used. Ternary modulation could be based on 0, $\frac{1}{2}$, or 1 levels of the optical source, or a higher number of levels could be used. For a simple comparison between binary and ternary modulation, consider an optical source which is peak power limited and assume a thermal noise limited detector. In this case the difference between adjacent power levels has been halved so that the optical attenuation which can be tolerated for a given error rate at the receiver has fallen by 3 dB. As against this the information per pulse is $\log_2 3 = 1.58$ bits, which means that the symbol rate can be reduced by a factor of 1.58 for the same information rate. The reduction in bandwidth will recover 1 dB of the original 3-dB penalty. Thus, the choice can be marginal, the final choice depending on many detailed practical considerations. Perhaps ternary modulation will mainly be used in shorter repeaterless systems where the cost of interface equipment is significant.

REPEATER SPACING

For many applications the key economic factor is the achievable fibre path length between the optical transmitter and the receiver. For a repeatered system this determines the number of repeaters needed in the system. In some situations enormous savings are possible where optical transmission allows one to avoid the use of intermediate repeaters altogether, since this does away with power feeding, housing, environmental, supervisory, and maintenance problems, apart from the sheer cost of the equipment.

In general the path length will be attenuation limited at lower bit rates, and fibre bandwidth limited at the higher bit rates (although the latter is complicated by system trade-offs, e.g. choice of single-mode fibres at the expense of poor coupling, the low fibre bandwidth can be partially compensated for by equalization, if there is spare optical power). In order to give an idea of the quantities involved, the shaded regions of Figure 178 indicate the way in which available path length falls off with increasing bit rate. Regions are shown both for current technology and as projected for next generation equipment based on consolidation of the most favourable laboratory results. (The region above this, for example 100 km at 100 Mbit/s, must be considered hightly speculative, but laboratory results have already demonstrated that such performance is fundamentally feasible.)

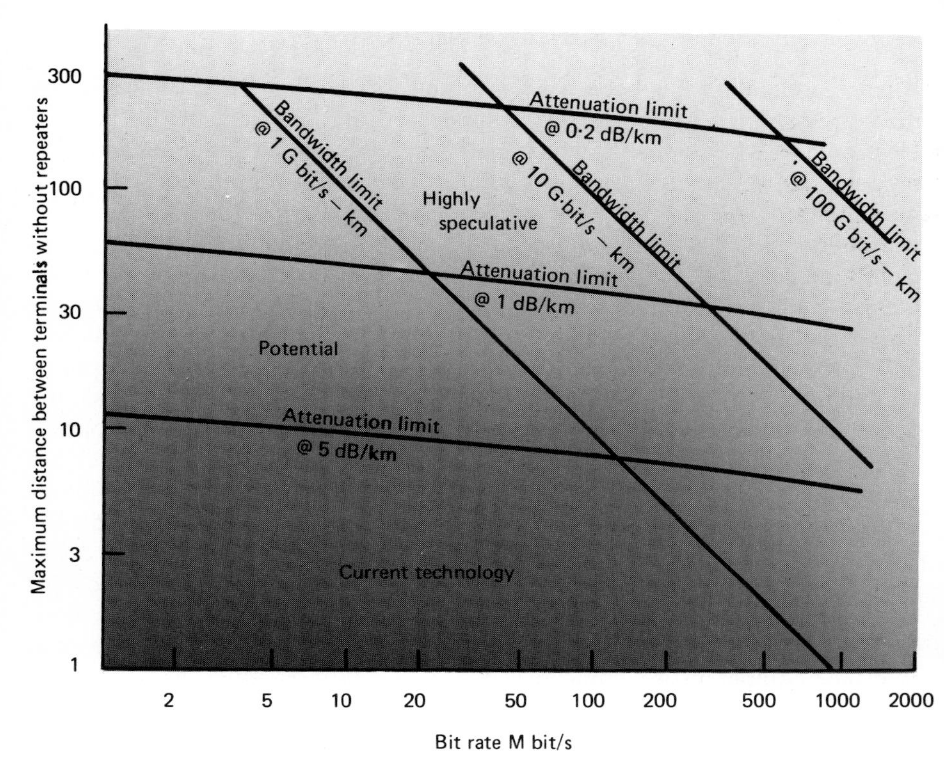

Figure 178. Maximum distance between terminals without repeaters

The explanation of the curves is as follows. The attenuation limits are based on 5 dB/km, 1 dB/km, and 0.2 dB/km total fibre loss. When it is remembered that splices (or connectors) are included, and that the figures apply to the routine production of cabled fibres, it can be seen that 5 dB/km is a conservative, but not a highly pessimistic, figure. The allowed loss in the optical path is taken as 60 dB at 1 Mbit/s, which allows for additional coupling losses at terminals, and has been taken as falling off due to reduced receiver sensitivity corresponding to a fixed number of photons per timeslot. The bandwidth limits are drawn for fixed products of bit rate and path length having values 1 Gbit/s·km , 10 Gbit/s·km and 100 Gbit/s·km. The lowest figure is a conservative one based on graded index fibres and the curve shows that the bandwidth limitation becomes noticeable around 100 Mbit/s, above which it begins to dominate. This is dependent not only upon the optical properties of the fibre, but upon the optical wavelength and linewidth of the laser. The 10 Gbit/s·km line gives the order of magnitude based on exploitation of laboratory results. Even the 100 Gbit/s·km line by no means represents the fundamental limitations, but it does correspond to considerable advances in the technology.

Figure 178 is intended to apply to PTT-type systems, where the handling and environmental requirements can be benign compared with some military applications. In the latter case, which may include highly rugged demountable connectors and special requirements which could limit the choice of components, the maximum path length could be much smaller than those indicated. Fortunately, for such systems long distances are often not required.

POWER FEEDING

In conventional line systems the coaxial cables normally carry d.c. power for the repeaters as well as the signals. In optical transmission systems it may be necessary to provide copper conductors alongside the fibres for this purpose. While the cost of these does not destroy the economic advantages of the system, it is high enough to warrant serious attention. Reducing the power requirements of repeaters is an obvious step. The cost of copper is further reduced by maximizing the line voltage, which both reduces the current and increases the allowed voltage drop. The only restriction here is the safety code.

ANALOGUE TRANSMISSION

Transmission of analogue signals can also be accomplished by optical means and will have its part to play in future communication networks in situations where the cost of digital terminal equipment is not justified. Perhaps the most obvious example is in cable television (CATV) systems, where many TV channels need to be carried for distances not exceeding a few kilometres, and with many branches, drops, etc. In the telephone local network, analogue Frequency Division Multiplex transmission may continue to be preferable, while in microwave transmission systems, links at the 70 MHz or 140 MHz IF can be carried by optical means using Frequency Modulation of a subcarrier. In addition there is an enormous requirement for video links of relatively short distance carrying one or a few channels, in the multiplicity of information systems, CCTV systems, etc. now being installed.

SYSTEMS CONSIDERATIONS

Figure 179 shows the essentials of an analogue transmission system. In the terminal equipment there may be several levels of modulation. The first level will be applicable where there is more than one baseband channel to

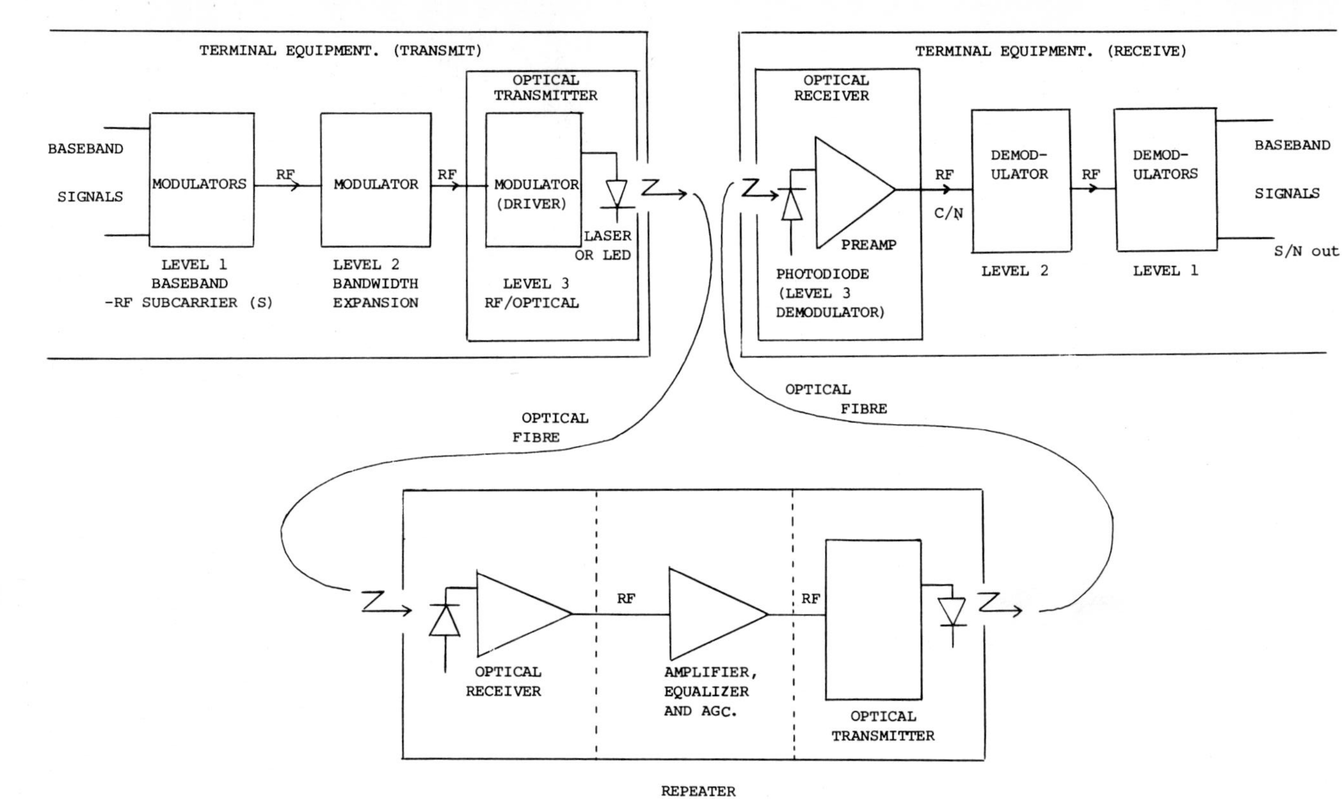

Figure 179. Analogue optical system

be transmitted, i.e. it is usually a frequency division multiplexing process. The second level will be applied to gain some transmission advantage, e.g. expanding the transmission bandwidth to gain signal/noise advantage (or to allow a lower carrier/noise in the optical link) or to overcome some component limitation, e.g. phase or frequency systems such as Frequency Modulation, Phase Modulation, Pulse Frequency Modulation, or Pulse Position Modulation will not require amplitude linearity in the optical source, at least if amplitude/phase modulation conversion is negligible. The third level of modulation is from an RF subcarrier to an optical carrier. At the present state of the art this is almost invariably a modulation of the source intensity, since phase and frequency are not well enough defined at optical frequencies to be successfully modulated. This situation may change in the future.

At an optical receiver the photodiode effectively performs the optical/RF conversion, and in an intermediate repeater this will be the only demodulation. (Alternatively, the RF–optical–RF conversion can be regarded as an electron–photon–electron conversion.) The RF signal will be amplified, level corrected, and possibly have some transmission distortions equalized and then the resultant signal will drive another optical source. At the terminal equipment, demodulators will finally recover the baseband signals as shown.

Although not shown in Figure 179 in order to avoid complicating the diagram, the terminal equipment will also contain gain blocks and level regulating equipment, probably using reference pilot tones, so that the signal levels can be controlled to keep intermodulation distortion within design limits. A considerable amount of frequency generating and filtering equipment may also be present.

Analogue optical systems can thus vary from the simplest, e.g. transmission of one television channel at baseband, where levels 1 and 2 of modulation and demodulation would not be used, through to the most complicated, e.g. 960-channel telephone transmission via FM at the second level. Table XI summarizes the essential parameters of a selection of possible systems, analysed on the basis of required (peak–peak signal)/(r.m.s. noise) at the optical preamplifier output, whatever the signal format. Normalizing the requirements to peak-to-peak signal removes confusion between standards for various systems. For example, in baseband IM TV systems, the peak–peak signal/r.m.s. noise ratio is a direct measure of unweighted output signal/noise. For systems using FM of a sinusoidal subcarrier, the r.m.s. subcarrier/r.m.s. noise is most often quoted. Evidently, in this case one must make a −9 dB correction from the peak–peak value. Similarly, for other systems (e.g. multi channel telephony) knowledge of the peak factors involved enables the peak–peak value of the overall signal, which must be handled by the optical components, to be derived. If the source can be fully modulated from zero light to a peak power of twice the mean value, the mean received optical power will be half the peak–peak

Table XI
Some possible analogue systems

Baseband signals	Level 1 modulation Baseband-RF	Level 2 modulation Bandwidth expansion	Level 3 modulation RF-optical	Bandwidth and centre frequency required in optical systems	Number of repeaters in line	Typical peak–peak signal to r.m.s. noise (C/N) required at the output of the optical receiver	Typical peak–peak optical power required at each repeater input	Source operating mode	Overall performance
Single video channel	None	None	Intensity Modulation (IM)	DC-6 MHz	0	44.5 dB	−30 dBm	c.w.	S/N (weighted) = 52 dB
Single video channel	None	None	IM	DC-6 MHz	9	54.5 dB	−22 dBm	c.w.	S/N (weighted) = 52 dB
Single video channel	None	FM	IM	±15 MHz typically 70 MHz	0	39 dB	−35 dBm	c.w.	S/N (weighted) = 52 dB
Single video channel	None	FM	IM	±15 MHz typically 70 MHz	9	49 dB	−28 dBm	c.w.	S/N (weighted) = 52 dB

Single video channel	None	PPM	IM	10–400 MHz	0	20 dB	−41 dBm	Pulse duty cycle ~7%	S/N (weighted) = 52 dB
Single video channel	None	PPM	IM	10–400 MHz	9	30 dB	−32 dBm	Pulse duty cycle ~7%	S/N (weighted) = 52 dB
12 video channels	AM-VSB	None	IM	>72 MHz typically in band 50–300 MHz	9	61 dB	−5 dBm	c.w.	$S/N > 40$ dB at synchronization peak
60 telephone channels	AM-SSB	None	IM	312–552 kHz	0	51 dB	−46 dBm	c.w.	Test tone signal/noise 40 dB
60 telephone channels	AM-SSB	FM	IM	±3.62 MHz typically 70 MHz	0	24 dB	−56 dBm*	c.w.	Test tone signal/noise 40 dB
960 telephone channels	AM-SSB	FM	IM	±20 MHz typically 70 MHz	0	59 dB	−16 dBm	c.w.	10 pW/channel over link length

*This figure may be made worse by noise from dark current in practical avalanche detectors

power. For modulation index m of less than unity, the peak–to–peak optical power is $2m\bar{P}$ (where $\bar{P}$ is the mean power).

In deriving Table XI it is also assumed that perfect linearity of the components can be achieved. In practice this is far from being true, although as will be seen the situation is improving. Nonlinear effects such as intermodulation, cross modulation and harmonic generation are most likely to occur in the optical source and its drive amplifier. At the receiver, the PIN diode detector has a very high linearity (distortion products below shot noise level) while avalanche diodes may be more limited (second order products −60 to −70 dB).[207] However, the major limitation at present is with sources.

Nonlinearity will cause a variety of effects, and amplitude or phase nonlinearity will be of dominant importance according to the modulation system. In telephony systems the effects appear as additional noise, while in television systems they may appear as beat patterns and/or changes in hue and saturation of colours. In any case, the practical effect is to cause a transmission impairment that can only be reduced by improving the system linearity.

The other assumption in the table is that the fibre transmission properties are ideal across the frequency band of interest (constant attenuation, linear phase). Where this is not true, equalization will be needed and more power will be required to offset the equalization penalty. Each example assumes a near-optimum receiver, but not necessarily a completely optimized one, e.g. in terms of avalanche multiplication, lowest noise PIN receiver, etc. However, accepting the limitations of the table, important conclusions can still be drawn from it.

Considering first the single channel video systems, the baseband–IM transmission is seen to require about −30 dBm peak–peak optical power at the photodiode input for a repeaterless system. Assuming full modulation of the optical source, the mean power will be half this (−33 dBm). If, however, the source cannot be fully modulated due to nonlinearity (of which the laser threshold is an extreme example) there will be a penalty of m^2 where m is the modulation index. This applies to all other systems also, and evidently the penalty rapidly gets severe as m falls. Eventually the signal may become dominated by shot noise in the relatively large mean photocurrent. There is thus a necessity for sources where m can approach 1. (The corresponding parameter in digital systems is the extinction ratio.) Knowing the power launched from the source (about −10 dBm from a high radiance LED and 0 dBm from a narrow stripe laser onto a 50-μm core fibre of 0.2 NA) the path loss can be calculated. With LED sources, larger diameter, larger NA fibres may be preferred, but these may have bandwidth limitations. If the distance to be covered necessitates repeaters, the optical power to each repeater must be increased to maintain the signal/noise performance, because noise accumulates in analogue repeaters. To avoid d.c. transmission, a double sideband reduced carrier system can be employed with an RF subcarrier and can give similar performance.

If greater path loss between repeaters is necessary a modulation such as FM can be used and will give some 5 dB advantage. The advantage is not great because bandwidth expansion is relatively small. Pulse systems such as PPM and PFM have similar signal/noise performance and can employ greater bandwidth expansion[208,209] and can use the advantage of high peak power from a source such as a sawn cavity laser. However, the penalty is modem cost and complexity, and also fibre bandwidth limitation will prevent full advantage being gained even with graded index multimode fibres—except for short distances where the attenuation capability is unlikely to be needed. In addition, jitter problems can dominate in the modems and unwanted beats between signal and sampling frequencies can be a limiting factor.

For twelve-channel video, a typical requirement for CATV in North America, bandwidth requirements, desire for least modem complexity, and for system compatibility renders AM–VSB subcarrier transmission almost mandatory. Since the effective modulation index for each channel is low, high overall carrier/noise and received power are necessary and the system tends to be shot noise limited. Very linear laser sources are a necessity, since with twelve or more channels, intermodulation is likely to be the controlling factor.[210] Such systems, for fibre optics, are still in the future, but sources such as the narrow stripe laser bring them nearer.

An alternative to transmitting twelve channels on one fibre would be to transmit, in the limit, one channel on each of twelve fibres—this is termed space division multiplex (SDM). Major problems with this concept are incompatibility to and from baseband at the terminals, the necessity for twelve optical connectors at each interface and, of course, the requirement for a twelve-fibre cable. It seems likely that technological advance to the point where a compatible single fibre system is possible will be preferable. However, the relative economies of fibres and sources versus modems, etc. may make SDM an applicable technique in some fields in a mature fibre-optics technology.

Turning now to telephone transmission, to transmit sixty telephone channels of low quality (e.g. for a single repeaterless link) low received power around −46 dBm suffices. Advantage can be gained by low deviation FM but calculations indicate that the dark current in practical avalanche detectors may prevent this advantage being obtained in practice. In contrast, 960 channels to a specification sufficient to satisfy the requirements for feeder 'tails' for microwave systems, require a linear laser source to give an output power of at least −16 dBm, and a high quality receiver tending towards shot noise limited operation, even employing wide deviation FM, if sensible distances are to be covered. Feasibility experiments are, at the time of writing, being carried out on such systems, in particular to assess the penalties due to source nonlinearity and fibre bandwidth limitations.

To summarize the conclusions from this brief examination of some systems of interest, the most obvious point is the diversity of requirements for optical power, bandwidth, and linearity. Sources with high modulated power capability, with output at low NA so that the power can be effectively launched and propagated, are of interest for all systems, and an absolute necessity for some. Very high orders of amplitude and phase linearity are also essential. In addition, modal noise (considered in a later section) is a phenomenon which may influence the choice of source, particularly where high carrier-to-noise ratios are essential.

SUBSYSTEMS FOR LINEAR OPTICAL TRANSMISSION

The fundamental properties of sources and detectors have already been covered in Chapters III and VIII and discussed in relation to digital systems in this chapter. This section will only emphasize those points that are relevant to analogue operation, particularly noise and linearity.

Analogue Transmitters

Both LEDs and lasers are applicable to analogue systems, with modulation of the input current leading to modulation of the optical power output. LEDs are suitable for short links where a low received power only is required, since a maximum of about −10 dBm mean can be launched into a graded index fibre. See Figure 180.

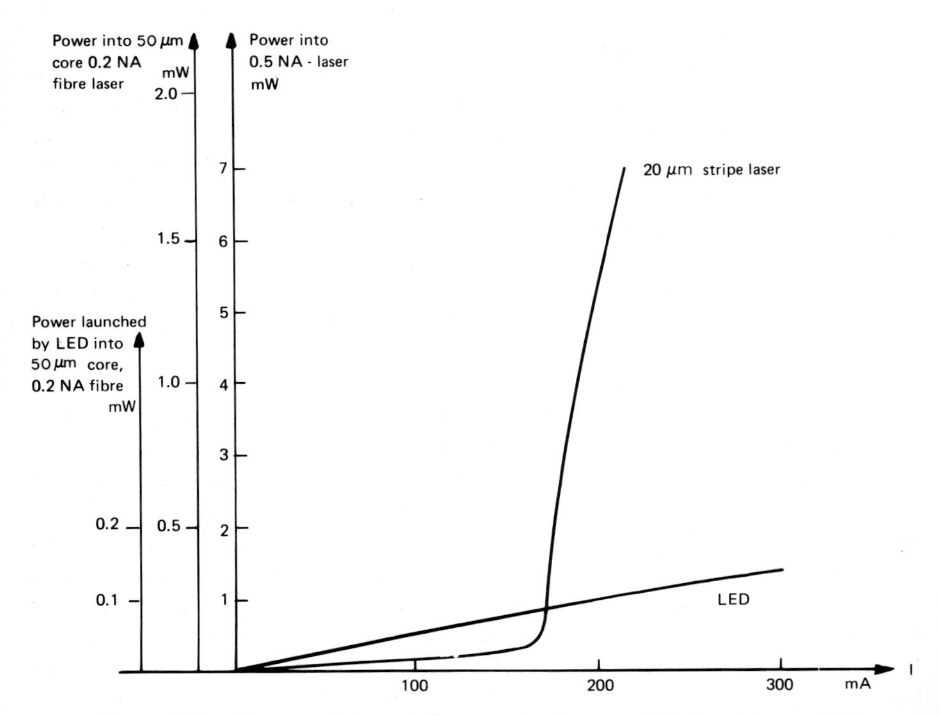

Figure 180. Characteristics of 20-μm stripe laser and high radiance LED

A disadvantage of the LED is the high modulation current required—up to 300 mA—and also the relatively slow response speed of high radiance devices. For example, for the LEDs made by Bell Northern Research, the high radiance device has a 3-dB frequency of 44 MHz, while for a sacrifice of 4.8 dB in power, a 3-dB frequency of 150 MHz can be obtained. Other LEDs also exhibit similar trade-offs.[211] Linearity is not good enough for high quality single channel video links, but linearizing can be achieved by predistortion of the electrical drive current.[212] There are, however, some doubts about this technique with ageing and wide temperature range operation of the device; also, it must be adjusted for each device. More useful techniques are likely to be feedback linearization, where a part of the optical signal is tapped off and used for negative feedback[213] and feedforward linearization. The disadvantages of feedback are frequency response problems in the loop, and loss of power in the optical tap, while feedforward[214] has the disadvantage of needing two closely matched sources. At present, this is a very significant cost disadvantage.

Lasers hold more promise for future systems and, as has been seen, for some applications their higher launched power is a necessity to achieve satisfactory system performance. Frequency response extends beyond several hundred MHz at full power. The light current characteristic of a 20 μm stripe laser is also shown in Figure 180. The device requires a relatively small modulating current, and because of threshold with temperature and device–device variation, practical operation requires mean power feedback stabilization of the operating point.[215] The characteristic exhibits marked nonlinearity, however, and device–device variations are large. Harmonic and intermodulation measurements have been made with a 1-mW peak–peak modulated light coupled to an attached fibre tail of 0.2 NA, and results are shown in Figure 181. It can be seen that high distortion levels are produced which would necessitate unrealistic levels of feedback linearization, etc. for useful operation. In addition, this device has a very narrow optical spectrum (see Figures 123 and 129 in Chapter VII), operating in a single longitudinal mode and, as will be made clear in the next section on modal noise, excessive noise may be generated by the use of narrow spectrum sources in multimode fibre systems. The use of the 20-μm stripe laser in analogue systems is not recommended.

Laser sources which promise extremely good linear performance have, however, been developed. They employ a stripe width of about 3 μm, and a typical characteristic is shown in Figure 182. Their main disadvantage is a relatively high light output at threshold, leading to a maximum modulation index m of about 0.5. Above threshold very satisfactory linear characteristics are exhibited, and Figure 183 shows the results of harmonic and inter-modulation measurements with 1-mW peak–peak modulated light from an attached fibre tail. Third order inter-modulation products are about 60 dB down, showing good potential for high quality systems. In addition the devices are well behaved and do not exhibit large device–device variations. They do have a noise output which is somewhat higher than the

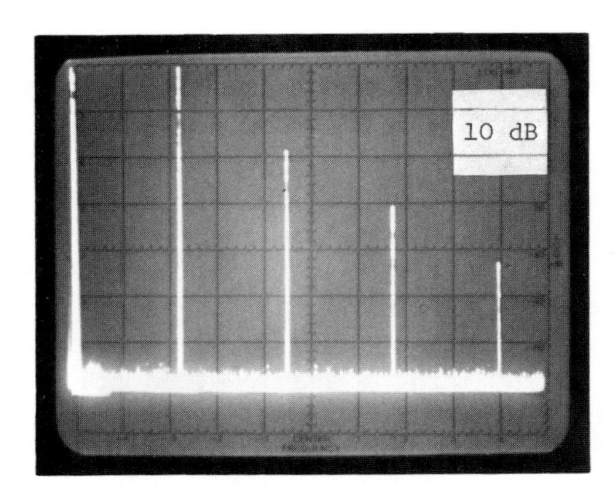

Stripe laser 20 μm wide
RB480
Harmonic distortion
1 mW p-p single tone 4.5 MHz
launched into 50 μm tail

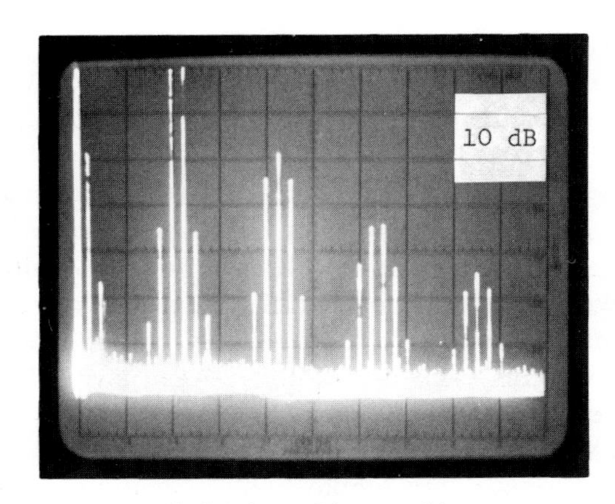

Stripe laser 20 μm wide
RB480
Intermodulation distortion
1 mW p-p two tone 4 & 4.5 MHz
launched into 50 μm tail

Figure 181. Distortion levels produced by 20-μm stripe laser

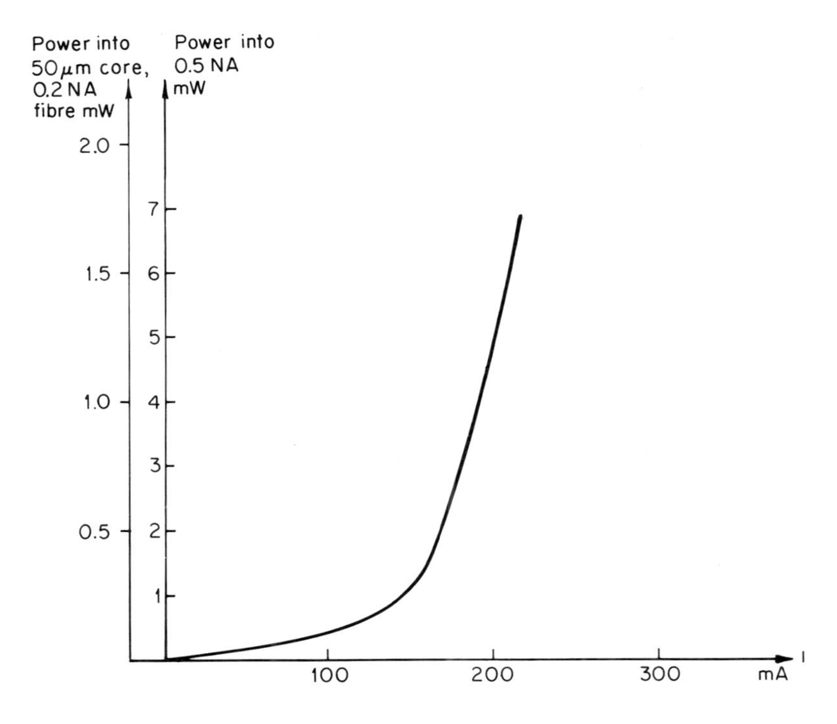

Figure 182. Characteristics of 3-μm stripe laser

quantum noise limit. Modal noise effects are much reduced but not entirely absent.

In operation, 3-μm stripe lasers still require mean power feedback stabilization of the operating point to allow for temperature variations. They exhibit good potential for further linearization by signal feedback, but it is not yet known whether devices will be sufficiently closely matched for feedforward techniques, although these have greatest bandwidth potential. With feedback or feedforward, one might hope to gain 20–40 dB of improvement of linearity, according to frequency and many other factors.

For pulse systems such as PPM, significant system advantage is not achieved unless the source can launch a high peak power. Low duty cycle operation is permissible. A source which has been used is the sawn cavity laser, but its lifetime is limited. A fully satisfactory source for this class of operation has yet to emerge.

Packaging of any of the devices is most important for high frequency linear operation. Lead inductance must be minimized and in addition output for mean and/or signal feedback must be available. A package has been developed which allows use of the light from the rear of the c.w. lasers for feedback. For signal feedback this light must be representative in distortion

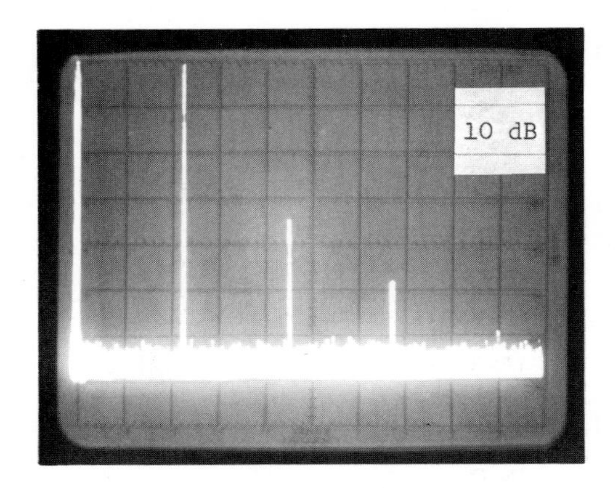

Stripe laser 3 μm wide
RB604
Harmonic distortion
1 mW p-p single tone 4.5 MHz
launched into 50 μm tail

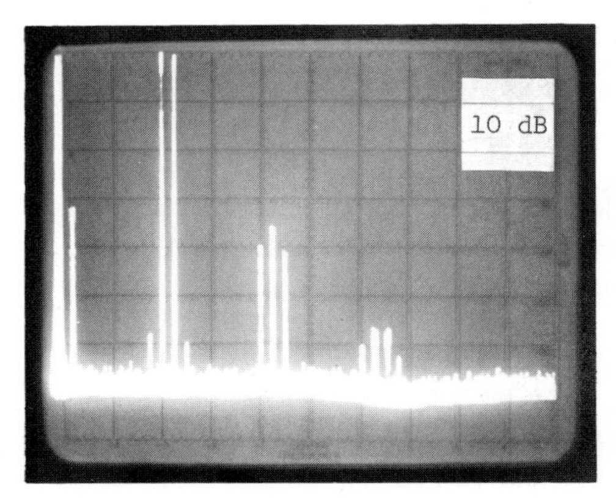

Stripe laser 3 μm wide
RB604
Intermodulation distortion
1 mW p-p two tone 4 & 4.5 MHz
launched into 50μm tail

Figure 183. Distortion levels produced by 3-μm stripe laser

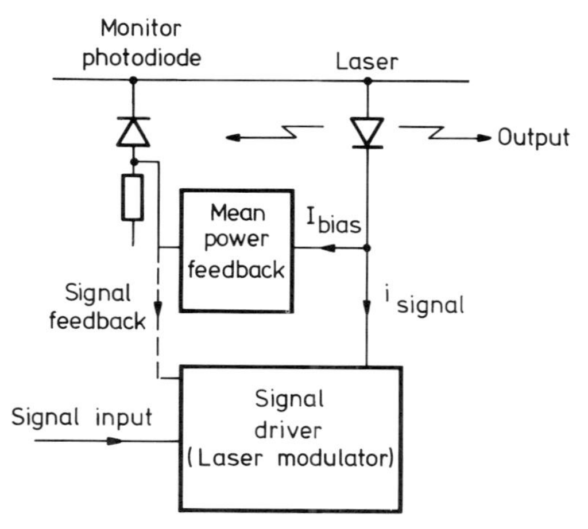

Figure 184. Outline of optical analogue transmitter circuit

of that from the front, and it is highly desirable that coupling efficiency from the back face to the feedback detector be high.

In practical systems both LEDs and lasers require mean bias current, which must normally be feedback controlled for lasers and also signal current. A block diagram is shown in Figure 184. High speed linear operation, particularly with signal feedback, needs full use of high frequency techniques such as short leads or strip lines for connections.

Analogue Receivers

For linear receivers, PIN and avalanche photodiodes are both usable. The choice between them depends on the carrier-to-noise ratio required, from system considerations, at the photodiode amplifier output. Factors such as ease of use and the possibility of gain control with avalanche photodiodes also need consideration. The performance of receivers is summarized in Figure 185 which shows the normalized power/bandwidth required for a wide range of carrier-to-noise ratios. It is similar to Figure 141 but, while that figure was drawn for the specific case of sine wave modulation with modulation index $m = 1$, this representation is in terms of peak–peak signal and is thus independent of waveshape and modulation index, unless the shot noise limit is approached. The abscissa of the graph is the optical receiver output requirement (see table XI) defined as follows

$$\frac{\text{carrier}_{(\text{p-p})}}{\text{noise}} = 10 \log_{10} \frac{i^2_{(\text{p-p})}}{\sum i^2_{\text{n}}}$$

where $i_{(\text{p-p})}$ is the peak–peak signal current and $\sum i^2_{\text{n}}$ the total mean square noise current referred to the same point in the detector/preamp chain. The ordinate of the graph is the quotient of peak–peak power and system

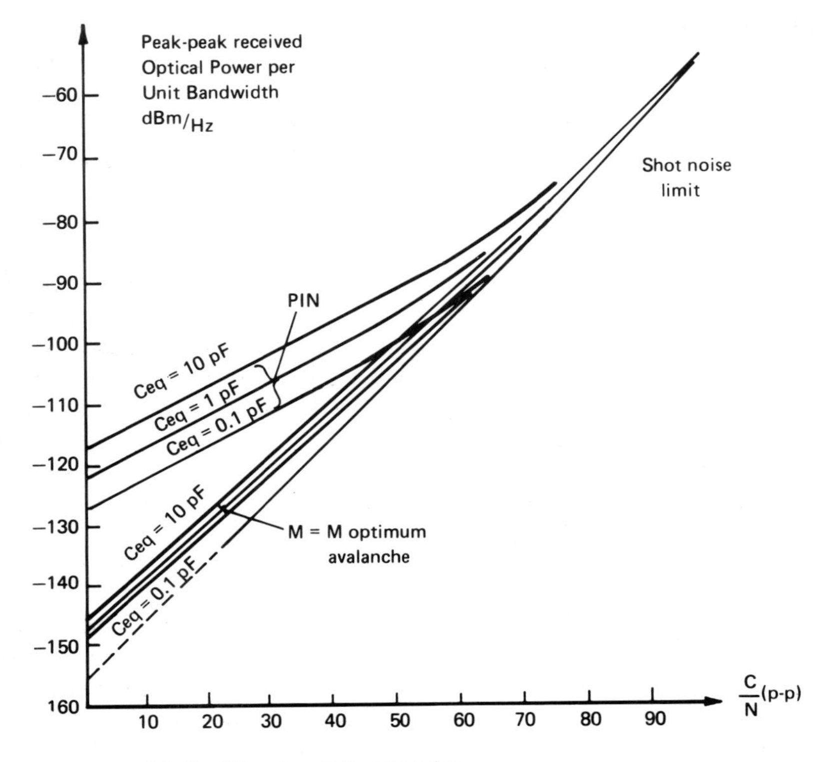

(1) Conditions to which graph refers.

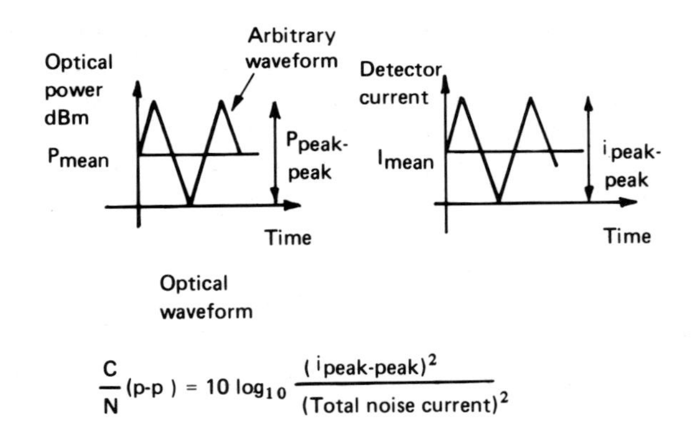

$$\frac{C}{N}(\text{p-p}) = 10 \log_{10} \frac{(i_{\text{peak-peak}})^2}{(\text{Total noise current})^2}$$

(2) To obtain required optical power, calculate $10 \log_{10}$ (Bandwidth) and subtract from value obtained from vertical scale.

(3) For assumptions, see text.

Figure 185. Optical detector power requirements

bandwidth, expressed in decibels relative to 1 mW optical and 1 Hz. Assumptions made in deriving this figure are as for Figure 141 in Chapter VIII, and are as follows:

(i) dark current and background light current are negligible;

(ii) for avalanche receivers, excess avalanche noise is proportional to $M^{0.3}$; and

(iii) All thermal noise arises from an equivalent resistor and the bandwidth is limited by capacitance across that resistor.

Hence, this capacitance dominates the receiver performance and curves are shown for a range of capacitance for receivers using PIN photodiodes, and similarly for avalanche photodiodes at optimum gain. Also shown is a line representing the mean optical power necessary for a given C/N ratio due to shot noise in the mean photocurrent, since this is the ultimate limitation. This has been calculated for $m = 1$, which gives the minimum mean photocurrent for a given peak–peak photocurrent: if the source cannot be fully modulated due, for example, to linearity considerations, then the shot noise limit will be less favourable. It also depends on wavelength and detector quantum efficiency. The line has been drawn for a wavelength of 850 nm and efficiency of 73 per cent.

Several things are evident from these curves. First, at C/N ratios of the order of 50–60 dB, the advantage gained from APDs is negligible. Secondly, the shot noise limit necessitates a certain level of optical power to attain a given C/N ratio even with a perfect detector, and at high C/N ratios it is possible to approach the limit closely with practical detectors. APD excess noise factors are being improved with new devices.

In the C/N region of 30–50 dB it is possible that the APD/PIN decision will be made on the practical grounds of the need for high voltage supplies, the fact that gain control, etc. is possible with APDs against the simplicity of the PIN receiver. Also, as frequencies increase above about 50 MHz it becomes more difficult to produce PIN–preamplifier assemblies near the performance shown, and avalanche operation becomes more attractive since the thermal noise of amplifiers with a 50–Ω input, as used at high frequencies, is high. With the possible addition of poorer linearity with APDs, the system designer's choice may not be an easy one.

Modern commercially available APDs and PIN diodes extend in operating frequency to about 500 MHz, making possible most linear systems foreseen at present. Experimental devices have been made operating up to several gigahertz.

The basic features of optical receiver circuits covered in the digital systems review are also applicable to the analogue receiver. In general, the transimpedance circuit (current feedback amplifier) is preferred for relative ease of design and good dynamic range performance.[216,217] For both PIN and avalanche receivers, the electrical signal output increases by 2 dB for every 1 dB increase in optical signal. Thus, if the system demands that large variations in fibre attenuation, source power, etc. be taken into account,

dynamic range problems can be severe purely on the grounds of signa
handling. If, in addition, linearity limits of the circuitry are approached a
high optical input power, then effective AGC with the preservation o
inter-modulation performance and C/N ratio may mean that optical attenuation is needed to bring the operation into a favourable signal range.

FIBRE PROPERTIES RELEVANT TO ANALOGUE TRANSMISSION

For analogue transmission, optical fibre response must be considered i
the frequency domain, rather than the time domain as is appropriate fo
digital systems. The situation is different for each of the three basic types o
fibres; step index multimode, graded index multimode, and single mod
fibres. These will be considered in turn.

Step Index Multimode Fibres

The bandwidth in this type of fibre is very limited and may be as little a
10 MHz·km. Fibres having a large NA and core diameter are best suited t
LED transmission because of high launch efficiency, but are most restricte
in bandwidth. The fibre frequency response is of low pass filter form, unlike
for example, coaxial cables.[218] In general, this type of fibre is likely to b
applicable only to short systems and/or bandwidths not exceeding a singl
baseband video channel, but some fibres having a bandwidth of 30 MHz·km
could be more widely applicable.

Graded Index Multimode Fibres

Bandwidths of about 300 MHz·km are predicted from Fourier transforma-
tion of the pulse response, but measurements so far carried out in th
frequency domain give more limited performance. The frequency respons
is again of low pass filter type. At these bandwidths, laser sources are mos
applicable, and it is known that results depend on launch conditions. It i
likely that the situation will be resolved in the near future and tha
definitive information for system designers, probably on certain souce–fibr
combinations, will become available. Graded index fibres will then b
applicable, at least for the shorter distances, to the widest bandwidt
systems.

Single Mode Fibres

For longer distances of the widest bandwidths, single mode fibres will b
necessary. The core diameter is about 5 μm, so that launching power from
source is less easy. Again, source–fibre combinations are likely to emerg
giving GHz·km capability in the near future. Single mode fibre may also b
necessary for systems needing high carrier/noise ratios, because of moda
noise.

THE PHENOMENON OF MODAL NOISE

The noise at the output of an optical fibre system receiver is usually determined by a combination of thermal noise, shot noise, excess noise in multiplication, and source noise. However, when lasers are used in multimode fibre systems incorporating misaligned fibre-to-fibre joints or mode selective taps, a further noise mechanism may occur and even dominate. This we call 'modal noise'. It appears as unwanted amplitude modulation of the received signal and is very sensitive to physical distortion of the fibre. Although this phenomenon only becomes dominant in particular circumstances, it is fundamental to multimode fibre systems in which mode selective loss occurs, so it is important to understand the conditions which bring it about so that it can be avoided.

OBSERVED PHENOMENA

Figure 186 shows an 'eye' diagram of a 140 Mbit/s system in which one fibre-to-fibre joint has been deliberately misaligned. The experimental configuration for all these experiments is shown in Figure 187. This misalignment is sufficient to produce about 10 dB of loss, yet the system margin is so large that the error rate should be negligible. However, modal noise is

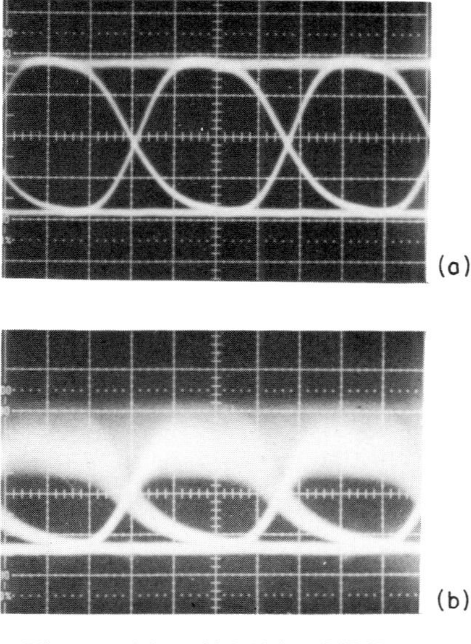

Figure 186. 140 M bit/s NRZ 'eye diagram' showing modal noise induced by misaligned connectors (2 ns/div). (a) Aligned connector. (b) Misaligned connector

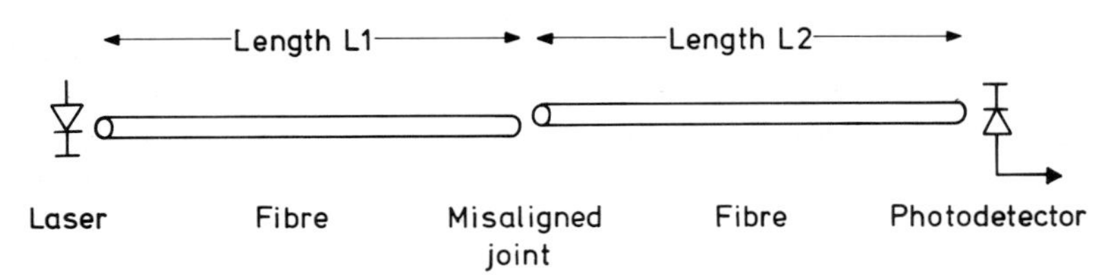

Figure 187. System configuration. Laser: 20-μm stripe, single longitudinal mode, low noise. Fibre: 50 μm core graded index, 0.2 NA, ~2 ns/km.

causing considerable 'eye' closure, resulting in a poor error rate. It is worth noting that it is modulation of the amplitude rather than the addition of noise, the effect being dominantly in the 'one' level (cf. shot noise). The sensitivity to mechanical distortion may be seen in Figure 188 which clearly shows the fluctuations in amplitude over a period of 200 ms caused by slight bending of the fibre before the misaligned joint.

The nature of modal noise can be conveniently demonstrated by modulating the laser with a low frequency square wave simulating, for example, a sequence of successive 'ones' in an NRZ (nonreturn to zero) code. This form of modulation is found to produce much more distinct repetitive modal noise waveforms (Figure 189(a), (b) and (c)). A high misalignment loss (10 dB) was used to make the effects more conspicuous but very much smaller values of excess loss can still produce problems, as will be seen below. A wide variety of traces can be produced simply by manipulation of the fibre at any point before the misaligned joint. Similar variations are obtained by adjustment of the joint itself (without necessarily changing the mean joint attenuation) as shown specifically in Figure 189(a) and (b). It

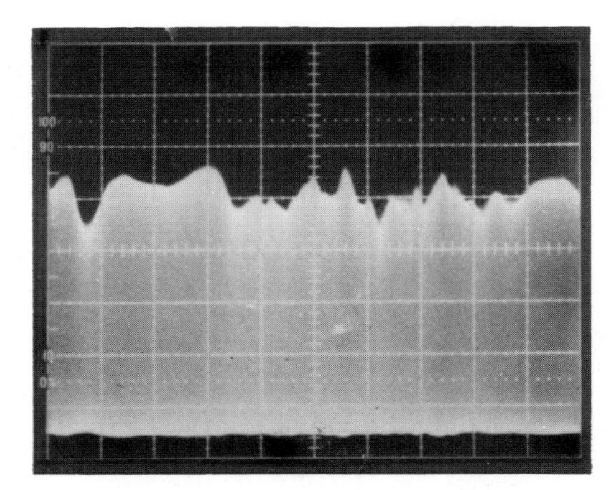

Figure 188. Slow scan of Figure 186 showing amplitude fluctuation of signal amplitude caused by flexing of the fibre before the misaligned connector

appears that the addition of these two waveforms would result in a reduction in the relative amplitude of the modal noise. The location of the misaligned joint has a significant effect on these waveforms. When there is only 1 m of fibre between the laser and the joint, fairly simple waveforms result (Figure 189(a) and (b)), whereas 1 km of fibre before the joint results in quite complex waveforms (Figure 189(c)). Observe in Figure 189 that the patterns appear to be bunched up at the leading edge of the pulse; this will be discussed in more detail below. The waveforms were found to be independent of the length of fibre following the joint.

We can describe the magnitude of the modal noise as the ratio of the peak-to-peak amplitude deviation in the one level x, to the mean amplitude y. The fibre may be manipulated before the joint to minimize or maximize the modal noise. If we measure the worst (i.e. maximum) ratio of x/y obtained for various values of misalignment loss, then we obtain curves as in Figure 190. Clearly, modal noise can seriously degrade digital systems, particularly with moderate values of longitudinal misalignment loss at joints near to the source.

Analogue optical transmission systems generally have a higher capacity per transmission bandwidth, but require high receiver power levels when compared with digital systems. For this reason the greater source power available from lasers compared with LEDs makes them quite attractive (provided that the linearity and other problems can be overcome). If we consider the implications of modal noise on analogue systems, then it will be quite clear from Figures 188, 189, and 190 that a simple intensity modulation system would be impracticable with these lasers (generally, nonlinearity already precludes this for 20-μm stripe lasers). The various pulsed analogue modulation systems (e.g. PFM, PPM, PIM) seem more practical for non-

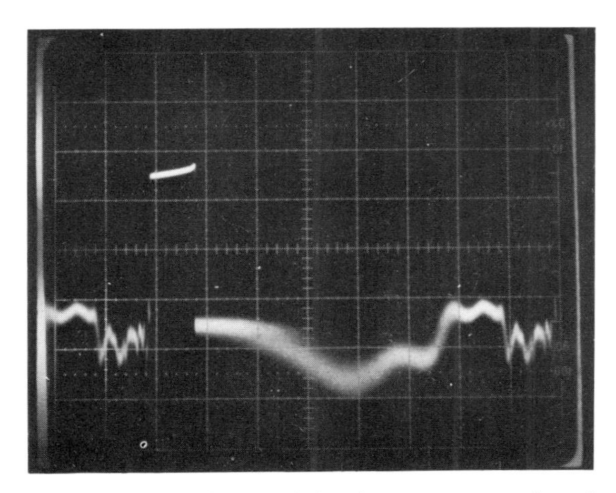

Figure 189. Repeating modal noise patterns on low frequency square wave (60 KHz) with gross misalignment of connector (2 μsec/div)

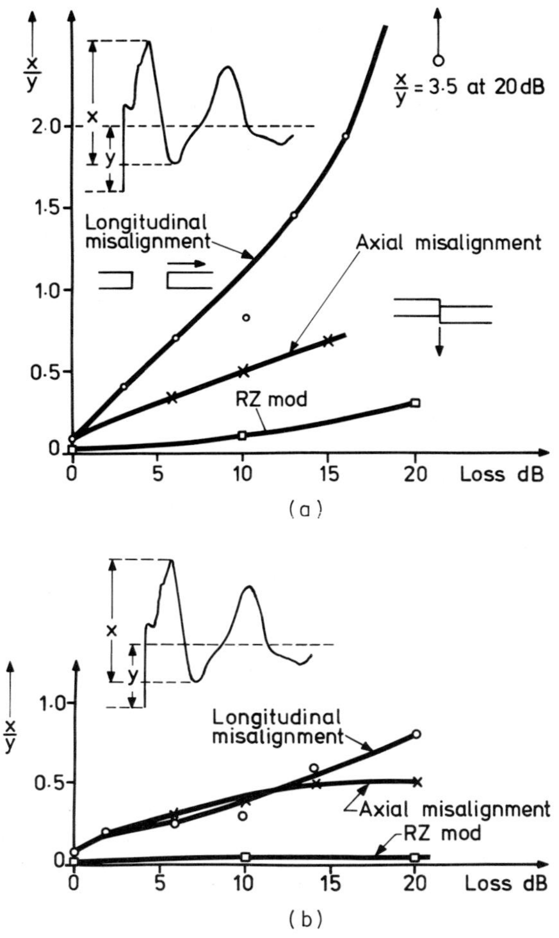

Figure 190. Maximum unwanted amplitude modulation vs connector loss. Measured for longitudinal and axial misalignment at L (see Figure 187). (a) $L_1 = 1$ m; (b) $L_1 = 1$ km

linear lasers, so we performed measurements of the signal-to-noise ratio obtained with 70 MHz 50 per cent duty cycle modulation, biased above threshold (to minimize nonlinearities resulting from turn-on delay which is dependent on the inter-pulse period). Figure 191 shows the results which, as with the previous experiments, are extremely dependent upon the precise physical position of the fibre before the connector and the connector itself. In general, only the measurement of the lowest C/N is satisfactory for system design, but it is interesting to note the range of values obtained and that C/N increases steeply as the aligned condition is approached. These curves show clearly that there would be very large C/N penalties produced by modal

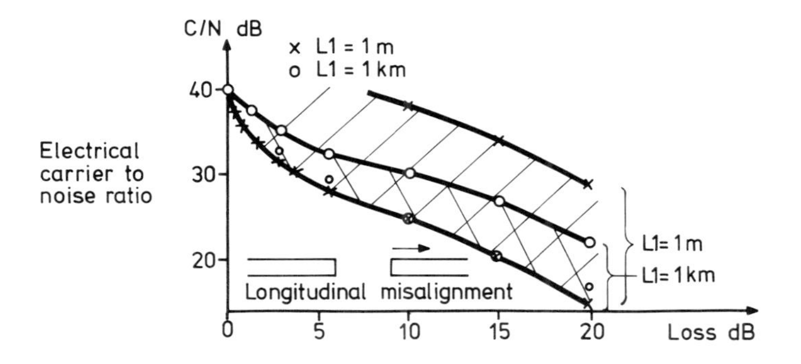

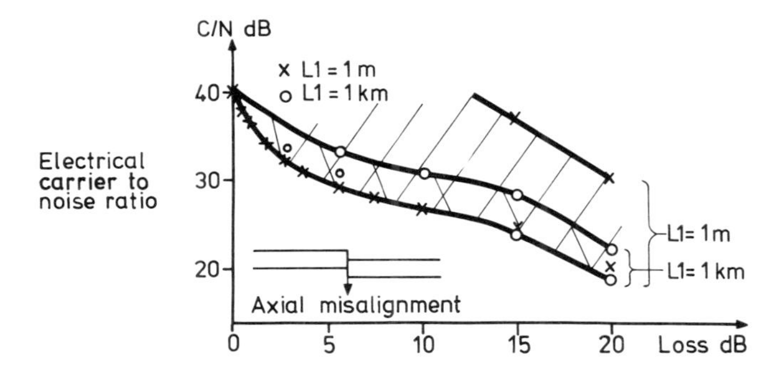

Figure 191. R.m.s. carrier/noise ratios versus connector loss for 70-MHz modulation and 40-MHz BW. Discrete points plotted correspond to the maximum and minimum C/N values. Hatched regions indicate the range of values obtained with $L_1 = 1$ m and cross hatched show the range for $L_1 = 1$ km

noise in such a system with only slightly misaligned joints. It is clear that if modal noise is to be eliminated then the mechanism by which it is generated must be understood.

THE EFFECT OF INTERFERENCE PATTERNS ON FIBRE JOINT LOSS

When light from a monochromatic source such as a HeNe laser is launched into one end of a length of multimode fibre, then the light emerging from the far end will have near- and far-field intensity distributions, of which Figure 192 is a typical example (the fibre was 50-μm core, graded-index). These 'speckle' patterns are produced by interference between the various propagating modes, each of which is subjected to slightly different delays

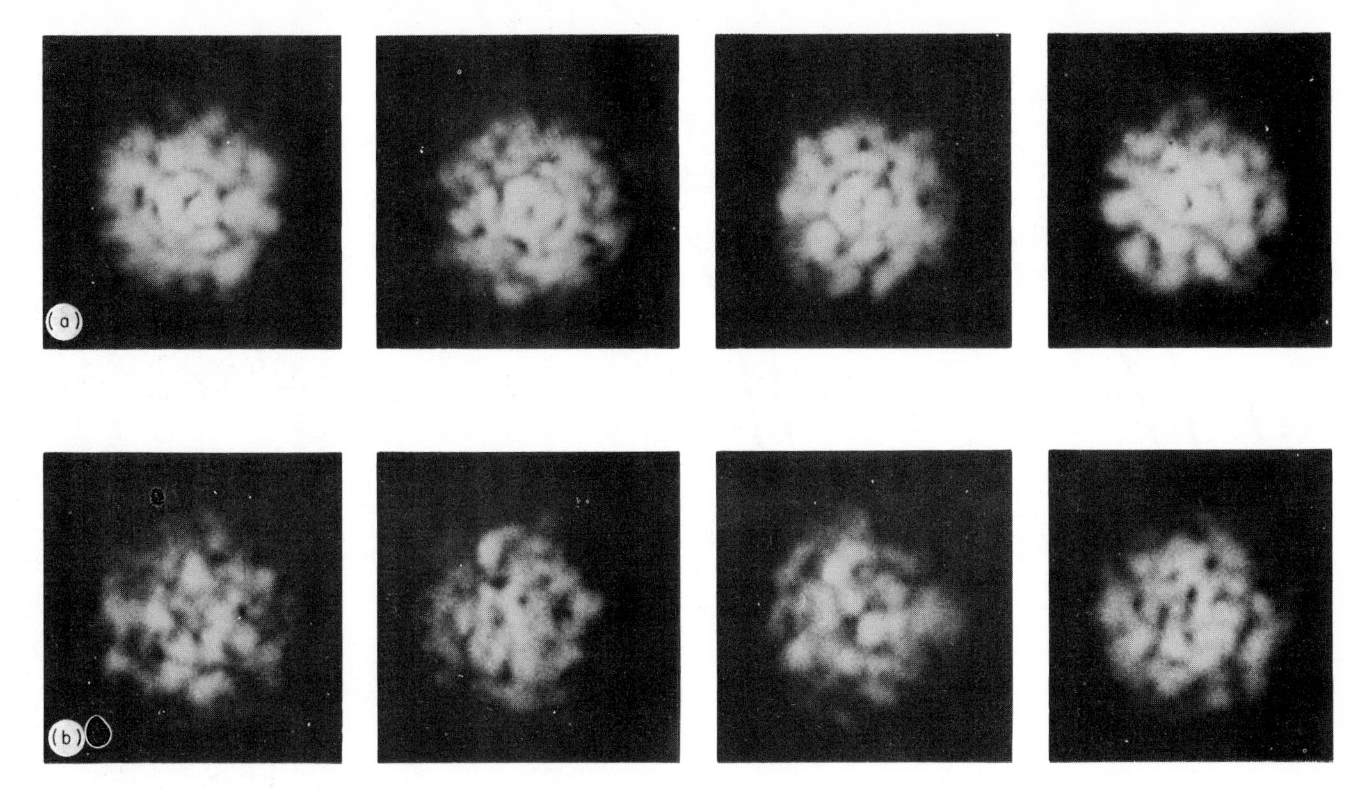

Figure 192. Some near-field and far-field speckle or interference patterns: (a) near-field and (b) far-field

through the fibre (mode dispersion). Readers interested in the analytical treatment of these patterns are recommended to study the work of Crosignani, Daino, and DiPorto[220–222] who have measured modal dispersion from the near-field speckle patterns of fibres. The position of the individual speckles is extremely sensitive to physical distortion of the fibre, due to changes in the local refractive index, and this effect has been studied and made use of in an optical fibre data collection highway by Davis and Kingsley.[223,224]

The four near-field and four far-field patterns in Figure 192 show the variations in near- and far-field patterns produced by slightly different physical bending of part of the fibre. Similar pattern changes may be caused by very small changes in the source wavelength. It is interesting to note the similarity between the near- and far-field patterns. The approximate number of speckles in either of these patterns is proportional to the number of modes propagating in the fibre. Therefore, a large core will have more speckles than a small core; a step index fibre will have more than a graded index fibre; and a high NA fibre will have more than a low NA fibre (if the numerical aperture is filled). Compare Figure 193, the near- and far-field patterns for a 350-μm, 0.4-NA step index fibre, with Figure 192 for a 50-μm core, 0.2-NA graded index fibre. All these patterns scale with λ.

If a fibre is joined to another similar fibre, but with axial misalignment, and monochromatic light is launched, then only the speckles within the circle drawn on Figure 194 will be coupled to the second fibre. Alternatively, if there is a longitudinal alignment error, then only the speckles within the circle drawn on Figure 195 will be coupled (the size of the pattern with respect to the fibre end will be proportional to the separation and the numerical aperture). With extreme misalignment only one speckle might be coupled. In practice there may be a combination of near- and far-field effects.

The precise number of speckles coupled to the second fibre will vary as the speckle pattern varies. Thus, a joint with fixed misalignment will have a loss

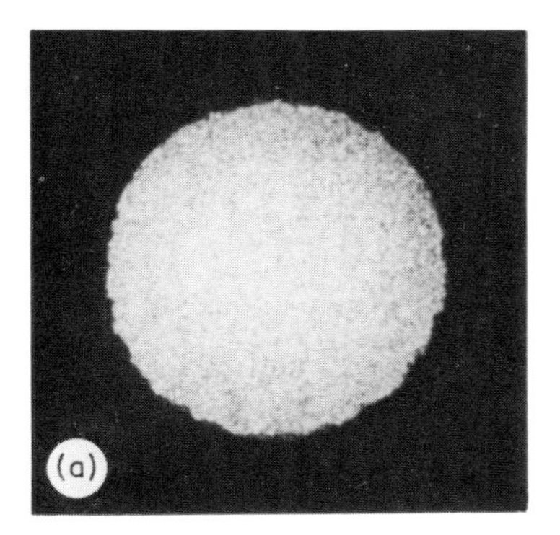

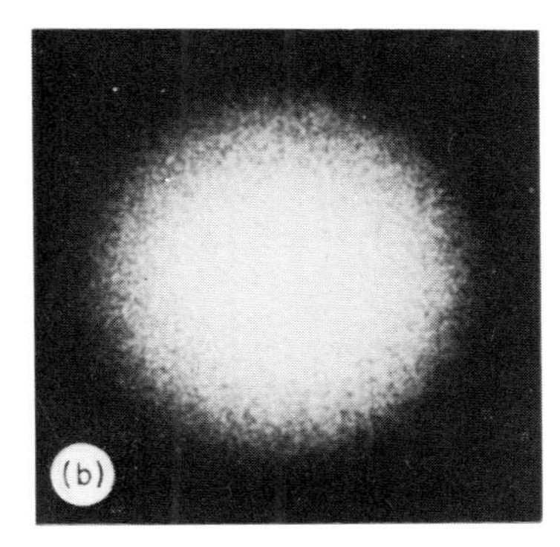

Figure 193. Speckle patterns for 350-μm core, 0.4 NA, step index fibre: (a) near-field and (b) far-field

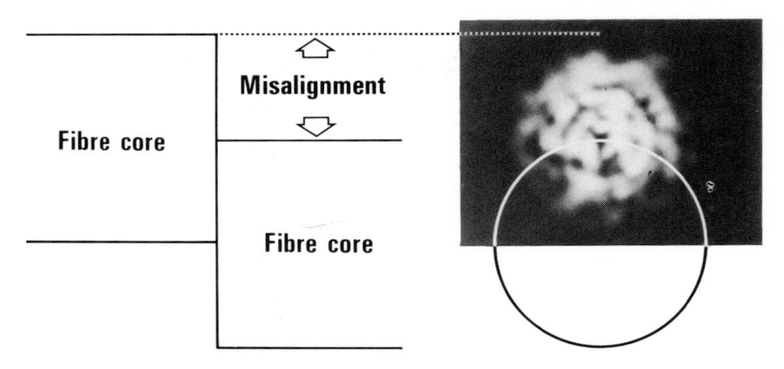

Figure 194. Diagram showing near-field speckle pattern and effect of axial misalignment

uncertainty and the loss will vary with changes in source wavelength and with physical distortion of the fibre before that misaligned joint.

The foregoing discussion of interference effects assumes a monochromatic source such as a HeNe laser, but injection lasers can be far from monochromatic when directly modulated. To help understand the effects of the source spectra on modal noise it is worth considering what conditions will produce a smooth pattern as opposed to a speckle pattern. If the fibre is flexed, it is possible to produce an average of several superimposed speckle patterns. Figure 196(a) is an individual speckle pattern of a 350-μm core plastic clad silica fibre under static conditions showing little evidence of surface damage. Figure 196(b) is the average of six successive exposures, each with different physical distortion of the fibre, and it may be noted that some surface damage is now apparent. Similar results would be obtained by simultaneous illumination with six monochromatic sources (or one source with six narrow lines in its spectrum). Figure 196(c) was a long exposure of the continuously varying speckle pattern produced when the fibre was subjected to a slowly varying distortion. Nearly all trace of the speckle pattern has disappeared, leaving a fairly reasonable image of the fibre end and the damage can now

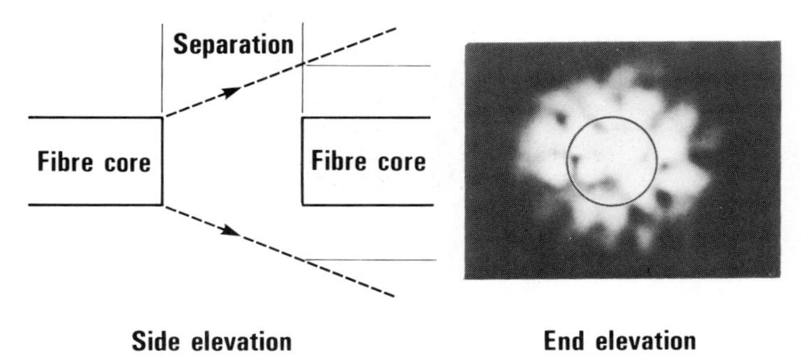

Figure 195. Diagram showing far-field speckle pattern and effect of longitudinal misalignment

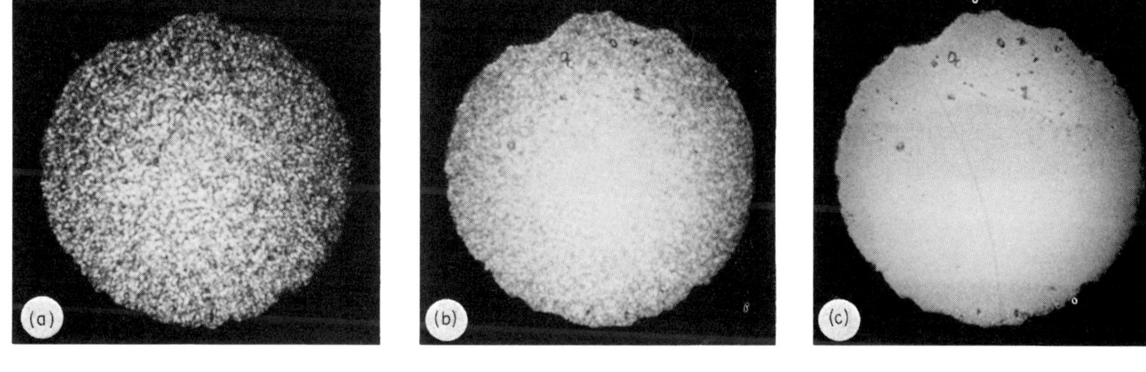

Figure 196. Near-field images for 350-μm core fibre. (a) Single speckle pattern. (b) Six superimposed speckle patterns. (c) Long exposure with varying speckle pattern

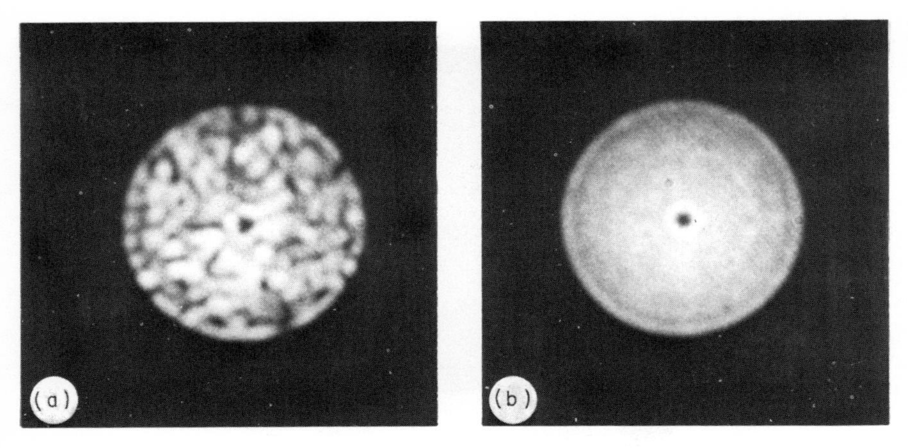

Figure 197. Near-field image of 50-μm core, step index fibre. (a) Single speckle pattern. (b) Long exposure with varying speckle pattern

be seen more clearly. Similar results can be obtained by increasing the spectral width, either by sweeping the source wavelength during the measurement interval or by phase modulating the modes, which is what occurs when the fibre is flexed.[223]

Figure 197 shows the comparison of a simple speckle pattern (a) with the time averaged speckle pattern from a 50-μm core, 0.2-NA fibre (b). This has a much coarser speckle pattern owing to the fewer modes propagating (the remaining dark spot in the centre is due to a refractive index dip in the centre of the profile).

COHERENCE TIME

The finite spectral width of light sources may conveniently be expressed as a 'coherence time' (which is simply the coherence length divided by the propagation velocity). This is the time up to which the wave may be delayed, but yet still correlate with the undelayed wave. The important concept to grasp is the *there can be no interference between two rays with a difference in delays greater than the coherence time, τ_c, even though they are from the same source.*

When light is launched into a multimode fibre exhibiting mode dispersion, the power is split between several modes, each propagating at slightly different velocities along the fibre; thus, then we measure dispersion in the time domain, a narrow launched pulse is seen to increase in width due to the spread in arrival times. Figure 198 shows a fibre system in which there are two misaligned fibre-to-fibre joints. If we consider the action of two modes only, it will help us to understand the interaction of dispersion and coherence time.

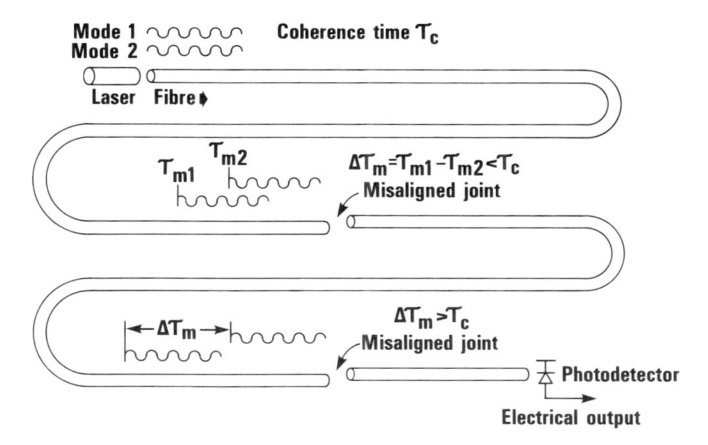

Figure 198. Simplified explanation of dependence of modal noise on source coherence time, mode dispersion, and position of joint. Illustrated for two modes only

At the launch end of the fibre the two propagating modes, 1 and 2, are completely coherent, i.e. no differential delay has occurred because there is insignificant delay. The light propagates along the fibre to the first misaligned joint with a total delay of τ_{m1} and τ_{m2} for modes 1 and 2, respectively. There is a difference $\Delta\tau_m$ between their arrival times, and at the first joint in the figure this is shown to be less than the coherence time τ_c of the laser. Near and far field interference patterns will therefore be produced at this joint and any misalignment will result in an uncertainty in the attenuation; furthermore, any change in the relative phase of these two modes will change the interference patterns and hence change the loss. Large phase changes will occur for small changes in source frequency due to the large difference in the number of wavelengths traversed in the fibre by the two modes, and large changes of differential phase occur for very little physical distortion of the fibre due to localized changes in the refractive index.[223] Either of these will result in a change of joint attenuation and a variation of these with time will produce amplitude modulation of the signal received at the photodetector, i.e. 'modal noise'.

In a practical system the attenuation of such a misaligned joint will continuously vary due to vibration of the fibre and to small changes in the source wavelength. Furthermore, the frequency of this unwanted amplitude modulation will be proportional to the product of $d\lambda/dt$ and the inter-modal delay $\Delta\tau_m$ (dispersion).

Let us now consider what effect a misaligned joint will have if the path length and dispersion are sufficient for the differential delay $\Delta\tau_m$ to be greater than the coherence time τ_c, as is illustrated at the second misaligned joint in Figure 198. There can no longer be any interference since the light waves in the two modes are now incoherent with each other and hence no modal noise can be introduced at this joint. Consequently, if there are no misaligned joints for $\Delta\tau_m > \tau_c$ (i.e. if the first joint in Figure 198 is perfectly aligned) there will be no modal noise. Any technique which reduces the effective modal dispersion (e.g. using successive joined lengths of under- and overcompensated profile fibre to optimize the mean profile) could lead to increased modal noise from a distant misaligned connector relative to a near one.

This is, of course, a gross simplification: a practical fibre might support several hundred modes, each capable of interfering with any other mode provided that their differential delay is less than the coherence time. Clearly, at any point along the fibre some pairs of modes will be capable of interfering, others will not, so we can consider the near- and far-field patterns to be the superposition of an interference pattern (Figure 197(a)) on a smooth illumination pattern (Figure 197(b)). As the light travels further along the fibre the ratio of power in the interference pattern to power in the smooth pattern will decrease. Therefore, there will be no sudden disappearance of modal noise after a given length, but rather a reduction in the effect with increased length. For analogue systems modal noise will still be significant at a greater length due to the higher *S/N* requirement. This discussion does not take into account mode conversion in the fibre or at misaligned joints.

CONDITIONS GIVING RISE TO MODAL NOISE GENERATION

Before looking at means of prevention, let us reappraise the necessary conditions for modal noise generation.

(1) Narrow spectral width (long coherence time).
(2) Modal or spatial filtering.
(3) Time variation of either modal or spatial filtering.

All three conditions must be present for 'modal noise' (a system carrying information at d.c. might be troubled by conditions (1) and (2) alone, e.g. joint attenuation or coupler coefficient measurement).

Narrow spectral width is obtained with good injection lasers,[219] particularly with the single longitudinal mode lasers described in Chapter VII (see Figure 123). Modal or spatial filtering can occur when splices or connectors are longitudinally or axially misaligned, by dirt on either of the fibre ends, or by any mode selective loss mechanism, e.g. mode selective couplers. It can also occur at the coupling from fibre to detector, but this is unlikely with practical detector sizes. Variations of filtering are caused by physical distortion of the fibre preceding the 'filter' or by changes in source wavelength. Such changes may be induced by changes in laser temperature (due to the low frequency content in the modulating signal current) and may be influenced by variations in reflections from adjacent external optical cavities.

All modal noise patterns are determined by the conditions before and at the connector. They are generally unaffected (ignoring fibre bandwidth limitations) by the conditions following the joint, each successive joint adding its own amplitude modulation of the signal. The frequency of the amplitude modulation is proportional to the rate of change of wavelength and to the inter-modal dispersion.[220]

It is worth noting that in Figure 190(a) the noise produced after 1 m is significantly greater than after 1 km. This may be because the large changes in wavelength (produced by the low frequency content of the modulating signal, below the thermal time-constant) make the effective coherence time (when measured with a 20 MHz receiver bandwidth) shorter than much of the inter-modal delay. Figure 191, however, shows minimum C/N ratio is the same for 1 m and 1 km but also shows that C/N cannot be maximized to the same degree by manipulation of the fibre. This is because the interference patterns are quite stable close to the laser but highly unstable after 2 ns of mode dispersion.

The 70-MHz modulation is well within the thermal time-constant of the laser so no large wavelength variations occurred. (No deviation was applied to the 70-MHz signal at the time of the experiment.) The different results of these two experiments highlight the need to consider carefully the specific modulation conditions before predicting the presence or absence of modal noise in a particular system. It also shows the necessity of measuring the coherence function of the laser under actual operating conditions.

MEANS OF PREVENTION OF MODAL NOISE

Since all three of the conditions are necessary for modal noise, prevention of any one will solve the problem.

Source Considerations

If we can use a broad spectrum source then we can avoid modal noise; but how broad a spectrum? The simple answer is that it must be sufficiently broad to eliminate interference effects at any modal or spatial filter. There are two ways in which the spectrum can be usefully broadened. The first is simply by increasing the width of the single longitudinal mode, i.e. decreasing its coherence time. The second is by increasing the number of lines (longitudinal modes) without necessarily decreasing the coherence time of the individual lines. (In practice multiple longitudinal mode lasers will usually exhibit short coherence times for each mode or line.)

For single longitudinal mode operation the coherence time should be shorter than the inter-mode dispersion time at the point of the first modal or spatial filter, i.e. the lower the dispersion then the shorter the coherence time must be. When the source has several longitudinal modes, then the effect is rather different. In this case we have a superposition of several interference patterns and these will tend to average out (see Figure 197(b)).

(However, such lasers exhibit several coherence peaks at multiples of the effective cavity length thereby making the analysis rather more difficult.) Therefore, the greater the number of lines the less modal noise. We have found that the use of narrow stripe lasers (3–5 μm) exhibiting multiple longitudinal modes (and broader linewidth/shorter coherence time) considerably reduces modal noise, almost eliminating this problem in digital systems.

One advantage of using the coherence time rather than the linewidth to describe the spectrum is that the relevant linewidth must be the instantaneous linewidth measured by time-resolved spectroscopy. To illustrate this point, consider a single longitudinal mode laser with long coherence length whose wavelength is slightly modulated by the small temperature changes induced by the modulating signal. The spectral width when measured as a time average will appear broader than it really is instantaneously. In a PCM system the photodetector receiver averages the received photons over a bit period, so if we can sweep the frequency during each bit period we can effectively broaden the spectrum.

When injection lasers are modulated from below threshold the spectrum is initially broad, consisting of a number of longitudinal modes, and over a period of several nanoseconds the number of lines reduces until with a 'good' laser there may be just a single longitudinal mode. At the same time, the mean frequency and, to a lesser extent the frequencies of each of the lines, are moving as the laser temperature changes with the increased current. When such a laser is modulated from above threshold, it operates continuously in a single mode (see Figure 199(a)) and therefore this has previously been considered the ideal modulation condition since material

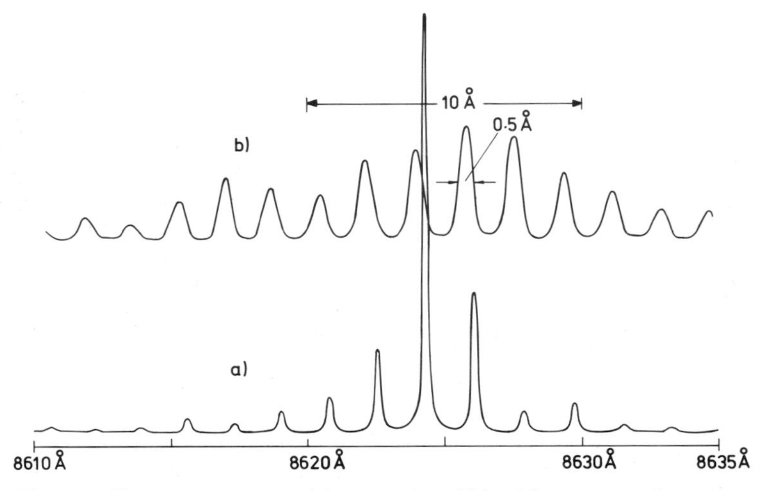

Figure 199. Laser spectra: (a) normal and(b) with spectrum broadening; with identical launched mean power

dispersion is minimized.[215] Also, with this method there is no turn-on delay. However, the occurrence of modal noise (as well as reduced extinction ratio and increased laser noise close to threshold) now makes modulation from slightly below threshold more attractive. It is vital that a return-to-zero (RZ) code is used to ensure that the spectrum broadening occurs for successive logic 'ones' and, of course, the pulse duration must not be long enough for the laser to settle back into single longitudinal mode operation.

Figure 199(b) shows the spectrum broadening induced by this method of modulation at 140 Mbit/s. (Note the increase in both the number of lines and in the individual linewidth (Figure 199(a) is monochromator, resolution limited.) For connectors close to the laser, the increased number of lines is probably much more useful than the increased linewidth in overcoming modal noise, the relative delay between modes being still quite small. Figure 200 shows the improvement in the 'eye diagram' produced by this spectrum broadening. Figure 200(a) shows the laser output and Figure 200(b) shows the 'eye' when normal NRZ modulation is used and the laser is biased to threshold during the zero level. The modal noise appears as gross 'eye'

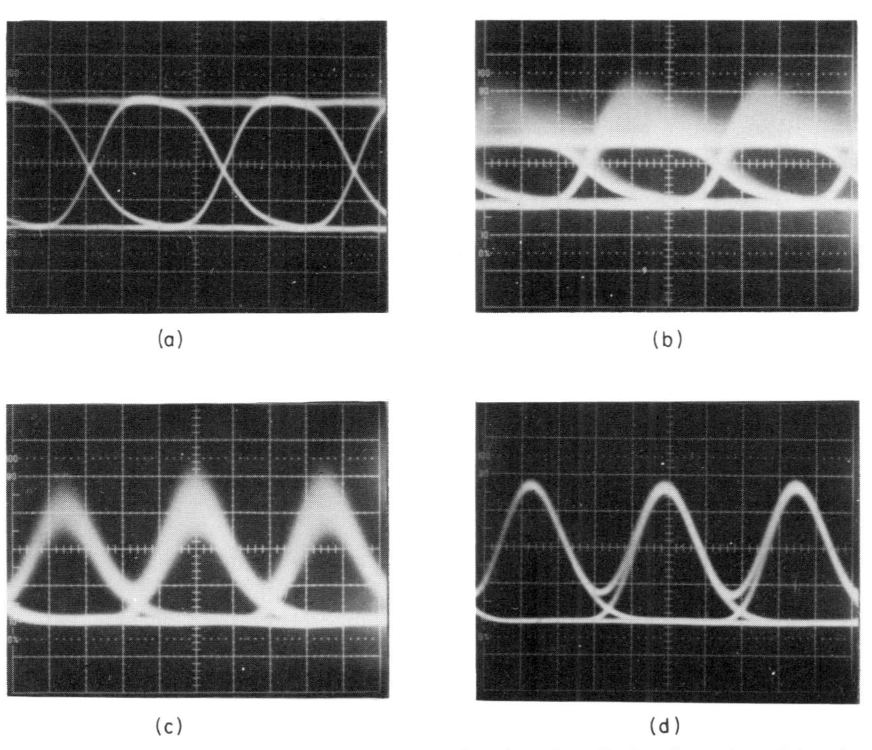

Figure 200. 140 M bits/s 'eye diagrams'—showing the elimination of modal noise by spectrum broadening. (Horizontal scale 2 ns/div). (a) NRZ laser output direct. (b) NRZ showing modal noise. (c) RZ showing modal noise. (d) RZ modulation from below threshold. No modal noise

closure. Figure 200(c) shows the effect of changing the modulation to RZ whilst keeping the bias the same. The 'eye' is still severely degraded. Figure 200(d) shows the effect of modulating RZ from below the threshold by increasing the modulation current. A good clean 'eye' diagram is evident despite the presence of a deliberately misaligned joint. If the same 140-MHz RZ modulation is applied to the low frequency square wave used in the tests recorded in Figure 190, then the improvement can clearly be seen in the curves marked RZ. Furthermore, the improvement is greater after 1 km (Figure 190(b)) than 1 m (Figure 190(a)) due to the much greater inter-mode dispersion time than coherence time.

It is possible artificially to reduce the coherence time simply by passing the laser light through an initial length of highly dispersive fibre. This is equivalent to ensuring that the first misaligned connector is never too close to the laser. The effect of dispersion is to reduce or, in the limit, eliminate the inter-mode coherence. The light emerging from the end of a highly dispersive n-mode fibre becomes equivalent to light from n coherent sources which are, however, mutually incoherent.

It remains to be seen whether any of these broad spectrum solutions will be satisfactory for future very high bandwidth analogue systems where there will be a compromise between modal noise and material dispersion. Perhaps a single mode fibre will be necessary.

Prevention of Modal and Spatial Filtering

There should be very few problems provided lossless connectors and splices are used! (though we have observed modal noise in a system with no joints, the mode selective loss being due to micro bending loss). However, if a misaligned joint should occur, a large system power margin may be ineffective in overcoming the increased loss because of the generation of 'modal noise' at that joint. The high S/N required in analogue systems makes them extremely sensitive to small joint misalignments.

It is worth reminding ourselves that the number of speckles in the interference pattern is proportional to the number of modes which are propagating in the fibre. Therefore, any increase in the number of modes will reduce the modal noise for the same relative misalignment. Also, if the increase in modes is achieved by use of a larger core diameter then the same alignment error will result in a smaller relative misalignment. Furthermore, high NA will give an advantage if only axial misalignment occurs.

It can be seen from Figures 190 and 191 that greater modal noise may be produced by longitudinal misalignment than by axial misalignment, especially for high attenuation. This is most easily understood if we consider separating the two fibre ends until only one speckle of the far-field pattern falls on the core of the second fibre (see Figure 195). Clearly, a small change in the interference pattern can now cause a dark region to be coincident with the fibre core, resulting in 100 per cent amplitude modulation of the signal. For the axial misalignment, however, the effect is of two overlapping circles, so even at the extreme misalignment case several speckles may be coupled. Also, remember that all the dimensions scale with wavelength so there will be a greater penalty with longer wavelength sources.

Mode selective couplers behave like misaligned joints. The modal noise generated will be greater the smaller the fraction coupled out, and furthermore the noise will be in the opposite phase. (This satisfies the requirement that the total power stays constant, i.e. no photons are lost.) Thus, such couplers will be unsuitable for feedback linearization of coherent sources and may present problems in multiterminal laser systems. For the coupler to be free of modal noise, identical speckle patterns must be coupled to each output port. Mode mixing in the coupler will not in itself eliminate modal noise since it cannot eliminate the speckle pattern, merely rearrange it. Beam splitter types will overcome this problem provided that loss is kept low (the loss will generate modal noise if it is mode selective).

Prevention of Time Variation of 'Filtering'

It does not seem practicable to prevent the slight physical distortion of the fibre which facilitates modal noise, especially when one considers the fibre as a distributed microphone. Even if this were possible, it would still be necessary to prevent any variations in source frequency and this would not be possible with direct modulation of the injection laser.

THE SIGNIFICANCE OF MODAL NOISE CONSIDERATIONS

The phenomenon which we call 'modal noise' is 'generated' at misaligned fibre-to-fibre joints when the source coherence time is longer than the inter-mode delay preceding the joint in question (mode dispersion). Digital systems are only seriously degraded by modal noise when both very narrow spectral width sources are used and when splices or connectors become misaligned. Some multimode analogue systems, however, may be quite impracticable with narrow spectral width sources, even with relatively low loss joints.

The best solution to the problem of modal noise appears to be to use as broad a spectral width source as is practicable for each system without introducing significant material dispersion. It also helps to use as high dispersion fibre as is practicable. Lastly, low loss connectors with little chance of accidental gross misalignment will help considerably.

Modal noise is a fascinating phenomenon since it involves and requires the understanding of so many of the components in an optical system. Some of the analyses of coherent propagation in fibres relevant to the study of modal noise have been published previously.[220,223] This discussion should help relate the phenomenon to practical systems design.

MULTITERMINAL SYSTEMS

The systems to be considered here comprise a number of terminals, all able to communicate with each other over some network of optical links. The system may or may not have a central switching complex that is analogous to a telephone exchange. After a general discussion, a particular example of a recently designed system will be described in more detail.

REPEATERED SYSTEMS

Probably the most straightforward way of designing an optical multiterminal system is to use simple point-to-point optical links to inter-connect electrical units. This can be arranged in a number of ways, for example Figure 201 shows a number of terminals connected into a ring by optical links. In this case it is necessary for terminals to act as repeaters for messages not intended for them. In some applications this could lead to problems of reliability since one terminal failure could result in failure of the whole network. Control of such a network could be on a 'distributed intelligence' basis or based on a single controlling master terminal. A second example is shown in Figure 202 where all terminals are connected to a central exchange-like unit. This unit could take the form of an electrical switching network or alternatively could simply mix the incoming messages and feed all outputs with this composite signal. In this latter case the central unit has become an active star transmission mixer and many of the design considerations mentioned below for similar passive devices would apply.

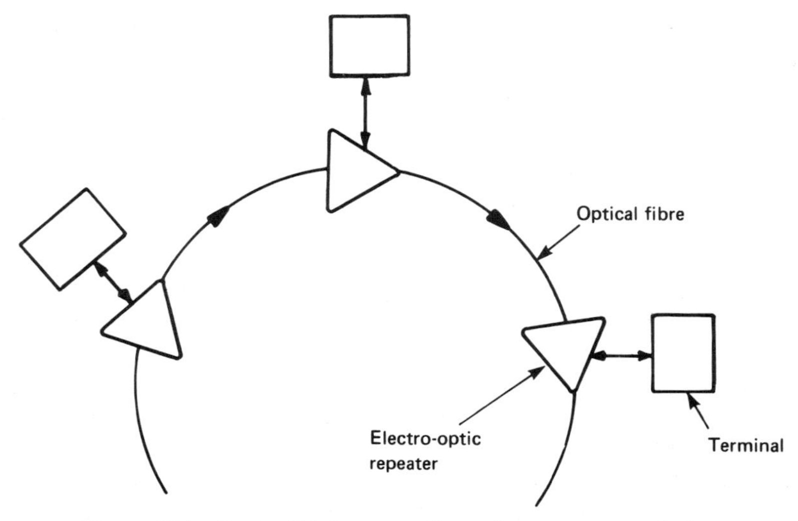

Figure 201. Optical highway configured as a repeatered ring

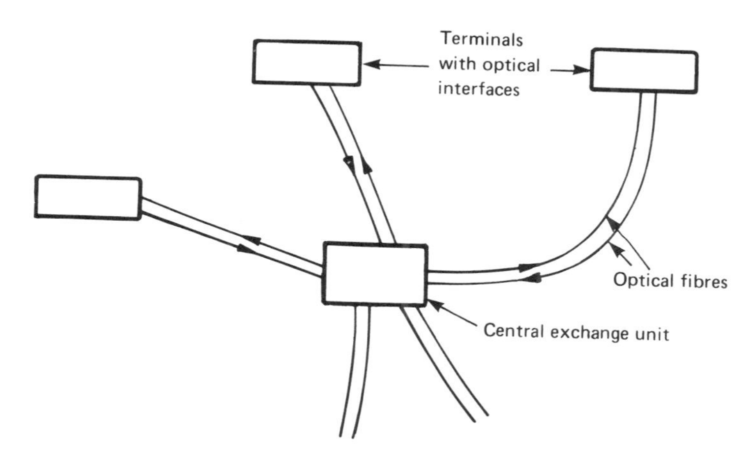

Figure 202. Optical highway with central exchange unit

The design of these systems involving only point-to-point optical links raises many interesting problems; however, they are not greatly influenced by the choice of fibre optics on the transmission medium and so will not be considered further.

SYSTEMS INVOLVING FIBRE-OPTIC HIGHWAYS

An optical highway is an arrangement for inter-connecting a number of terminals solely by optical means. In its simplest form this would be a number of optical fibres jointed together with splitters, combiners, and mixing components. These components, unlike many of their electrical equivalents, will introduce significant losses into a system, and this has to be given careful consideration in complicated networks. Some suitable optical components will be described below in the Optical Highway section.

Multiplexed Signals

With the use of an optical highway to transmit several channels of information some form of multiplexing is required, and several possible schemes are described below.

(a) TDM—time division multiplexing. TDM in various forms is the most common form of multiplexing used in electrical data bus systems, and is equally applicable for optical systems. Typically it is arranged so that each terminal's transmitter will have sole use of the highway for the duration of one message, and that during that message all other terminals will be receiving.

There are several possible arrangements for determining the order of transmissions, for example:

rotational—where order is predetermined to be a certain sequence;
controlled—where a central controller allocates time slots as the system demands require; and
contention—where transmitters use any empty slot.

An example of a TDM system designed for a fibre-optic highway is given below in the section Example Multiterminal System.

(b) FDM—frequency division multiplexing. FDM is also possible for optical systems. In this case the frequency concerned is not the optical carrier frequency but the frequency of an amplitude modulating subcarrier. Information would then be carried by the usual forms of modulation of this subcarrier (FM, AM, phase, etc). Separation of signals at the receiver would be carried out by electrical filtering following a broadband optical receiver.

Inter-modulation products could be a problem with this type of arrangement, but since each transmitting source handles only one frequency all inter-modulation would take place in the receiver, and very linear receivers have been demonstrated.

(c) Wavelength or colour multiplexing. Wavelength multiplexing is also a form of frequency division, although some optical sources are in fact narrow band noise sources in the frequency domain. The two most commonly used optical sources for fibre-optic systems, LEDs and semiconductor lasers, both appear suitable for this type of multiplexing.[225,226] The simplest form of demultiplexer is to use wavelength selecting filters or mirrors to select out the wanted signal which is then detected conventionally by a photodiode and an amplifier.

(d) Phase modulation. Phase modulation of a coherent optical carrier differs from the previous examples in that it is most suitable for systems with a single optical source, and many transmitters that can all-phase modulate this carrier.

One example of this, where a single laser launches an optical carrier along a fibre, and transmitters at intervals along the fibre phase modulate the carrier by using piezoelectric transducers to distort the fibre geometry,[223] has already been referenced in connection with the discussion of modal noise. This is a transmit only system, all the signals being received and demultiplexed at one point at the other end of the fibre.

The choice of one of the above multiplexing techniques must be made to suit the particular system requirements.

OPTICAL HIGHWAY

An optical highway may be constructed of purely passive components such as mixers and access couplers, or it may contain active components such as optical switches (relays) and amplifiers. The use of such active devices will have significant implications for the design of optical highways; however, as these components are at an early stage in development, and it is not clear what their ultimate capabilities will be, they will not be discussed further.

Optical Highway Configurations

In choosing a layout for an optical highway many parameters must be considered, and matched to the rest of the system, for example: total optical loss, differences in loss over different paths, delays caused by path lengths, and multipath effects and echoes caused by component reflections.

Two well known configurations for such a highway are the star or ring systems, illustrated respectively in Figures 203 and 204. The relative merits of these are well documented.[227]

Briefly, the star system has the advantages to the system designer of low optical loss, freedom from echo pulses, and can easily be designed to give loss and delay which is the same for all pairs of terminals. The restrictions on configuration, and the vulnerability to the star component itself are, however, unacceptable for some applications. The ring system suffers from high and variable loss, variable delays, and echoes, but the configuration itself is sometimes preferable, especially if access points can be added at will.

More complex configurations, which can be thought of as hybrids between the ring and the star, can readily be conceived to avoid the very high loss of ring systems or the vulnerability and inflexibility of the star configuration.[228] One example is shown in Figure 205.

A hypothetical configuration is shown in Figure 206 which combines several concepts for illustration. It shows that a hybrid can consist of a ring of star subgroups, or the reverse, and it also shows how two- or three-way fibre mixers can be used as couplers for a ring configuration. The use of extra optical paths is illustrated which provides redundancy and greatly

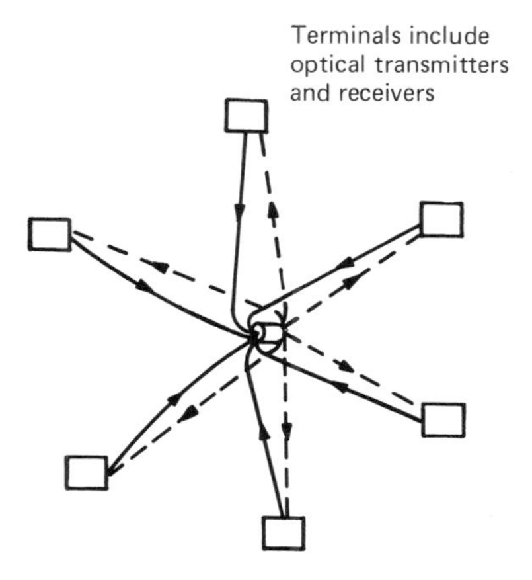

Figure 203. Multiterminal optical data system—star configuration

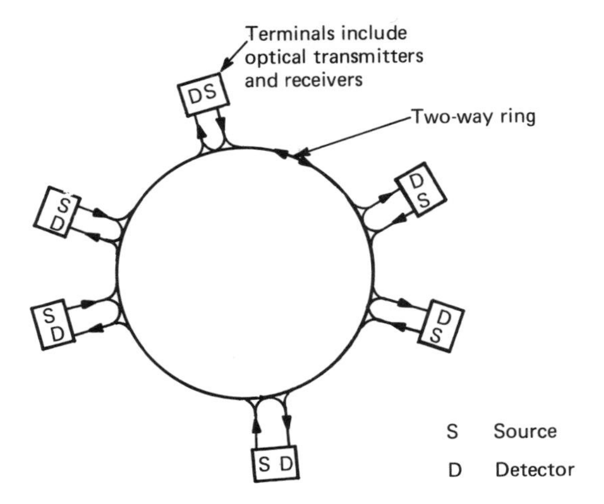

Figure 204. Multiterminal optical data system—ring configuration

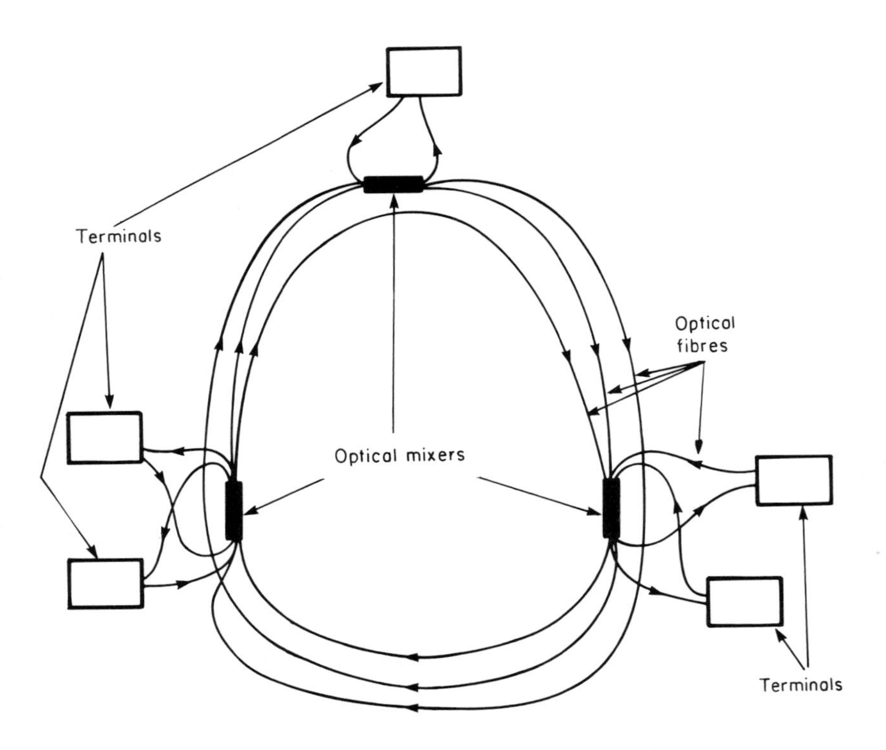

Figure 205. Hybrid ring and star configuration

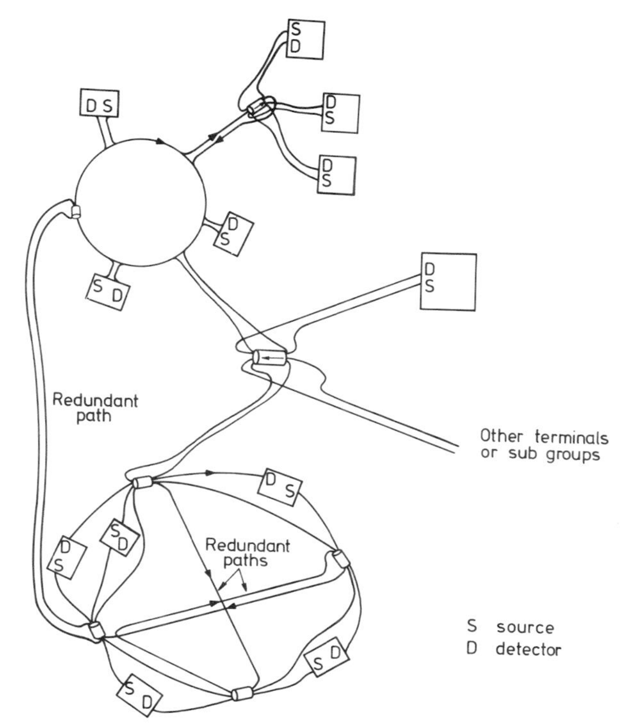

Figure 206. A hypothetical optical highway configuration to illustrate several concepts including redundancy and the use of a range of coupling components

reduces the loss between the most distant terminals. There are considerable opportunities for tailoring the configuration to meet constraints on layout, installation, and redundancy to suit the particular requirements but only if the terminals are well able to tolerate a wide range of optical loss, delay, and echo pulses. These considerations have guided the example terminal design described below.

Coupling Devices

Figure 207 shows an idealized three-part tee-piece (Y-junction) used for combining or dividing optical power. This representation indicates a wastage

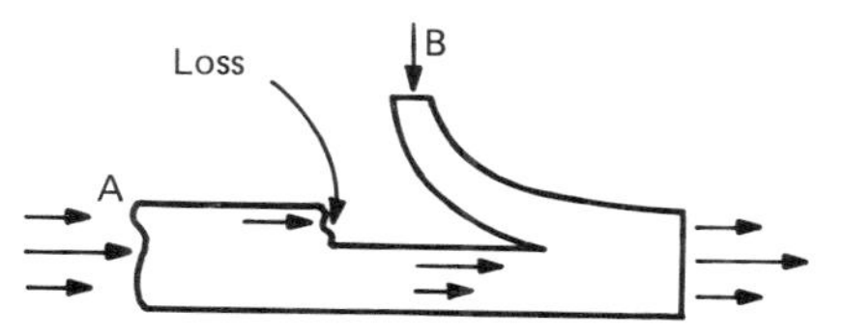

Figure 207. Three-port coupler used as combiner

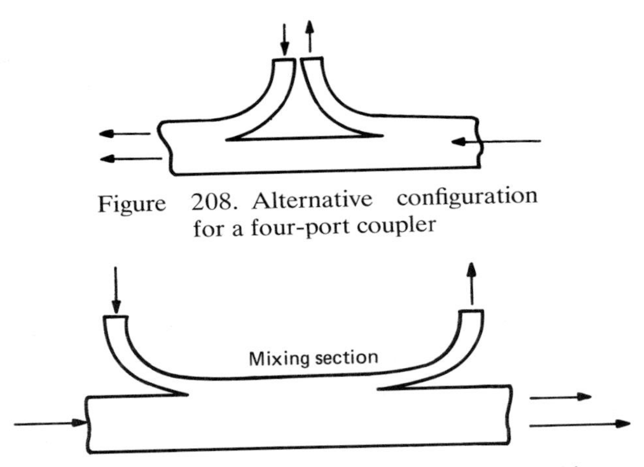

Figure 208. Alternative configuration for a four-port coupler

Figure 209. Four-port coupler used as both combiner and divider

of optical power, and it may be thought that this could be avoided by suitable design. However, under conditions of a uniform transmission fibre, the numerical aperture of which is always filled, such an improvement is not available and the loss is fundamental. This is simple to prove by considering optical sources of the same brightness completely filling ports A and B. If the loss indicated did not occur, the power at C would be greater than that at A, which would mean a greater brightness at C than at either A or B, contradicting the Law of Brightness (a corollary of the Second Law of Thermodynamics). The device shown can obviously be used as a divider or combiner, and the foregoing has shown that ideally the transmission loss between A and C is the same for a combiner as for a divider.

The four-port devices indicated in Figures 208 and 209 are applicable for transmit/receive access points, and in effect make use of the otherwise wasted power referred to above. A configuration for mixing device is shown in Figure 210. They can either be used for star couplers (e.g. in Figure 203) or can replace access couplers in bus or ring configurations, as shown in Figure 205. A photograph of half of a four-port mixer showing two of the fibres connected to the central mixing section can be seen in Figure 211. The same technique has been used to mix up to 32 ports. An encapsulated eight-port star mixer is shown in Figure 212.

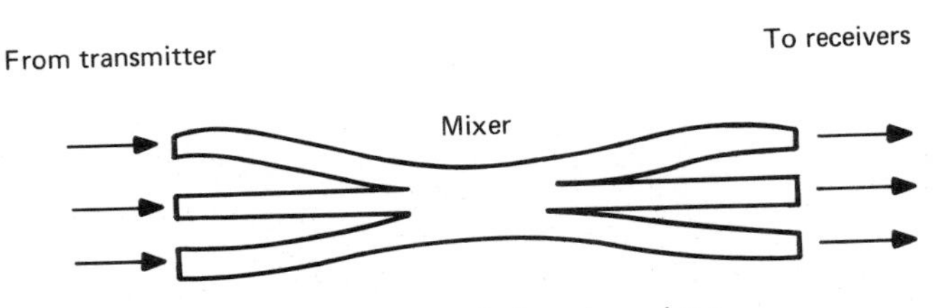

Figure 210. Transmission star mixer

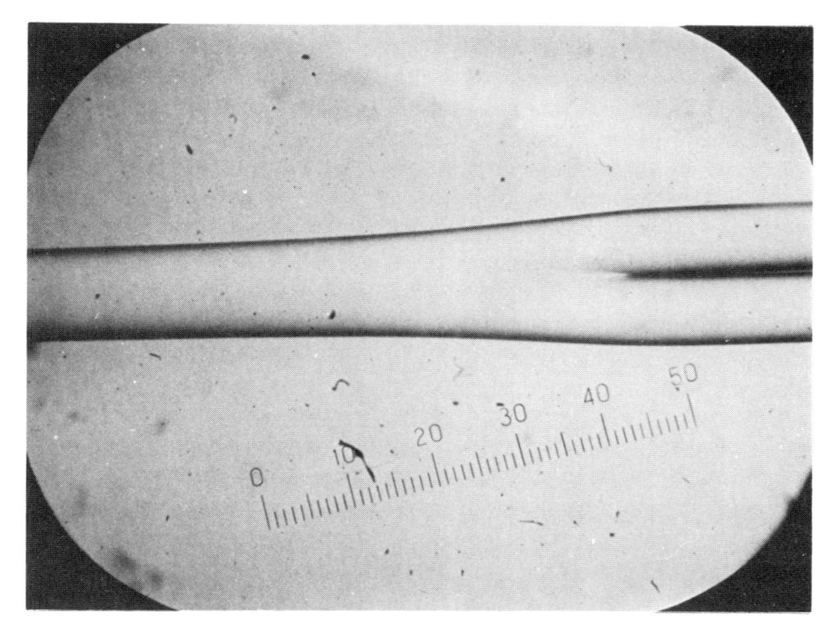

Figure 211. Half of a four-port mixer showing two of the fibres converging to the central mixing section

Figure 212. An encapsulated eight-port star mixer

EXAMPLE OF A MULTITERMINAL SYSTEM

The following system is described as an illustration of some of the design considerations mentioned above. It is a TDM system designed to cope with the problems of using a fibre-optic highway and takes advantage of the available bandwidth to have only predetermined transmission capabilities and no central controller. It describes a system that has been built as a four-terminal functional model at STL.

Synchronization

For a TDM system where only one transmitter is allowed to send information at any one time, the timing arrangements are of crucial importance. The most straightforward timing arrangement is to use a master terminal, where a single master unit has responsibility for absolute timing, and sends whatever information is needed to the other terminals to maintain the synchronization of the system. In the extreme, the master terminal can make all decisions about timing and send a short message to each terminal in turn to start its transmission. This approach can give a very flexible system, where time can be allocated by the master terminal in a way that varies with the amount of information needing transfer, and this leads to very efficient use of the data bus. Unfortunately, this arrangement gives a single vulnerable point in the master terminal where damage or a fault can disrupt the whole system. This situation can be improved by having several terminals ready to take over from the master if it should fail. This needs a system of priorities to define which terminal will take over first, and involves many problems in deciding whether the first unit has failed or not. To avoid this situation it may be preferable to have no master terminal. To achieve this each terminal must do its own timing and some means must be found to keep all the terminals synchronized. One possible means of achieving this is described on pages 276 to 279.

One consequence of this is that the data transmission capacity of the terminals must be predetermined, and cannot easily be changed to meet different network loadings. This approach is more acceptable for a fibre-optic network since the high bandwidth available means that excess data capacity can be made available at low cost.

Coding

The choice of coding in an optical system can considerably affect the rest of the design. For example, the choice of a PPM code would enable a low duty cycle optical source to be used, or a code with no low frequency content allows an a.c. coupled receiver. Because of the rapid changes of signal level expected at a receiver, it is desirable that the chosen code should allow fast AGC to be applied. Another requirement is that the data clock can easily be

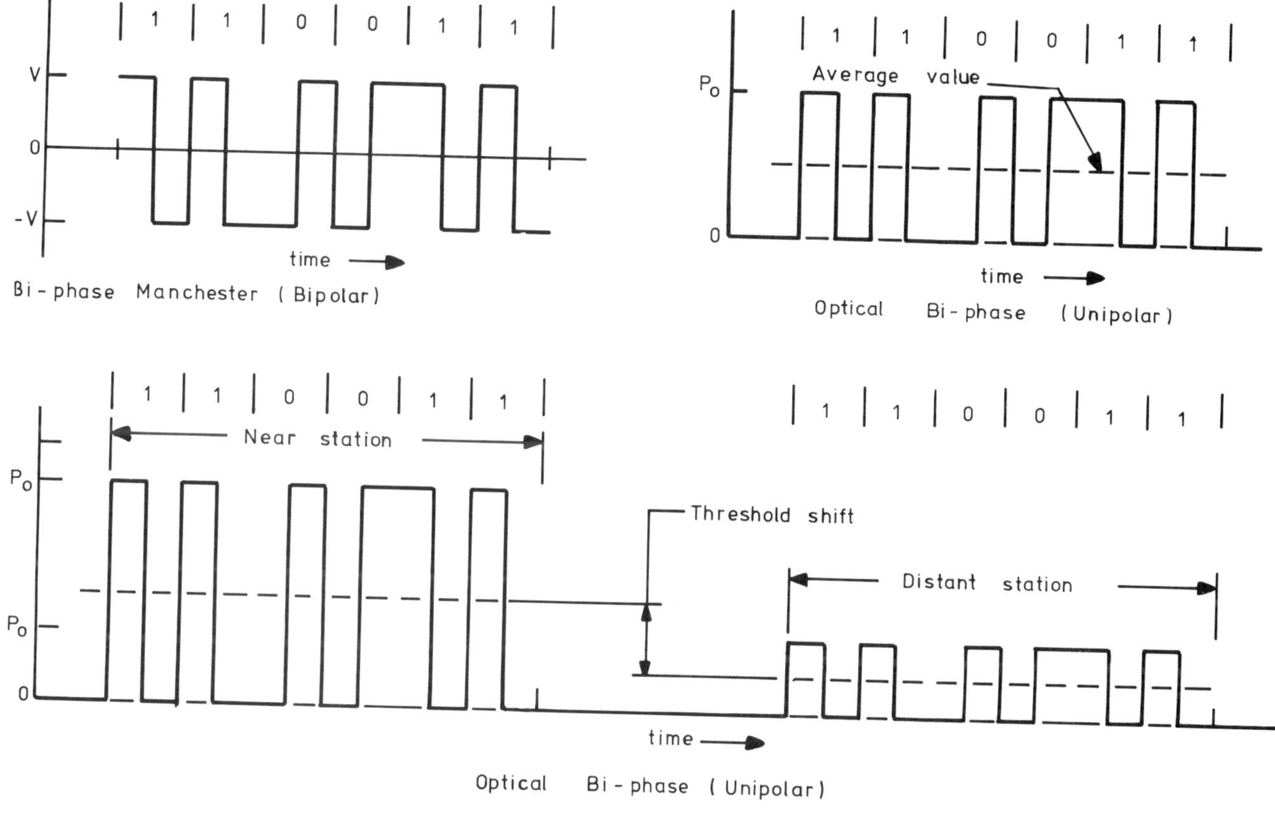

Figure 213. Bipolar and unipolar biphase

extracted from the data, preferably without the use of a phase locked loop. This is necessary because each received message will be at a different phase owing to the different path lengths involved. A short preamble can be included on each message to allow the receiver decoder to settle down, but if this is too long then the system becomes very inefficient.

The most appropriate code for the system being described was considered to be biphase, as seen in Figure 213, because this offers several advantages. There is a data transition at the centre of every clock interval which makes clock extraction easy, it only being necessary to time from one cycle to the next with a delay element such as a monostable. The biphase code is also arranged to have very little low frequency content, whatever the data being sent. This occurs because the code has a constant average value for ones and zeros. This allows fast AGC and a.c. coupled receivers to be used.

This code is most ideally suited to bipolar applications where differential signalling is employed, but unfortunately this is not possible with a single optical fibre, and so a unipolar version must be used. This has the disadvantage of needing a threshold shift if the signal strength changes, also shown in Figure 213, and ways of overcoming this difficulty will be discussed below.

A FLEXIBLE MULTITERMINAL SYSTEM

The Adopted System Design

A time division multiplex system approach was chosen for a system built at STL where each transmitter in turn has complete use of the optical highway for a burst period.[229] It sends a fixed length message in that period, complete with an address code to specify the terminal that is to receive it (see Figure 214). The transmitters send in strict rotation and if a terminal is inoperative, or if no information is available for sending, then that period is either left empty or a dummy message is sent. Each terminal contains its own clock oscillator and divider chain which provides all the timing information within the terminal. These clocks must be kept approximately in step so that the transmission periods do not overlap. This was achieved by making each transmitted message contain the address of the sending terminal. This is received by all other units and used to correct small timing errors that have developed. Special attention has been given to the problems of synchronization from switch-on, and to the effects of a single faulty terminal.

The block diagram of a terminal design for this system is shown in Figure 215. It consists of three main sections: the receiver, transmitter, and timing.

The Receiver

The optical signal is detected by a photodiode and the signal is then amplified. For biphase signals the amplifier can be a.c. coupled and limiting

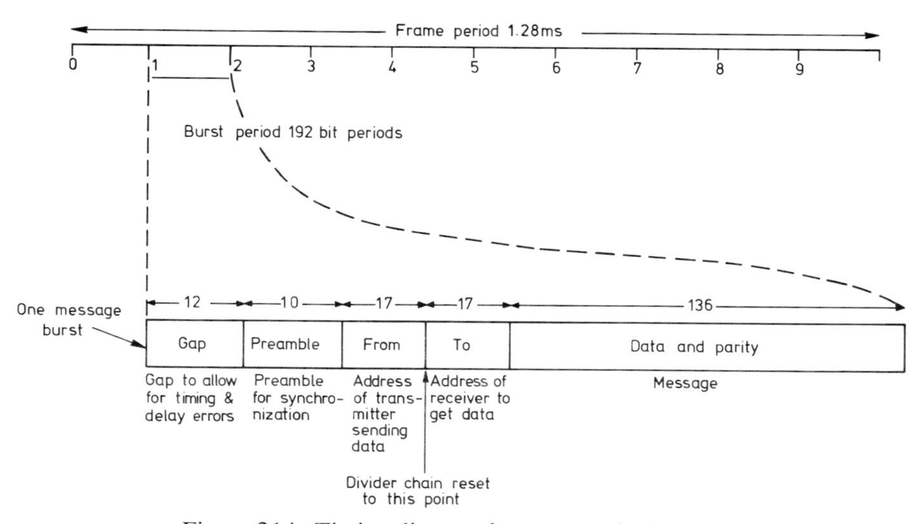

Figure 214. Timing diagram for ten-terminal system

thereby reducing the amount of AGC necessary in the following circuitry. This amplifier can be disabled during the terminal's own transmission period, if this is desirable. This received signal is decoded from biphase and then fed via a serial-parallel converter to two comparators. These act as word recognition circuits and detect the transmitter address and receiver address in the message. These addresses are protected by redundancy so that they are not falsely decoded from data or each other.

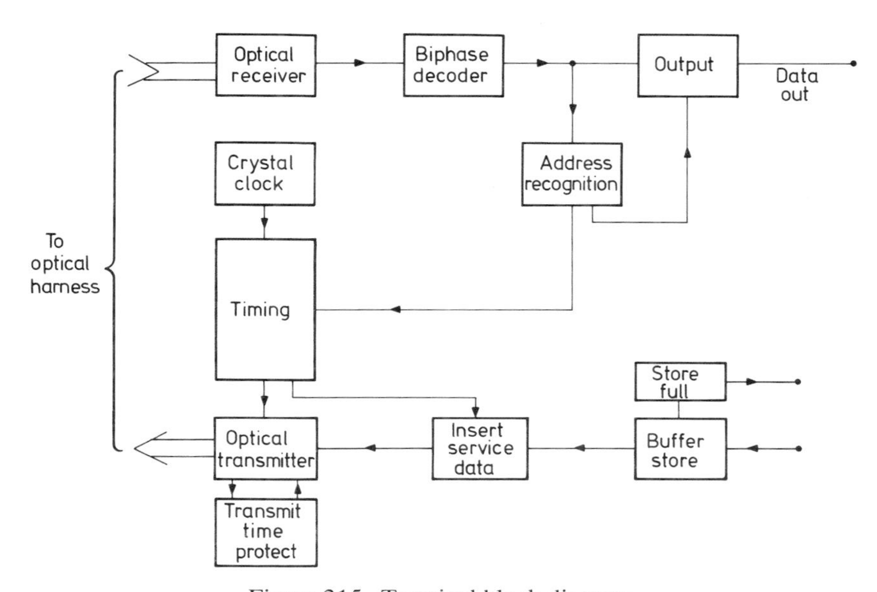

Figure 215. Terminal block diagram

When a comparator recognizes the terminal's own address in the received message the following data is gated into the terminal's store for output to the external equipment.

The Transmitter

Data to be transmitted is clocked to the transmission store by the external equipment. During the transmission burst period, defined by the timing circuit, the data is clocked out of the store and transmitted as a biphase signal at a data rate fixed by the terminal clock. Before the actual data certain service data is inserted into this message. This service data consists of:

(1) a preamble to enable the received decoder and AGC to settle;

(2) an address code for the transmitting terminal for checking and timing purposes; and

(3) the address of the terminal to receive the message.

This part may be preset, or may be part of the input data.

The optical transmitter must handle the 50 per cent duty cycle biphase signal, and so a high duty cycle source must be used.

The Timing Section

The timing for the terminal is carried out by a crystal clock oscillator and divider chain. The various waveforms needed in the terminal, such as the transmission burst period, can be derived from this divider chain by simple gating.

Each time a transmitter address is received by the word recognition circuit, it is compared with the divider chain and the difference used to correct errors that may exist due to signal delays or clock oscillator differences.

Large discrepancies may occur at switch-on before all the terminals have reached synchronism or if a terminal develops a fault in its timing system; both of these possibilities should be catered for. To do this a type of majority voting has been used. If large errors are consistently detected between the internal clock and the received transmitter addresses, the terminal's transmitter is disabled.

Dynamic Range

An important parameter for a receiver in a system such as this is the dynamic range that can be tolerated from one message to the next.

The received biphase signal has a d.c. component that is proportional to the average received power, and in a simple receiver this will be removed by the a.c. coupling before the signal is passed to the level detector (see Figure 213). This process will limit the dynamic range since the recovery after a large signal is exponential. The time available for recovery is limited to the inter-message gap and the preamble before each message, and if these are lengthened a greater change in message amplitude can be tolerated. For an inter-message gap of 10 bit periods, a dynamic range of greater than 40 dB can be expected.

The capacity to tolerate a large dynamic range for received messages would only be needed for certain optical harness configurations. If necessary these configurations could be avoided in practice; however, more flexible and redundant harnesses should be possible if they are allowed.

Data Input and Output

Data for transmission is supplied from external equipment and stored in a buffer store until the transmission burst period. Only if there is a complete message in the store is the transmission made. For synchronization purposes, however, it is still necessary to send the service data, but with a dummy receiver address. The true data section of the message is left blank.

The receiver address to which data is to be sent may be preset or may be taken as the first word of the input data. This would allow the external equipment to select the destination of the message.

Received data may also be stored in a buffer store to allow the external equipment to take the information whenever it is ready. Unfortunately, the terminal has no control over when data is going to be received and so no action can be taken to prevent store overflow. For this reason it may be preferable to output received data as it arrives.

COMPROMISES IN DATA FORMAT PARAMETERS

Consider the data format described above and shown in Figure 214. The peak bit rate, B_m, which is required to be carried by the optical fibre depends primarily on the mean data rate, $\bar{B}$, per terminal, and on the maximum number of terminals to be accommodated, M. Ideally, $B_m = M\bar{B}$, but in practice B_m will be higher because there are periods where useful data cannot be sent. These consist of the inter-burst gap periods and the periods where service data is being sent.

A further increase in B_m will occur when redundancy is built into the message, but this is considered to depend on the user and will not be considered here.

The significance of these periods is clearly reduced as the message length in each burst is increased, but this requires increased data storage capacity and increases the update period for each terminal.

These compromises will be considered in broad outline here. Let T_0 be the interburst gap period which depends on maximum fibre length; n be the number of bits of service data per burst period; N be the number of message data per burst period; and T be the frame period = update period for each terminal.

Then, considering the uniform data output of any one terminal

$$T = N/\bar{B} \tag{44}$$

Alternatively, by summing the elements of a frame

$$T = M\left(\frac{n}{B_m} + \frac{N}{B_m} + T_0\right) \tag{45}$$

Eliminating T we have

$$\frac{B_m}{\bar{B}M} = \left(\frac{n+N}{N}\right)\left(\frac{1}{1-(M/N)\bar{B}T_0}\right) \tag{46}$$

A plot of this function for some typical parameters is shown in Figure 216.

For the experimental model we chose $N = 128$, $n = 52$, $\bar{B} = 100$ kbit/s, $M = 10$, and $T_0 = 8\ \mu$s.

Equation (46) leads to

$$\begin{aligned} B_m &= 1.406 \times 1.071 \bar{B}M \\ &= 1.50 \bar{B}M \\ &= 1.50 \text{ Mbit/s} \end{aligned}$$

The factor 1.071 occurs because T_0 is much larger than it need be in an optimized system. For example, in order to tolerate an optical delay in 200 m of fibre the gap period need only be 1 μs, and the factor on the r.h.s. of equation (46) would only be 1.004. Except for much larger systems than

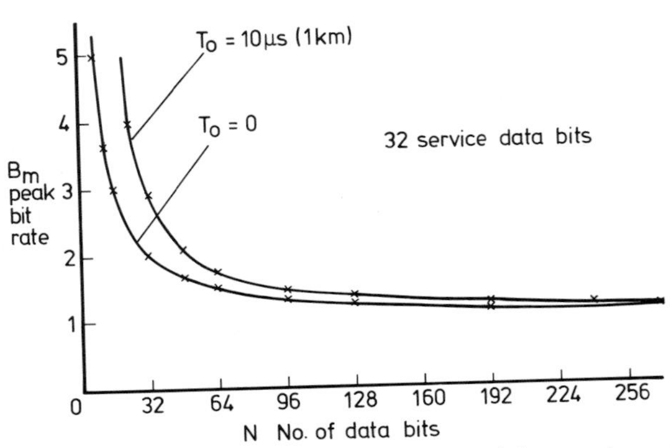

Figure 216. Peak bit rate as a function of the number of message data bits per burst period for a 1 M bit/s system

this, or when the burst length N is reduced, this factor can be neglected and equation (46) approximated by

$$\frac{B_{\mathrm{m}}}{\bar{B}M}=\frac{n+N}{N}$$

This represents simply the ratio of total to useful message data.

The frame period T is also the update period for each terminal, and represents a message delay between any two terminals. In the experimental model $T=1.28$ ms from equation (44). This can be reduced if a lower value of N is chosen, at the expense of requiring a higher peak bit rate B_{m}. A reasonable alternative design for $M=10$ and $\bar{B}=100$ kbits/s would be $n=32$, $N=16$, and $T_0=1\ \mu$s, which leads to a frame period of $T=160\ \mu$s and peak bit rate of $3.2M\bar{B}=3.2$ Mbit/s.

APPLICATIONS OF OPTICAL MULTITERMINAL TECHNIQUES

The particular system that has been described here has proved to be a very flexible approach, easily, modified to fit a wide range of applications. It is insensitive to optical harness imperfections, and so should be suitable for use with a wide range of harness configurations, including hybrid star/ring combinations. It is, however, only one example of many possibilities for optical fibre multiterminal systems.

Optical transmission can be applied to multiterminal systems but the limitations of practical coupling devices govern the total number of terminals that can be connected by a possible optical highway.

FUTURE POTENTIAL OF OPTICAL TRANSMISSION SYSTEMS

The technical problems in optical fibre transmission systems have been solved to the extent that there is no longer any doubts that they will go into service, earning their place on economic advantages. Before this happens, much more engineering work will be carried out to optimize designs, demonstrate reliability, choose standards and specifications, work out installation and maintenance procedures, and so on. The tasks that remain no longer rely on further major scientific breakthrough for their success and will be solved by development techniques.

In parallel with this work, however, the R&D activity continues and there is room for much further extension of the scope of optical fibre systems. In the case of repeaters for PTT applications, where economic advantages are paramount, the economic improvement (at any given bit rate) is measured by the reduced cost per repeater and the increased repeater spacing. In general one can expect a trade-off between these factors. The advantage of reducing the cost of the repeater circuits and components tends to be swamped by standing costs such as housing, power supplies, and maintenance. For this reason it is usually preferable to choose the highest quality components to maximize repeater spacing. However, if repeaters could become not only less costly, but also very compact and economical on power, installation and maintenance procedures, etc. there could be an economic advantage in accepting a reduced repeater spacing. Ideally, the repeater would be built into the cable and be pulled into the duct or ploughed into the ground with the cable, thus requiring no maintenance. Work on integrated optics may lead to an all-optical repeater with no reversion to electrical signals, but the feasibility of this is still to be proved.

Of more immediate significance are the steady advances in optical sources and photodiodes. New structures and processes are being developed for better performance, reliability, yield, temperature range, and low cost mass production. Further improvement can be obtained by the monolithic integration of several transmitter or receiver functions on a single semiconductor chip. As the optical receivers improve, the required number of photons per time slot will become so low that the designer will have to think in statistical terms.[231–235]

As already mentioned, there is a growing interest in operation at wavelengths above 1 μm to take advantage of the lower fibre attenuation and material dispersion. Both lasers and LEDs are being successfully developed for operation at these wavelengths. With silicon photodiodes there is a material limitation (low absorption coefficient) which results in poor efficiency. However, at wavelengths above 1 μm germanium photo diodes can be used and devices using quaternary semi-conductors are being developed. It is likely that these sources and detectors will be available with excellent linearity. These advances will enable the system designer to take full advantage of the low attenuation and wide bandwidth of the cables at longer wavelengths, for analogue transmission as well as digital.

If there is an increase in the requirement of bandwidth, perhaps augmented by widespread use of video facilities, very high speed systems will be of interest. For these applications the single mode fibre has the advantage not only of great bandwidth but of having a frequency response which is independent of launching and coupling conditions. In this sense it is a cleaner transmission medium than the graded index multimode fibre, but it presents more difficult splicing problems. Single mode fibre systems will require sources capable of launching adequate optical power, and narrow stripe lasers with improved emitting geometry are now emerging together with several efficient coupling techniques such as cylindrical microlenses. To take full advantage of the single mode fibre it is necessary to minimize the optical linewidth of the source, and this appears to be emerging as a by-product of general improvements.

The use of colour multiplexing (combining signals of various wavelengths onto one fibre) is a possibility where it is desired to increase the capacity of each fibre still further. Similarly, it will be possible to use a single fibre for bothways transmission. These will require a range of new devices at the repeaters, such as filters, combiners, and splitters, but it is too early to predict whether it will be economically attractive to increase repeater complexity to reduce the number of fibres in the cable.

Looking still further into the future, coherent optical techniques may be harnessed and only then will it be possible to realize the ultimate bandwidth capability of the fibres.

The success of this work will determine not only whether optical transmission can be used for the highest capacity routes, but also whether it will be economically competitive for low capacity applications.

CHAPTER X

Potential Applications of Optical Fibre Communication Systems

INTRODUCTION

Information transmission along glass fibres was originally proposed for wideband long haul communication systems[1] in the trunk networks of PTT administrations, in view of the enormous bandwidth capability. Since then the situation has radically changed for two reasons. First, the performance criteria postulated, particularly fibre attenuation, though considered excessively optimistic at the time, have been handsomely bettered. The resulting wider repeater spacings make many more types of PTT systems economically viable. Secondly, it has been realized increasingly that optical fibres have much more to offer than low-cost bandwidth and now appear suitable for many applications which are either very difficult (and hence expensive) or impracticable by conventional means.

In this chapter the many advantages and special features offered by optical fibre technology are summarized in the light of the detail given in preceding chapters. Some of the applications which are emerging in the civil telecommunications, military, and industrial fields as a result of these advantages are identified. Many user organizations already have experimental systems installed or planned—a necessary step towards a full understanding of the new technology's potential.

ADVANTAGES AND SPECIAL FEATURES OF OPTICAL FIBRES

To provide a background against which the wide range of emerging and potential applications can be appreciated, it is useful to discuss the main advantages and special features offered by optical fibres.

High Bandwidth

Optical transmission systems based on single fibres are capable of carrying very high information rates. Even simple multimode fibres have bit rate times distance products of a few Mbit/s. km, while graded index fibres have been demonstrated above 1 Gbit/s. km. Single mode fibres are capable of transmitting about 40 Gbit/s. km at 0.85 μm and about 400 Gbit/s. km at the longer wavelengths. These high bandwidths can be exploited in two main ways:

(a) where the volume of traffic justifies, the transmission capacity can be increased with the objective of reducing the cost/channel; and

(b) the high bandwidth can be used in a way which would be considered expensively wasteful on ordinary wire lines, to achieve simplicity or some performance or convenience benefit.

Very Low Loss

Cables containing multimode fibres with losses below 4 dB/km at 0.85 μm are already currently permitting long distance applications with wide repeater spacing and short haul applications with no repeaters. Single mode fibre cables having characteristics such as those shown in Figure 40 further extend the system capabilities. Fibres have been produced with losses down to 1.5 dB/km at 0.85 μm and 0.5 dB/km at 1.2 μm[238] and even lower losses have been demonstrated at longer wavelengths; when such performance is realized routinely in production cable, extremely long repeater sections or unrepeated links will be feasible.

Small Size and Weight

As will have been evident from Chapters III and IV optical fibres and cables are extremely compact. Bearing in mind also the high bandwidth capability (permitting multiplexing) it is clear that considerable savings in volume are possible over copper cables for a given overall transmission requirement. This is important in the utilization of PTT underground cable duct space and in many other applications. Since fibre cables are also less dense overall, the weight saving is even more marked, especially important for applications in aircraft and guided missiles, and for aerial cable installations.

Ruggedness and Flexibility

In spite of earlier fears, it has proved possible to manufacture coated fibres routinely with very high tensile strengths. Furthermore, some very simple and effective cable structures have been evolved which have already led to compact, flexible, and extremely rugged practical cables. Apart from obvious deployment advantages, the small size and permissible bending radius enable much smaller reels to be used, and lead to storage, transportation, handling, and installation benefits.

Low Cost

Although at present the better quality fibres are relatively expensive, they are made from material which is not a scarce resource (unlike copper) and moreover require such small quantities that there seems little doubt that rapidly improving manufacturing techniques will produce significant cost advantages over copper cables. Even before this happens overall system cost savings will be possible from reductions in electronic equipment such as repeaters and simplifications made possible by features discussed below.

Electrical Isolation

Optical fibres are inherently electrical insulators and can now be readily incorporated into cables which have no conducting components. This feature makes them ideally suitable for use in hazardous environments. For example, large potential differences between the ends of the cable can be tolerated without protective circuitry, there are no earth loop or EMP problems, and no arcing or spark hazards from short circuits or abrasion.

Immunity to Interference and Absence of Radiation

Problems commonly encountered in electronic systems are radiated electromagnetic interference (EMI) from switching transients, electric motors, etc. and radio. Being insulators, fibres are naturally immune to EMI or RFI problems. In addition it is easily ensured that they are completely immune to optical interference, so there are no cross-talk problems.

As the converse of the previous point, it follows that it is easy to ensure that fibres do not radiate, thereby giving a high degree of signal security. This is achieved by the design of the waveguide, the splices, the connectors, and by using opaque materials in the cable manufacture. It is very difficult to achieve similar results with metallic cables.

PTT AND SIMILAR APPLICATIONS

General

In PTT transmission networks those special features of optical fibres which lead to lower overall system costs are of main interest. Metallic conductor cables are capable of providing excellent service, VF pair, HF pair, and coaxial types having become well established and optimized for a wide range of requirements. Millimetric waveguides[239] have been developed for very wide bandwidth applications. Thus, a prerequisite of optical fibre systems development for PTT network applications is a study of their economics in relation to metallic cable systems, although the possibility of novel applications where optical fibres may be particularly suitable must not be overlooked.

It is desirable to plan optical fibre systems which are compatible with the existing network structure and can make use of existing facilities, at least in the earlier years. This implies a family of systems covering the required capacity range, interfacing with existing types of terminal or exchange equipments, and using existing cable ducts and manholes. At the present stage of development, digital transmission is most suited to the optical fibre technique. A range of systems of bit rates corresponding to the European PCM hierarchy is shown in Table II, Chapter I, with the probable repeater spacings for first generation optical fibre systems compared with existing or planned metallic cable systems. For US conditions an equivalent table might extend from 1.544 to 560 Mbit/s, and for Japan from 1.544 to 400 Mbit/s.

Cost Considerations and Comparisons

Deriving future costs and price levels for a new product is always difficult, but estimates have to be made to permit realistic planning. Some economic comparisons based on classical present value of annual charge (PVAC) calculations[240] are presented in Figures 217–219. They assume large scale production of cable a few years hence, cable installation techniques modified to benefit from the small size and weight, volume production of reliable optical components, and systems engineered on the basis of current developments. A survey of current forecasts, published and unpublished, indicates future price levels for low loss graded-index fibres in the range 10–20 cents per metre, and the calculations have used values in this range. Although these are of necessity only approximate, they do permit tentative conclusions to be drawn.

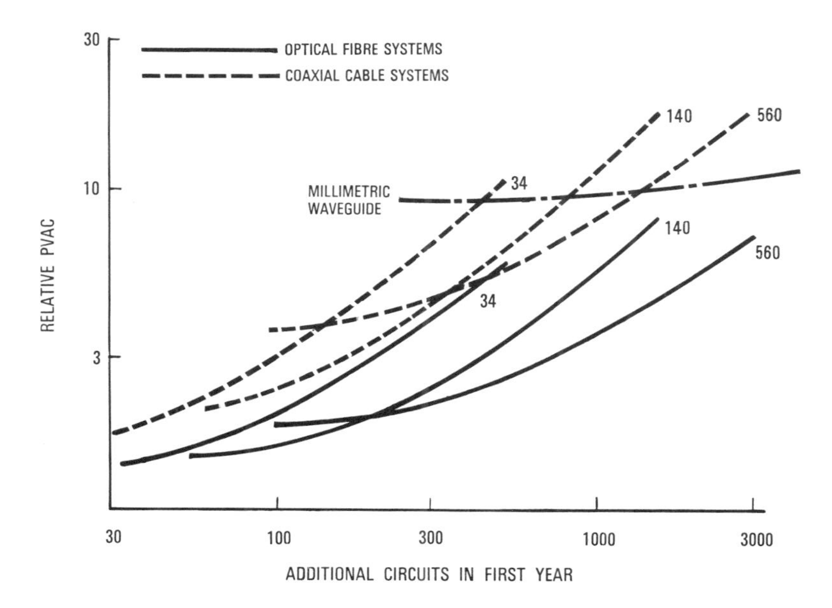

Figure 217. PVAC comparison of medium-high capacity systems

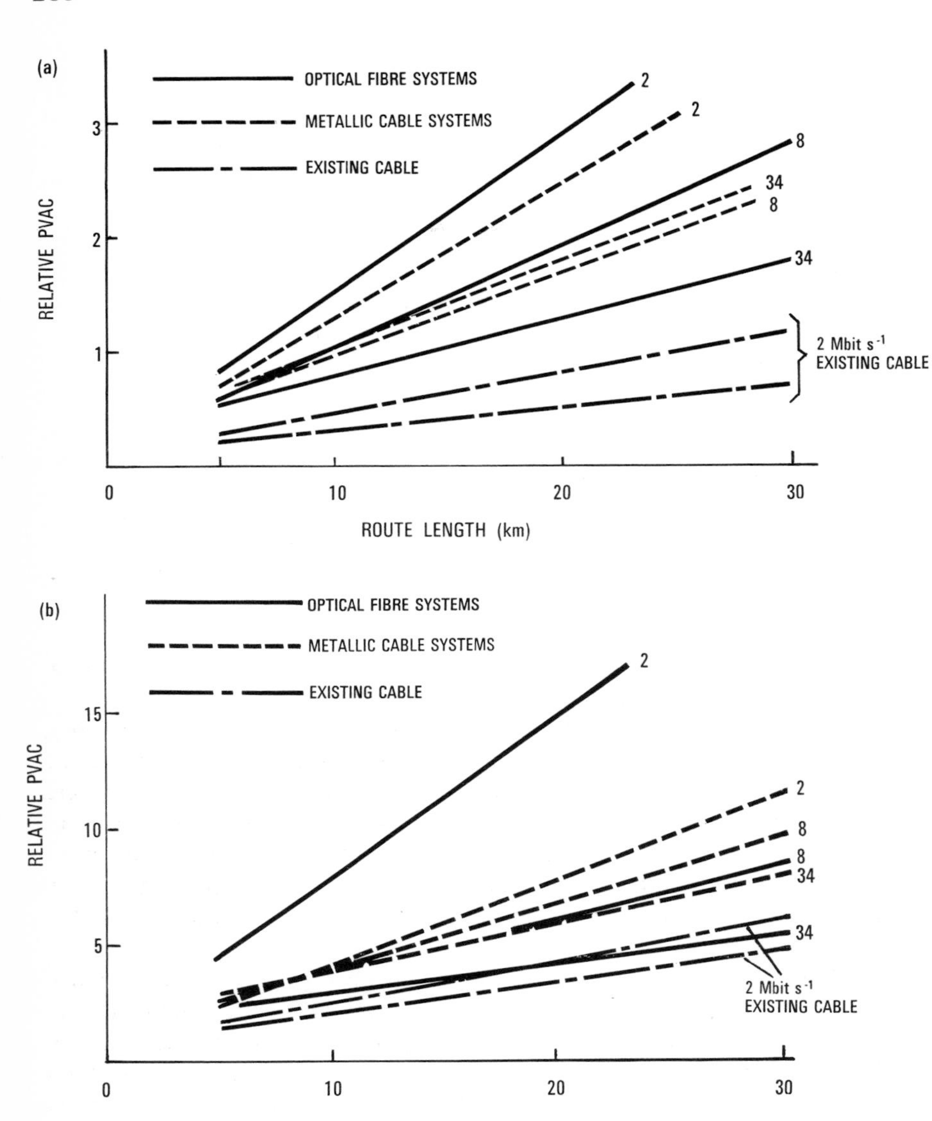

Figure 218. (a) PVAC comparison of low-medium capacity systems (20 added circuits first year). (b) PVAC comparison of low-medium capacity systems (120 added circuits first year)

Figure 217 is a comparison of 34–560 Mbit/s medium-to-long haul optical fibre and coaxial cable systems, and a millimetric waveguide system. High order multiplex is included down to a level of 34 Mbit/s for all systems, and the route length is approximately 100 km. It is seen that the three optical fibre systems have lower cost than the corresponding coaxial cable systems, the cost advantage being greater at higher capacities. The millimetric

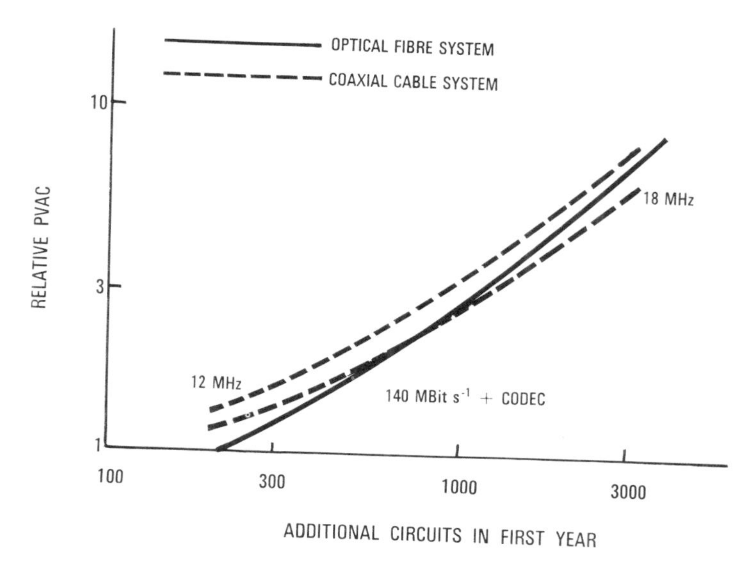

Figure 219. PVAC comparisons for FDM circuits

waveguide system cost is shown higher than that of any fibre-optic system, except at extremely high growth rates.

Figures 218(a) and 218(b) compare 2, 8, and 34 Mbit/s short-to-medium haul systems. Route length is of importance as multiplex costs can be the controlling factor on very short routes. Two growth rate situations are shown; higher order multiplex is included down to a common level of 2.048 Mbit/s. These figures show fibre optics to be attractive at 8 or 34 Mbit/s, compared with new metallic cable systems, and comparable with applying 2.048 Mbit/s to existing voice-frequency cables at higher growth rates.

Long haul voice circuit transmission on coaxial cables is normally considered of lower cost in analogue (FDM) form compared with digital (PCM), although of course the latter can show considerable economies when integrated with digital switching. Thus, first generation optical fibre systems (digital) are unlikely to have great economic advantage over high capacity analogue coaxial systems. However, Figure 219 indicates that 140 Mbit/s optical fibre systems in association with very high speed codecs, e.g. one super master group of 900 channels encoded at 70 Mbit/s, can have a cost for analogue traffic comparable with 12 or 18 MHz coaxial systems, whilst providing flexibility for initial analogue and subsequent digital applications.

Administration Plans and Applications

A number of PTT administrations are showing active interest in trial systems within the above range, and are planning exploitation in their networks.

A tertiary system (34 Mbit/s or its equivalent) offers attractive cost saving, is well within the capabilities of graded index fibres and components, and has adequate capacity for many junction and short haul toll applications. Several experimental systems are in progress or planned. ATT have described[241] an experimental installation of 44 Mbit/s systems on multifibre cable under simulated field conditions, the system being aimed at inter-office trunk applications in metropolitan areas, followed by a 2.5-km, 24-fibre cable trial installation now providing service in Chicago. In Japan, a trial system at 32 Mbit/s system with 8-km repeater spacing has been installed by NTT who propose to make widespread use of such systems on short and medium haul routes by the mid-1980s.[242] A study by the Italian organization SIP[243] concludes that 34 Mbit/s systems are economic for their network. 34 Mbit/s systems produced by several German manufacturers have been installed for Deutsches Bundepost trials over a 4-km route in Berlin.

Higher capacity systems compare even better with their metallic cable counterparts. With the introduction of 120 Mbit/s coaxial systems already in progress and a 140 Mbit/s coaxial and microwave radio network planned, the British Post Office are confident that 140 Mbit/s optical fibre systems will become a major feature of their toll network in the second half of the 1980s. STL and STC have co-operated in a field demonstration of such a system with power fed repeaters, installed since July 1977 in Post Office ducts on the Stevenage–Hitchin route and used to carry live traffic. This is described in detail in Chapter XI. The British Post Office now plan to install commercial 140 Mbit/s systems in the early 1980s with a repeater spacing up to 10 km. Other European administrations show interest in 140 Mbit/s systems. In North America, Alberta Government Telephones have ordered a field trial of 274 Mbit/s systems. This probably is the economic limit for graded index fibre; 560 Mbit/s and above on monomode fibre will surely follow, but probably not before the late 1980s.

Although the economic advantages are not so pronounced, secondary rate systems (6 or 8 Mbit/s) are likely to be in full service in the early 1980s, linking exchanges up to 10 km or more apart without intermediate repeaters. Field experiments or trials are in progress in the UK, France, the US, and Japan, and many administrations world-wide have similar plans or show great interest. The UK Post Office plan to have commercial 8 Mbit/s systems in service by the early 1980s; their own experimental system is on a 13-km duct route between Ipswich and Martlesham,[244] and STC have tested one on the Stevenage–Hitchin route. SIP are planning an 8 Mbit/s trial between two exchanges in Rome. Aerial cable installations, making use of existing open-wire pole routes, are likely to prove particularly attractive in developing countries.

With the cost assumptions made to derive Figure 218, primary PCM systems (1.5 or 2 Mbit/s) show no economic advantage since they considerably under utilize fibre capacity. Most current interest and trials are reported from independent telephone companies in the US, but they may find application in the relief of overcrowded cable ducts where only modest capacity growth is required. However, if low loss step index fibres become available at a cost of a few cents per metre, optical cable prices similar to those of VF copper pair cables could be achieved and primary fibre-optic PCM systems would be attractive for many routes.

In summary, optical fibres may be expected gradually to replace coaxial and high frequency pair cables for new installations in junction and toll networks, and it now appears that administrations are abandoning plans for millimetric waveguide systems in favour of fibre optics. The British Post Office have already ordered commercial systems at 140, 34 and 8 Mbit/s capacities from the main UK manufacturers. Current studies in CCITT, the world-wide technical authority for telecommunications, reflect the great interest of PTT administrations and their desire to achieve standardization for international operation.

Other PTT Applications

The subscriber area is potentially of great importance because of the very large amount of cable employed. Optical fibres may not compete with conventional low cost copper or aluminium subscriber loops, but as digital carrier techniques, and later broadband services, penetrate the subscriber area, optical fibre will become of great interest. There are already trial systems of this type, for example the Hi-Ovis scheme in Japan.[245]

Submerged systems are mainly long haul, and relatively large diameter coaxial cable is used to gain wider repeater spacing. Optical fibre systems offer wide spacing with low cost cable and should become very attractive, although power feed, component, and cable strength problems have to be solved economically, and PTTs must be expected to await operational experience on land and, perhaps, the availability of long wavelength sources and detectors. Already experimental work is reported in the US, Japan, and the UK.

Whilst most PTT transmission systems operate in a relatively benign environment, interference from tramways and railways, and from electromechanical exchange noise, is not uncommon. Also lead-in or tail links for hill-top microwave radio sites can be exposed to lightning. Optical fibre systems will be especially useful in such cases.

MILITARY APPLICATIONS

General

With military systems, although economics are important, there are normally other, often overriding, considerations such as weight, deployability, high integrity, and survivability under the most severe conditions. The special features offered by optical fibres can have an impact in the military area in two types of situation.

(1) Where a particular job can be, or is already being, done by conventional techniques but fibres could do it with less complication, to a better specification or simply with more operational convenience.

(2) Where fibres open up the possibility of achieving functions which are impracticable using conventional approaches.

Even now, when the technology is relatively new, a large number of possibilities suggest themselves and it is difficult to treat them in a nicely ordered fashion. For convenience, potential applications are grouped largely by system length.

Short Distance Systems

There are a number of very short distance applications, up to tens of metres, which usually make only modest demands on the technology. These comprise links between fairly closely spaced items of electronic equipment, for example in an operations room. They include internal computer links and links carrying information such as raw radar data to processors and display units. Fibres offer wide bandwidths, freedom from interference, compatibility, and ground loop problems, and have the ability to bridge very high voltage drops without protective circuitry. They also make equipment less susceptible to damage from accidental misconnections, since short or open circuited fibres cannot harm the terminating circuitry.

Fibre bundles provide the obvious medium for these applications. However, for some of the wide bandwidth applications, for example fast computer links, bundles are inadequate even for surprisingly short distances and single fibres must be used. Indeed, the future trend is likely to favour discrete fibres (which can be fairly simple and cheap large core types such as medium loss glass, plastic, or plastic clad silica fibres) as the price and availability of the essential components improve. Also, there will be many cases where a number of parallel paths are needed, which are most conveniently handled by multiway cables, but it is much more difficult to preserve high mechanical flexibility with bundle designs.

The above types of application merge into rather longer systems linking equipment sited in different parts of a building, involving distances up to a few hundred metres. Losses must be rather lower and unless the links can be laid in protected troughs, the cables must be more rugged to withstand installation stresses or continual handling. Thus, again for most systems discrete fibre cables will prove more suitable than bundles.

The immunity to interference and ground loop problems is even more valuable with these longer distances. In addition there are other advantages such as security, since fibres are much more difficult to tap.

Applications in Mobiles

This class of application, which includes aircraft, ships, and tanks, is presently one of the most promising for optical fibres. These mobiles are characterized by an extremely noisy electromagnetic environment which, as bandwidth requirements grow with the sophistication of modern equipments such as radar, navigation, weapons, etc., makes it increasingly difficult for wire systems to cope, even when expensive precautions are taken. The interference immunity property of fibres enables the wide bandwidth capability to be exploited, e.g. to multiplex many channels onto a common highway without excessive complication. Bandwidth is 'cheap' enough to affect overall system philosophy. For example, it can be traded to enable simpler and more flexible approaches to be used for data highway management.[246] However, the most obvious attractions are small size, since space is tight in virtually all military vehicles, ships, and aircraft, and also light weight in airborne applications. Modern aircraft carry so much electronic equipment that conventional wiring accounts for a substantial fraction of the total aircraft weight, and optical fibres would effect savings thereby significantly improving flight performance.

The immunity of fibres to lightning and EMP is particularly valuable in airborne applications, permitting routing through fuel tanks and magazines. Indeed, the optical medium has such advantages in this type of application that it makes sense to avoid the electrical regime where not actually necessary (see section on Future Prospects).

There are now many government programmes both in Europe and the US, following up the earlier projects such as ALOFT[247] in which an A7 aircraft was wired up with a fibre bundle system. So far bundles have been most used, for the usual reason, plus the additional one that the necessary star and branching couplers have been more easily realized in bundle form. However, these applications provide a major incentive for the development of single fibre branching couplers and in most cases single fibres are likely to provide the long-term solution.

Longer Distance Fixed Links

In the military field these cover fixed site communications over distances ranging from 100 m to many kilometres. Typically, they include links within headquarters complexes, bases, airfields, and dockyards. Bandwidth requirements range from very moderate, e.g. below 2 Mbit/s where only speech communication is involved, to many tens and even hundreds of megabits for video, radar data, and sophisticated sensor outputs. The larger distance, higher bandwidth links in particular make heavy demands on the technology, especially since a high degree of ruggedness and all-round 'survivability' (e.g. ability to continue operating after nuclear irradiation and structural damage to portions of the complex) are required.

Deployable Links

In the preceding applications the role of the transmission medium has been virtually static. However, in 'deployable' applications the cable is subjected

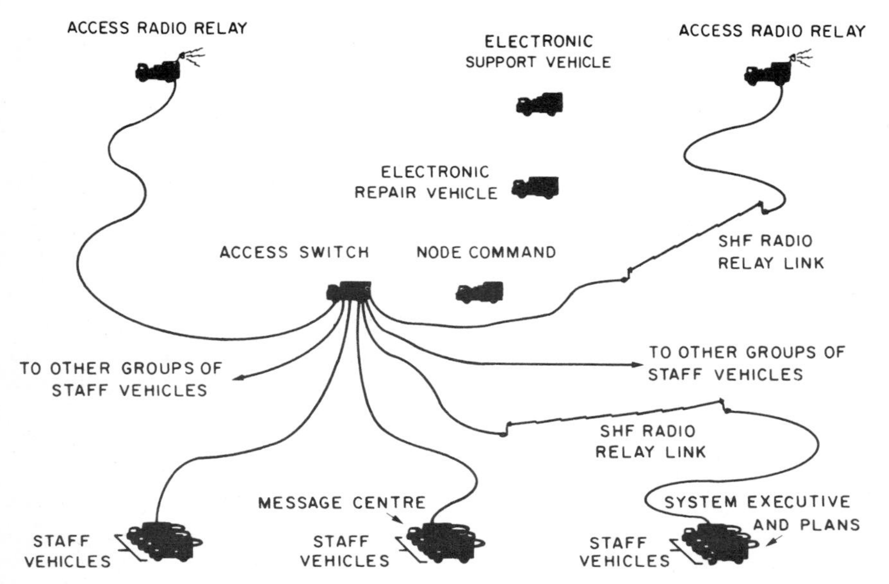

Figure 220. Military trunk node system illustrating tactical application of fibre-optic cable

to continual movement and handling and high strength and extreme ruggedness are usually required.

The Ptarmigan tactical communication system being developed for the British Army is a particularly promising application. HF quad cable was originally planned for the intra-node deployable links (see Figure 220). These handle up to 2 Mbit/s and can be as long as 2 km, necessitating the inconvenience of intermediate repeaters. The operational requirements call for frequent deployment and redeployment, thus exposing the cables and connectors to considerable handling and rough treatment under harsh environmental conditions. Somewhat surprisingly, in view of the early worries about fragility, optical fibre cables are now demonstrating greater ruggedness than conventional deployable cables and in view of their small size, light weight, high flexibility, and nonmetallic nature, appear ideally suited to this application. In addition, a 2 km range is comfortably achievable (with simple LED sources and PIN detectors) even making realistic allowances for connector degradation and system margin. A set of functional models (to enable the potential of such an optical system to be assessed under real operational conditions) is being built for the UK MOD.

Another example is a link to a tethered platform where, apart from the normal advantages, a fibre presents a much smaller 'radar cross section' than a comparable conducting cable.

Other deployable applications include various surveillance and espionage activities where fibres are sufficiently small to be unobtrusive and difficult to

Figure 221. Remote control of TV camera

detect by electronic means, while offering increased security over radio. Figure 221 shows a remotely deployed TV camera system using optical fibres for both the control and video signals.

Long Distance Applications

Long distance military applications promise to be quite different from the PTT type and will include cases where fibres offer the possibility of realizing some facility not practicable with conventional technology.

One example is information links between military vessels and towed sensor arrays. For safety, considerable distances (literally the longer the better) are desirable, requiring the lowest loss high bandwidth fibre.

Another marine application is torpedo guidance, where again very long distances are involved. Fibres here offer the great advantage that the link could literally carry video bandwidths, enabling the torpedo to be homed in on its target even if this were completely out of sight of the firing ship.

The ruggedness and small size of packaged fibres, combined with their low attenuation and high bandwidth, should make them an ideal transmission medium for missile guidance. Video bandwidths can be carried over tens of kilometres, providing the facility of accurate guidance beyond line-of-sight with complete security.

CIVIL AND INDUSTRIAL APPLICATIONS

General

Civil and industrial applications include the counterpart of some PTT and some military applications. Economics are very important in what is often a highly cost conscious market, but hazardous and other special environmental conditions will make optical fibres especially beneficial in some cases. It is convenient to discuss applications under the headings of long, medium, and short distances.

Long Distance Applications

Public utility organizations may provide their own communication transmission facilities over moderately long distances. Main examples are railways, pipe lines, and electrical power lines. They have the same interest as PTTs in low costs and, although capacity demands may be modest (e.g. 2 or 8 Mbit/s for a mixture of voice and data circuits), optical fibre systems can be expected to be attractive due to the high electromagnetic interference or hazardous nature of the environment. Standard PTT systems appear to be suitable for railway applications where increasing electrification and thyrister control gives a particularly harsh electrical environment; British Rail have an experimental 2 Mbit/s system installed over a 6-km route in Cheshire where the optical fibre is suspended between the gantries shown in Figure 222 which support the power lines. Proposals have been made for optical fibres to be carried within the steel/aluminium earth conductors of overhead electric power routes; in the UK an experimental optical cable has been suspended from the earth conductor by means of a plastic tube.[248]

Video applications, which conventionally require coaxial or special pair cable and closely spaced amplifiers, may be more economic using optical fibre. Closed circuit television (CCTV) links are provided up to distances of tens of kilometres, for example for remote surveillance of road traffic, and several developments giving analogue video transmission of adequate quality over distances up to 10 km without repeaters have been reported using transmission similar to that shown in Figure 221, where the optical link is used both for camera control and the video signal transmission. Common antenna television (CATV) trunk links are also a promising application, and an experimental optical link has been installed in the UK.[249] CATV distribution is a potential application requiring a suitable tee-off device; at this time the difficulty of multichannel VHF or UHF transmission, and adequately high launched power to feed many subscribers, suggests a combination of subscriber-controlled central switching with one or two analogue

Figure 222. Application of optical fibre for railway communication

channels per fibre.[252] Outside broadcast links provided by television authorities also are potentially attractive applications, in this case requiring high transmission quality, and taking advantage of the light weight and strengh of the optical cables.

Medium Distance Applications

Under this heading fall a variety of applications within a single operational site, with distances ranging up to say a few kilometres. Electricity generation and substation sites require reliable telemetry and control communication in close proximity to conductors carrying up to a few hundred thousand volts and with large earth-potential differences. In a typical substation site heavy multipair copper cables linking transformer and switching points to a central control may be replaced by single fibres. Several trial systems are planned by power companies, including those in the US, the UK, and Belgium, and some have already been installed in Japan.

Manufacturing and chemical processing sites present problems of high voltages, high electrical noise levels, and explosive or corrosive gases. Growing automation of process control makes reliable communication increasingly important, and optical fibres provide an excellent solution. A major oil refinery in the UK plans a trial installation, whilst the European nuclear research organization, CERN, have requested tenders for data and control links for evaluation. Modern hospitals present somewhat similar problems; X-ray and other electronic equipments generate high interference levels, whilst patient monitoring and records data must be communicated accurately. Computer–computer and computer–peripherals links are required in many other types of location, probably the earliest operational installation of this type being the system seen in Figure 223 which was provided several years ago by the ITT Optical Equipment Division of Leeds for a Police Headquarters in the UK to replace a copper cable link in a system which had been susceptible to damage from lightning strikes.

Opportunities are developing in the relatively new area of hydrospace, although challenging requirements for strong yet flexible underwater cable have to be met. An application of current interest is the communication link

Figure 223. Data terminal link by optical fibre in Dorset Police Headquarters

(television and data) for robot submersibles, where electrical interference exists from power conductors within the same 'umbilical' cable. Successful trials have also been carried out using fibres for communications between human divers and the surface, as shown in Figure 224. Two-way speech and multiplexed physiological data on the diver, which facilitate safer control in saturation diving situations, can easily be accommodated in cables adding very little extra weight and stiffness to the existing umbilical line.

This is one of the areas of widest current interest in optical fibre applications, but potential users are moving cautiously and plans are experimental, with conventional system back-up whilst experience is gained.

Short Distance Applications

These include communications within buildings and mobiles (aircraft, ships, road, and rail vehicles).

Applications within buildings overlap with the applications within sites discussed above. Additional examples currently under consideration include broadcast studios, where the move towards digitization of video and sound signals requires high bit rate transmission, and hotels where multichannel TV and sophisticated automated services are required for every room, all in close proximity to electric power wiring.

Mobile applications are similar to those described under military systems, although the internal communication needs of civil transport are considerably less than those of their military counterparts. Moreover, cost is all-important, and early applications appear unlikely.

FUTURE PROSPECTS

Most applications of optical fibres are in the future, but the technology is advancing so rapidly that most of those discussed above could certainly be realized in the very near future. In this section a few of the longer term prospects which arise out of some of the newer developments will be discussed.

Analogue Systems

At present the linearity of LED and laser sources is sufficient to enable a number of single channel or relatively uncritical systems to be demonstrated. However, both LEDs and c.w. lasers are beginning to show marked improvements in linearity as device behaviour is better understood. It is commonly assumed that lasers are inherently less linear than LEDs, which in turn are inherently less linear than transistors. However, as described in Chapter VII, remarkable linearity is being achieved with STL narrow stripe lasers. Further improvements are also being effected using elegant feedback

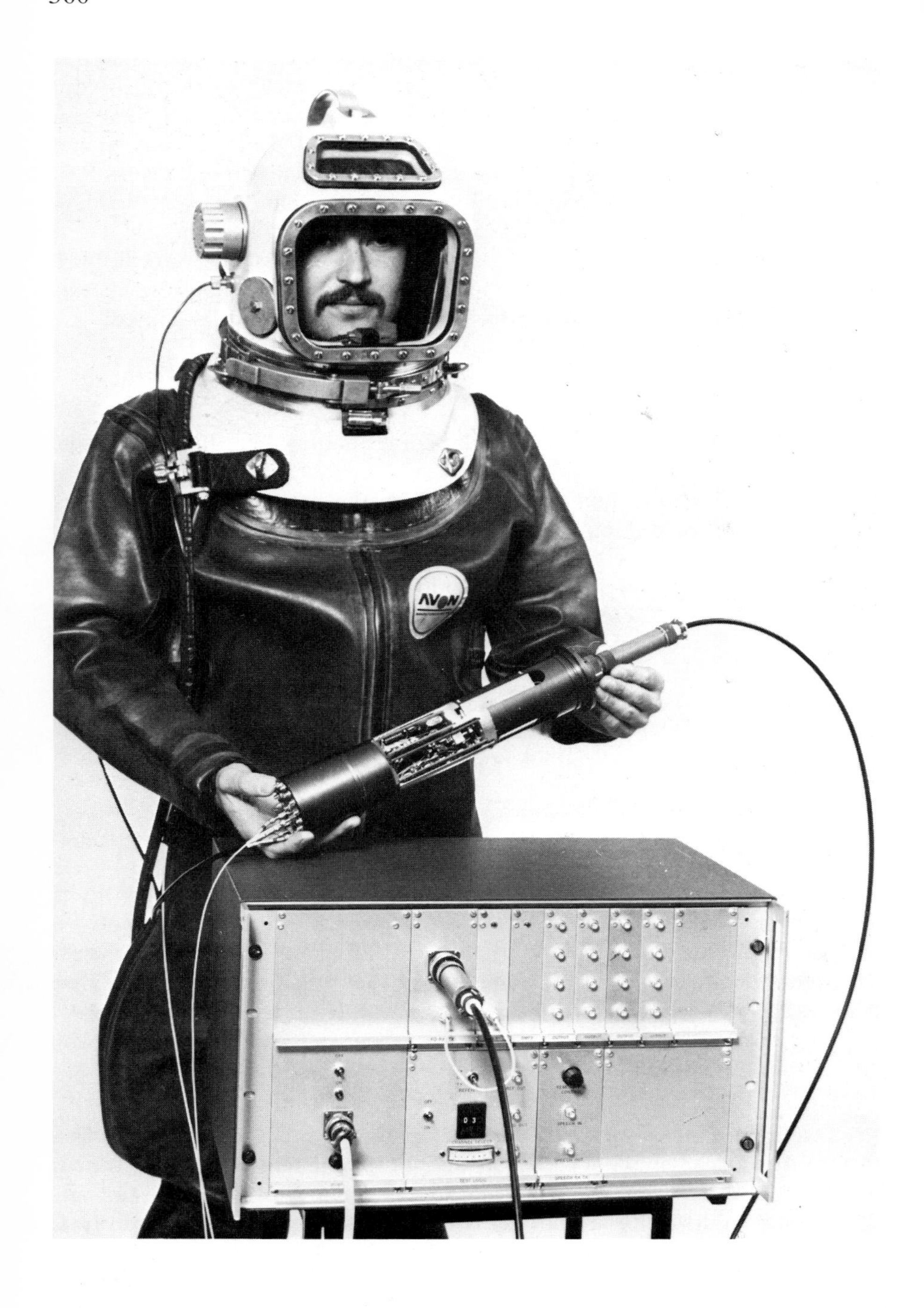

Figure 224. Optical fibre communication link for divers

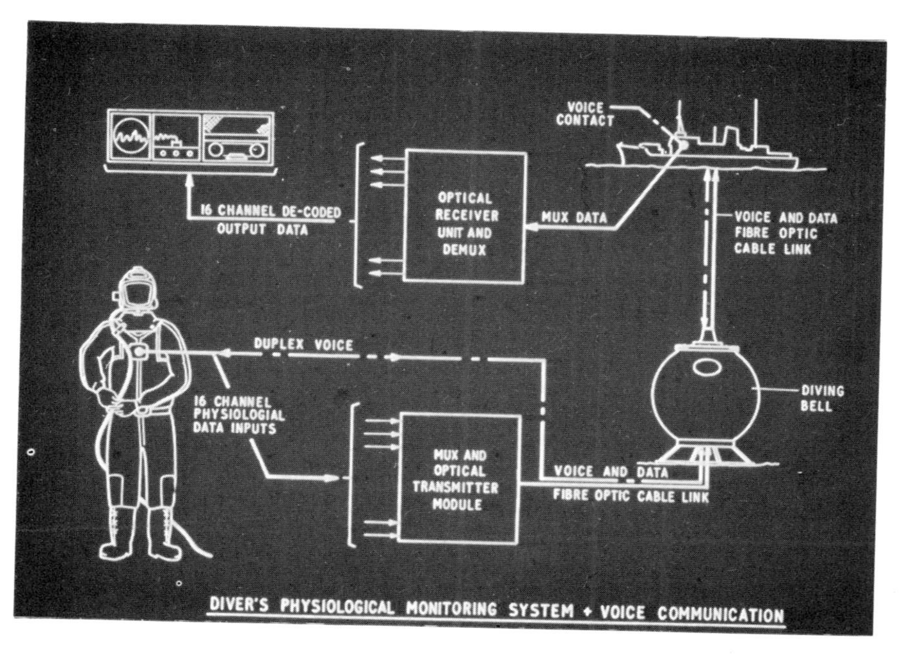

Figure 224(b)

techniques to control the source output. In principle assuming the development of optical power combiners, directional couples, etc. there is no reason why feedforward techniques should not be adopted also. Optical systems, which are inherently no more nonlinear than electrical systems, should prove capable of accessing many analogue applications, including some critical FDM systems.

Long Wavelength Systems

Although the lasers and LEDs based on gallium arsenide technology are advancing rapidly, and will certainly prove adequate for a very large number of applications, a new range of longer wavelength sources using quanternary materials such as gallium indium arsenide phosphide are starting to emerge. These sources are potentially important because achievable fibre attenuation is so much lower at longer wavelengths. For example, at 1.55 μm attenuations down to 0.21 dB/km have been reached compared with a best result at 0.85 μm of about 1.5 dB/km. Such low attenuations, if combined with high bandwidth, e.g. using a single mode fibre, obviously open up the possibility of quite enormous distances between repeaters. To exploit this potential fully, splicing losses must also be reduced or installation procedures adapted to enable long continuous lengths to be handled, when it seems likely that repeater spacings of greater than 50 km and possibly 100 km will prove feasible, with obvious application to submarine as well as land-based systems. For example, the NTT in Japan have disclosed laboratory results demonstrating unrepeatered transmission of 32 Mbit/s over 53.3 km of fibre.[253]

The possibilities of working at even longer wavelengths, for example in the 2–10 μm range, are also being investigated. At present this must be regarded as speculative, and although an attenuation of 0.01 dB/km attenuation at 5 μm in a thallium bromo-iodide core fibre was reported, this result proved erroneous. However, if suitable fibre, source, and detector materials can be developed at these wavelengths there would be significant advantages quite apart from possibly still lower attenuations. For instance, single mode fibres will have correspondingly larger cores, considerably easing connector and splicing problems, and the lower frequency implies lower quantum noise.

Direct Transducers

Some of the potential advantages of fibres, such as simplicity and immunity to electrical interference, are only partially exploited because at present the terminals still work in the electrical regime. This is not always necessary and work has been reported on transducers which directly modulate light. Thus, remote monitoring of basic parameters such as position, temperature, and pressure can be carried out without involving electronics, or indeed requiring any source of power at the remote points.

Another interesting approach has been demonstrated[223] which, though using electronics, enables information to be injected into an optical fibre highway at any desired number of positions without the need for individual optical transmitters. Briefly, a piezoelectric transducer is used to apply local deformations to a 'signal fibre' which cause phase modulation of the transmitted signal. A parallel 'reference fibre' carries an unmodulated signal derived from the same original source. By mixing these two signal at the detector at the far end of the highway the information imparted at each individual terminal can be recovered. This idea is particularly suitable for applications where many signals originating in different places need to be transmitted to a central point.

Integrated Optics

Integrated optics,[250] though not nearly so advanced as discrete components, are nevertheless clearly going to provide further scope for staying more in the optical regime. The feasibility of switches such as the one shown in Figure 15, modulators, and multiplexers have all been demonstrated and these, along with components such as distributed feedback lasers, will doubtless bring about the realization of all-optical repeaters in due course. This will free system designers from some of the speed limitations associated with conventional electronic circuitry and enable the enormous bandwidth potential of optical fibres to be fully exploited.

Wavelength Multiplexing

An additional facility which arises from the high frequency of optical radiation and the broad spectral regions of low attenuation in modern fibres is wavelength (or colour) multiplexing. The feasibility of transmitting independent information modulated onto different optical frequencies over a common fibre has already been demonstrated by a number of workers. Particularly impressive results have recently been reported[251] with multiple channel high speed transmission with near state-of-the-art performance on each individual channel.

This approach is very attractive for very high capacity applications, apart from the rare exceptions where a single very broadband channel is required. Very high overall capacities can be achieved without involving high speed (i.e. above integrated circuit capabilities) circuitry, and possibly with the continued use of a graded index multimode fibre instead of single mode. Also, the amount of electronic multiplexing equipment is reduced. The approach also provides useful flexibility to accommodate high future growth rates while minimizing initial expenditure.

Consumer Applications

The largest potential market for optical fibres is, of course, consumer applications, e.g. in the car industry. Although fibres will almost certainly make an impact there eventually, it should be remembered that traditionally these industries have proved immensely conservative and seldom adopt any technique which is not totally accepted and, above all, very low in cost. Progress in this type of application will depend heavily on the extent to which the earlier applications build up in volume and costs come down.

CHAPTER XI

A 140 Mbit/s *Optical Fibre Field Demonstration System*

INTRODUCTION

In recent years momentum has been gathering throughout the world on both research and development in the field of optical fibre communications. By 1975 much had been written on the performance in the laboratory of most of the constituent parts of a practical system and some laboratory based experiments had been described; however, there was very little experience of optical systems applied to realistically harsh environments. Consequently, a decision was made to launch work on a system which could be operated in a typical PTT environment.

Clearly the correct choice of system to demonstrate the viability of a new technology is complex but briefly the decision was made to develop a digital system since this was likely to be the most compatible with the PTT environment and the technology. A high bit rate was chosen on the basis that this would be more likely to expose any unexpected problems than lower bit rates systems. Also, because of the increased complexity of high bit rate systems relative to low bit rates it was judged to be a simpler transition to extend the range of systems at a later stage from the more complex to the simpler, rather than the other way round. In this chapter the work leading to the commissioning of one of the world's first major optical communications field trials is described.

OBJECTIVE OF DEMONSTRATION SYSTEM

The primary objectives of developing and building a demonstration system may be summarized as follows:

(a) to extend the optical system technology;

(b) to demonstrate the viability of optical systems, particularly of high bit rate systems, in a realistic environment;

(c) to demonstrate that optical communication technology had reached the stage of development where it could be installed with relative ease on a typical PTT route; and

(d) to develop the technology to the point where repeatable performance could be achieved both in the production of the individual component and in the performance of the submodules making up a repeater.

CHOICE OF SYSTEM PARAMETERS

A major factor in ensuring that the system achieved its objectives was the choice of system parameters. It is essential though to realize that at the beginning of 1975 when the parameters were selected that many of the targets set had not even been achieved in the laboratory and certainly not on a repeatable basis. Considerable effort was put into selecting reasonable targets to ensure that in spite of this fast moving technology they would still appear to be realistic. On the other hand it was essential that they were not so difficult that there was a high probability of failing to meet them.

A list of the major system parameters are given in Table XII, and in Table XIII the system power budget as planned in 1975 is given. This power budget is compared with the performance actually achieved, from which it is clear that in most areas the initial targets were exceeded. It is the overall system margin of 24 dB that enabled a repeater to be by-passed thus achieving unrepeatered operation over 6.2 km without significantly changed error performance.

Table XII
Major parameters for the field demonstration system

Information transmission rate	139.264 Mbit/s
Line transmission rate	147.456 Mbit/s
Line Code	Scrambled binary plus 1 parity bit after every 17 bits
Optical transmitter	c.w. double heterojunction stripe geometry GaAlAs laser (STL) using optical feedback
Optical receiver	Silicon avalanche photoiode, constant current biased followed by bipolar transimpedance amplifier
Terminal equipment	Full range of digital multiplex equipment to CCITT standards, from voice channels to 140 Mbit/s. Also includes standard 50-mA power feed circuits, engineering order wire and alarms connected to station alarms
Repeaters	Two intermediate regenerative two-way repeaters power feed via copper conductors in the optical cable. Power consumption of one-way repeater <4 W
Cable	Four fibre (two working, one spare and one filler) four copper conductors (0.5-mm dia.) stranded steel strength member, polyethylene sheath 7 mm over-all diameter

Table XIII
140 Mbit/s field demonstration

	Planned 1975	Achieved 1977
Power launched into fibre	−3.0 dBm	−3 dBm
Connector losses (2 at 1 dB)	−2.0 dB	−2 dB
Jointing losses (7 at 1.0 dB)	−7.0 dB	−2 dB
Cable losses (8 dB/km)	−24.0 dB	−15 dB
Dispersion penalty (7 ns/3-km dispersion)	−5.0 dB	−3 dB
Required receiver sensitivity	−41.0 dBm	−25.0 dBm
Achievable receiver sensitivity	−45.0 dBm	−49 dBm
Overall system margin	4.0 dB	24.0 dB

PROBLEMS REQUIRING A SOLUTION TO ACHIEVE A PRACTICAL SYSTEM

Early in 1975 when the demonstration system was being defined there were a number of problems which required technological solutions before a practical system, even for a field demonstration, could be achieved. These included:

(1) lasers having adequate life (>10,000 h);

(2) stabilization of laser output power against the effects of ageing and temperature;

(3) compensation of change of performance with temperature of the APD;

(4) manufacture of large quantities of fibre and cable having the required optical and mechanical properties;

(5) splicing fibres in the field; and

(6) a low loss demountable connector.

All the problems listed above, and others, were addressed and solved in bringing this demonstration to fruition.

Laser Life

To have a viable field installed demonstration system it was judged that it would be necessary to have lasers with life times of at least 10,000 hours. However, at the time that the development of the demonstration system started the first STL c.w. lasers having the prospect of significant lives (hundreds to thousands of hours) had just been made but progress had been such that the demonstration equipment incorporated devices that had been shown to have a 50 per cent probability life of >50,000 h at 18°C. (End of life is an arbitrary definition given by the threshold current increasing by 50 per cent.) Details of those lasers are given in Chapter VII.

Laser Control

A typical laser characteristic of the type used in the demonstration system is given in Figure 225 and, in common with all lasers of this type, the threshold current has a temperature sensitivity of approximately 2 mA/°C. Clearly if a fixed bias current and fixed modulation current is employed the optical output equivalent to the 'zeros' and 'ones' will vary according to the temperature. To overcome these problems special circuit techniques were developed[216] involving the monitoring of the light output from the back face of the laser to control the mean power launched into the fibre. Of the units employed in the field, one unit used a sampling technique to control the level of the 'zeros' and 'ones' independently and the remainder used the mean power feedback control circuit.

Under the conditions prevailing in the field experiment, where the terminal equipment temperature has ranged between 24 and 40°C and the temperature of the repeater equipment has ranged between 16 and 36°C, both types of control circuit have operated satisfactorily.

The sampling feedback control circuit, although not of full system bandwidth, has a bandwidth much larger than the mean power feedback circuit and therefore is more critical and consumes more power. Consequently, if the laser slope efficiency proves not to vary with long-term life the simpler, lower power consumption mean power control circuit is likely to be preferred.

APD Bias Control

To achieve the greatest receiver sensitivity at these bit rates it is necessary to employ avalanche photodiodes of which the reach-through type is probably

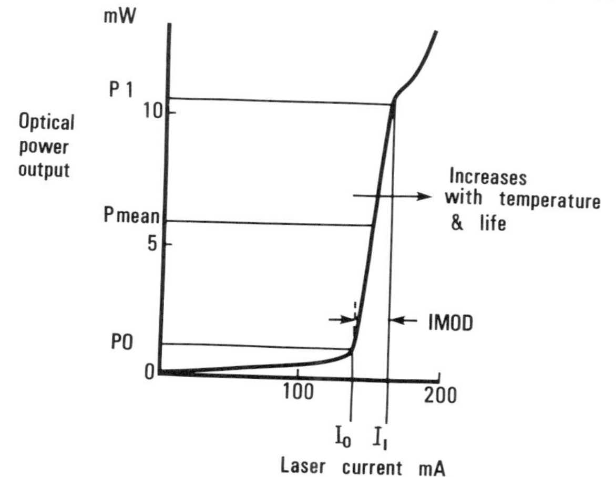

Figure 225. A typical curve of the light versus current characteristics of double heterostructure GaAlAs c.w. laser

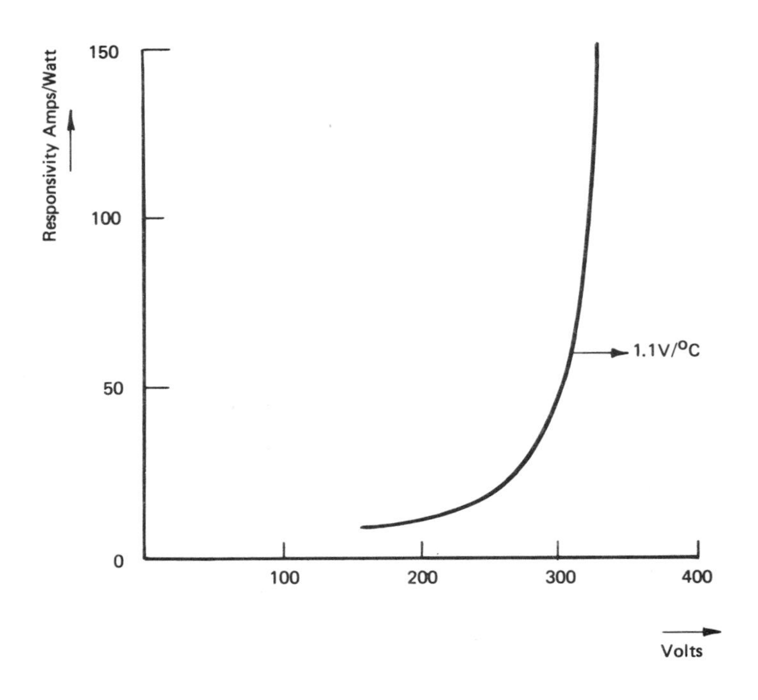

Figure 226. A typical curve of the responsivity versus bias voltage of an avalanche photodiode. A temperature sensitivity of 1.1 V/°C is indicated

the best. These devices exhibit useful internal avalanche gain factors of several hundred, resulting in low effective noise factors for the receiver and have fast response times. A typical gain characteristic is shown in Figure 226, from which it is evident that the devices have two disadvantages. First, they require a relatively high bias voltage at maximum multiplication and secondly the characteristic is temperature sensitive (approximately 1.1 V/°C) giving rise to large changes of multiplication at the highest gains (lowest optical powers) where the system will be least tolerant of these variations.

The very low current requirements of the APD enable very compact inverters to be built to provide the high bias voltage, thus minimizing the first disadvantage. To obtain stabilization of performance against changes in temperature, novel techniques have been adopted[216] in which the APD is arranged to have constant current bias over a wide range of optical input powers, from below threshold level and up to a power level where the voltage of the constant current bias has fallen to 200 V. A typical bias characteristic is shown in Figure 227 where it can be seen that the current is limited at high input powers to avoid damaging the device, and at zero input power the gain sets itself to such a value that the total dark current (i.e. surface current plus multiplied bulk current) is equal to the bias current.

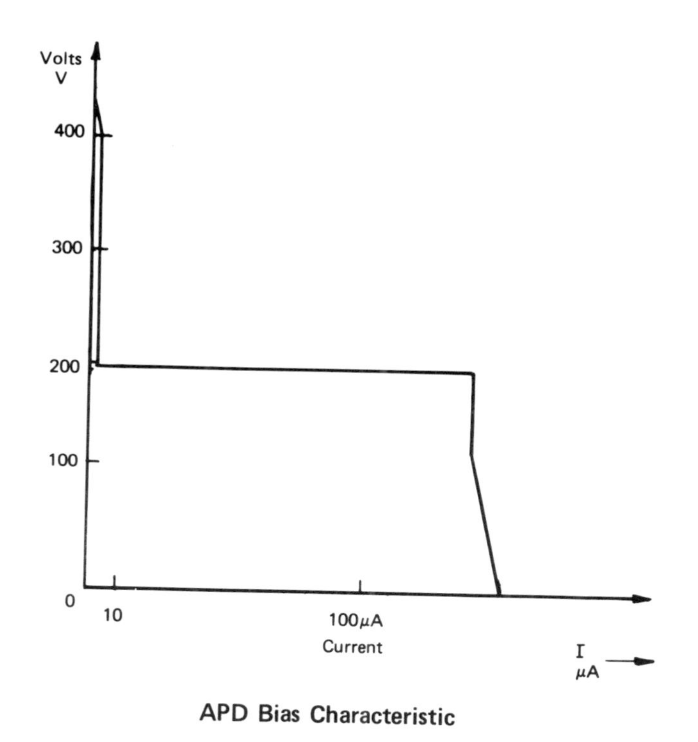

Figure 227. The characteristic of the avalanche photodiode bias circuit employed in the field demonstration equipment

Fibre and Cable

The production of a relatively large scale system requires the production of a large quantity of high quality fibre and cable in which all the required parameters are attained simultaneously. To achieve this it is desirable, if not necessary, to introduce a high degree of process control into the manufacturing processes.[254] Descriptions of fibre- and cablemaking processes are given in Chapters III and IV, where these process controls are described. Table XIV gives the average data of the optical performance and geometric characteristics together with the standard deviations. It can be observed that a closer tolerance of core diameters than outside diameters has been achieved. This is a result of attempting to obtain consistent core diameters at the expense of variation of outside diameters in order to minimize both splice and connector attenuation.

Field Splicing

Most splicing techniques up to 1976 relied on a V-groove technique or variation of that technique (e.g. square sectioned tube) for fibre alignment.

Table XIV
Field demonstration cable performance

	Average value	Standard deviation
Performance of all cables made for the demonstration twelve cables)		
Loss in primary coated fibre (dB/km)	4.1	0.69
Loss in cabled fibres (dB/km)	4.7	0.64
Disperson in cable (ns/km)	1.9	0.75
Fibre o.d. μm	100	4.2
Core diameter μm	31.8	0.75
Ellipticity	0.03	0.02
Performance of last nine cables made for the demonstration		
Loss in primary (dB/km) coated fibre	3.9	0.41
Loss in cable (dB/km)	4.6	0.58
Dispersion in cable (ns/km)	1.6	0.54

Vee-grooves have the disadvantage that if fibres of differing outside diameters are used their axes are offset giving rise to possibly even greater attenuation than the corresponding mismatched core diameters. Consequently an alternative solution to low splicing attenuation was sought; the collapsed sleeve technique[255] described in Chapter V was the solution adopted for this field demonstration.

To adapt the technique for field use it was necessary to exclude gross contamination and to achieve this a rigid transparent cover enveloped the splicing jig, as shown in Figure 64. Using this equipment, and making no attempt to match core diameters, an average attenuation of less than 0.5 dB per splice was achieved in spite of the fact that small core diameter (30-μm) fibre was used.

The collapsed sleeve splice·has now been superseded by other techniques such as electric arc fusion.[93] Nevertheless, it was an important step to demountable connectors to facilitate easy and quick replacement of the precautions having to be taken.

Demountable Connectors

Any practical system intended for a PTT environment must incorporate demountable connectors to facilitate easy and quick replacement of the repeaters in the event of a failure. The problem posed for this demonstration was the achievement of connectors having low and stable attenuation with a 30-μm core diameter fibre. To ensure a satisfactory solution within

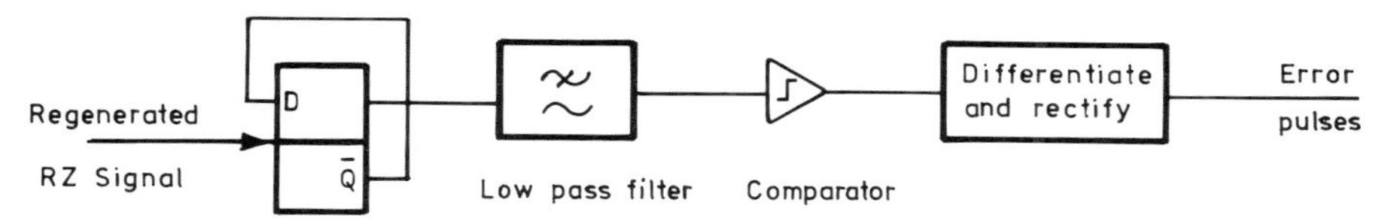

Figure 228. Block schematic diagram of the parity error detector incorporated in the dependent repeaters

the timescale an adjustable device was designed. A full description of the adjustable connector can be found in Chapter V.

LINE CODES

Calculation and experimentation showed that for optical fibre transmission the cost of repeaters was not significantly reduced by the use of nonredundant codes[256]. In addition, even in high bit rate systems the transmission medium bandwidth is not at as great a premium as in coaxial systems, for example. Accordingly, the line code chosen was scrambled binary with an added parity bit after every 17 information bits thus making the signal regenerable with the added facility of in-service error monitoring at both the terminal and at dependent repeaters. The error detection circuitry at dependent repeaters is very simple and requires only one high speed element, a toggle; the circuit is shown in Figure 228. The parity bit is used to control the

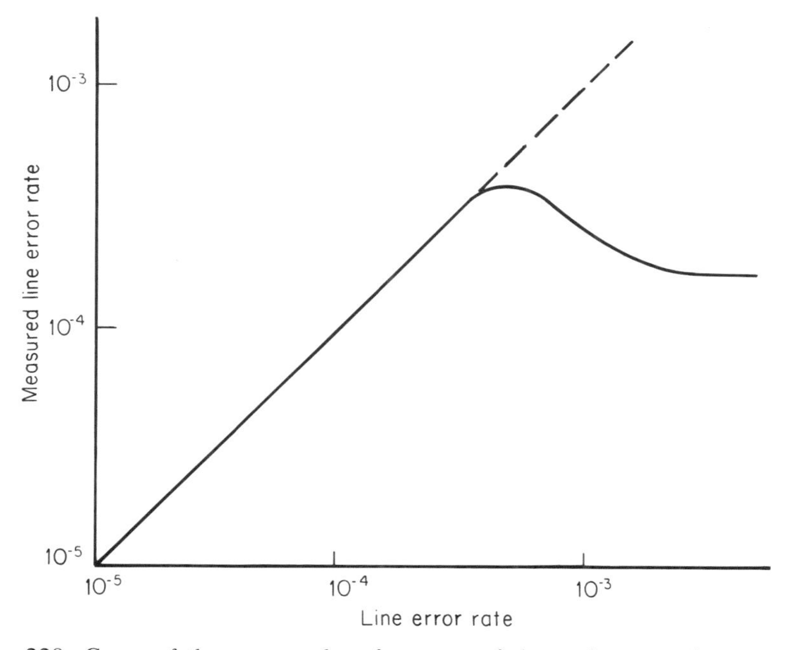

229. Curve of the measured performance of the parity error detector (measured line error rate versus line error rate)

state of the toggle every eighteenth time-slot in such a way that it is the same as the previous control instant. The output of the toggle therefore has two components, namely a d.c. component due to the parity bits and a random component due to the scrambled informant bits. The d.c. component can be distinguished from the random component to any desired degree of confidence by appropriate low pass filtering. A single error or odd number of errors in a parity frame causes the toggle to change state at the next and all subsequent control instants. The change in the d.c. level in the toggle output due to a transmission error is sliced, differentiated, and rectified to give error pulses. The performance of the circuit having incorporated a filter with a low pass cutoff of approximately 13 kHz is shown in Figure 229.

EQUALIZATION

The pulse dispersion of individual lengths of fibre had been shown by experiment to be almost directly proportional to length, indicating very little mode conversion. However, when different fibres were joined in the field, the overall dispersion was somewhat unpredictable; the law of addition of pulse dispersion varying between linear and square root dependence on length. The cause of the variation being in part attributable to the degree of mode selection or mode mixing at the splice and in part attributable to optical equalization effected by fibres having under and overcompensated refractive index profiles occasionally being installed in a random order.[117] This unpredictability made it essential to have either an adaptive equalizer or one which was easily adjusted in the field. For simplicity the latter course was chosen. A single tap transversal equalizer in conjunction with a variable roll-off filter proved adequate for all cases where dispersion incurred penalties up to 10 dB.

Half-width optical pulses were transmitted, but received as full-width pulses at the decision point of the following repeater. Half-width pulses were used for two reasons: (i) it gave nearly optimum theoretical S/N in equalizing for dispersion and (ii) it greatly reduced the effect of 'modal noise' by switching the laser off during each time slot, thus reducing the coherence time. (A full description of the modal noise phenomenon is given in Chapter IX.)

SYSTEM DESCRIPTION

To ensure that the final system would be as realistic as possible and that it would be possible to transmit telephone traffic, a full range of standard digital multiplex equipment was included at each terminal. Reference to Figure 230 shows that the terminals are situated 9 km apart. These are situated in British Post Office repeater stations, one in a small market town (Hitchin) and the other in a larger new town (Stevenage). The intermediate

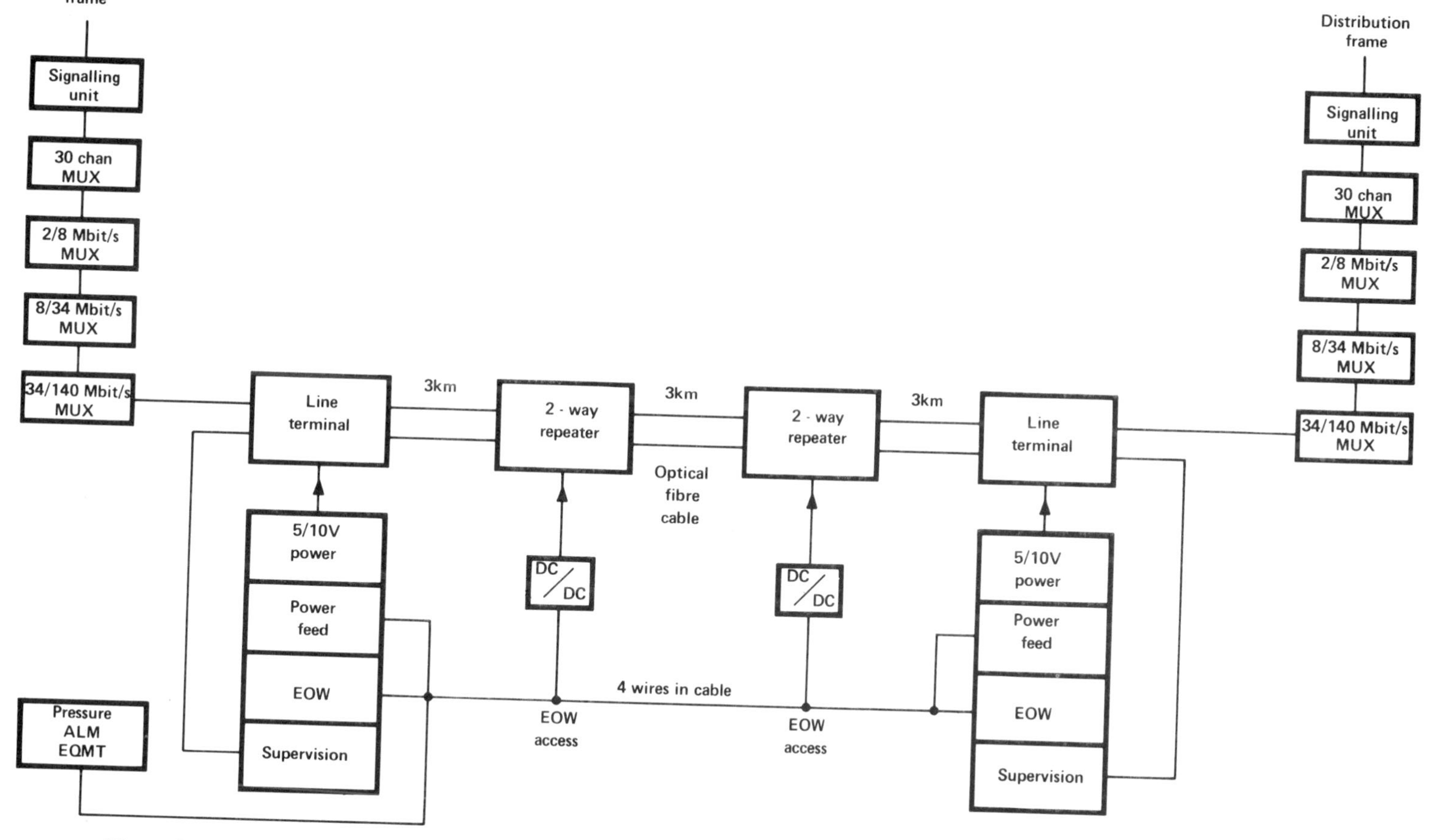

Figure 230. Block schematic diagram illustrating all the major constituents of the 140 Mbit/s field demonstration system

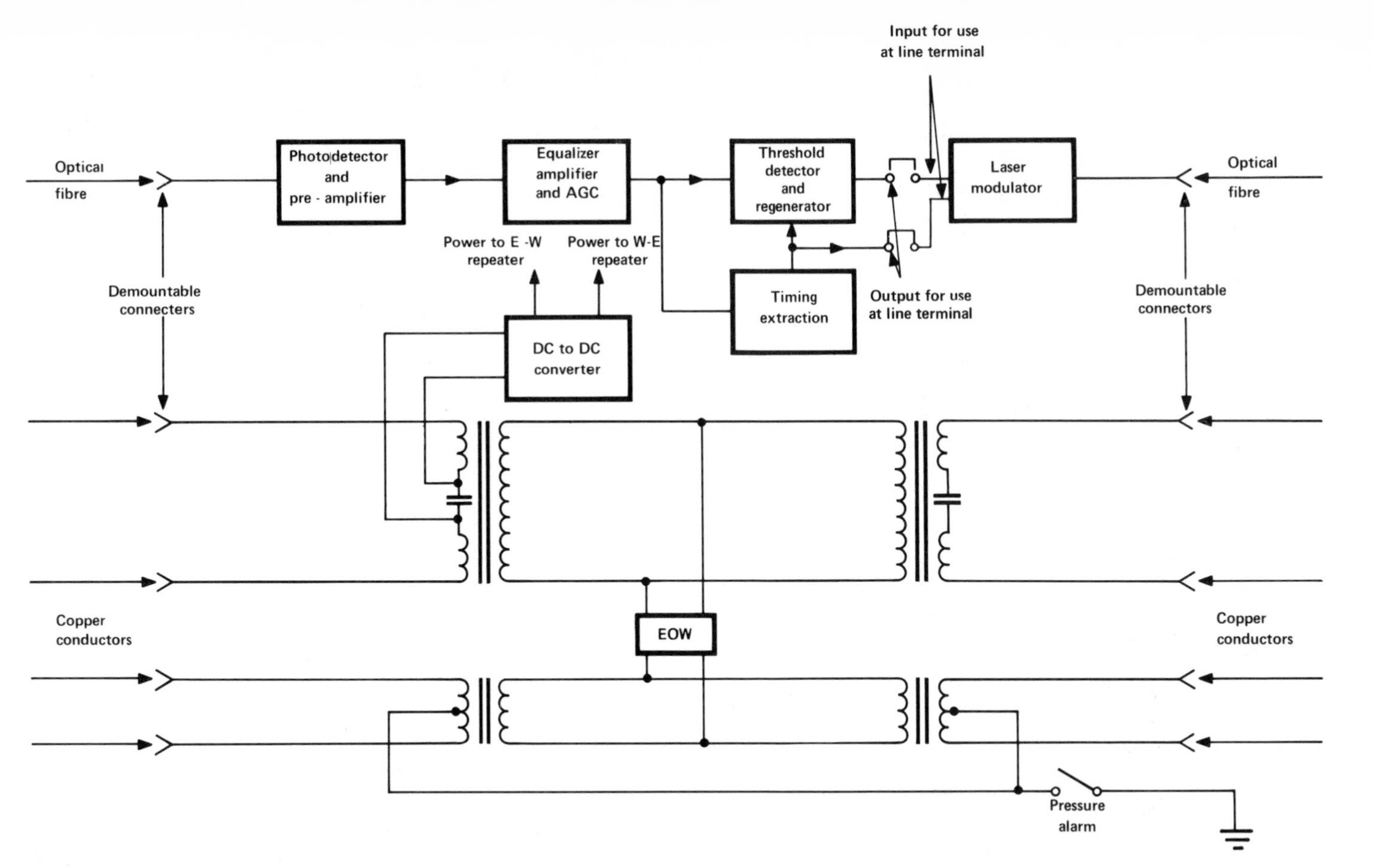

Figure 231. Block schematic diagram of the line repeater showing the major functional blocks together with the engineering order wire and power separation facilities

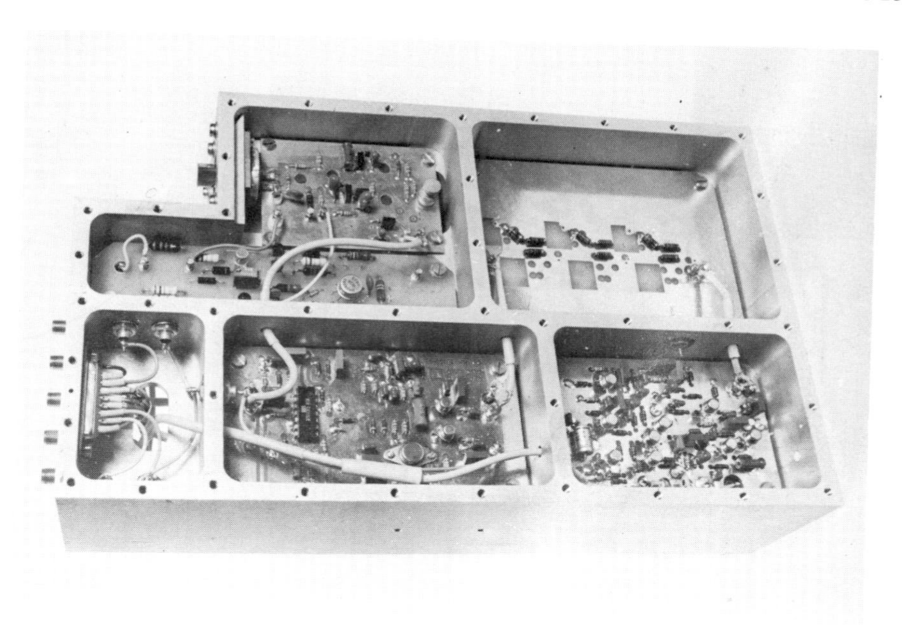

Figure 232. View of a repeater with the receiver boards exposed. The visible boards include the APD and preamplifer, APD bias circuit, AGC amplifier, and equalizer

repeaters are spaced by approximately 3 km and are housed in standard British Post Office equipment enclosures (CRE 8A) which are located alongside the roadway in standard footway boxes (JRF 10).

The repeaters (Figures 231 and 232) are power fed from the terminals through 0.5-mm diameter copper conductors contained in the optical cable. A standard 49-mA constant current power feed is used, but because of the current demanded by the laser (approximately 180 mA) and the necessary avalanche photodiode (APD) bias coltage (~400 V) d.c. to d.c. converters are used at the repeaters. By means of power separation circuits, shown in Figure 231, an engineering order wire facility was also provided over the copper conductors. In addition to the standard order wire units used at the terminals, other standard units such as pressure alarm units and power supplies were incorporated.

LINE TERMINAL EQUIPMENT

A block schematic of the line terminal is shown in Figure 233. The incoming CMI coded bit stream from the multiplex equipment is converted into binary. The signal is then scrambled in a 15-stage scrambler and parity check bits are inserted after every 17 information bits. This signal drives the laser modulator. The inverse processes are performed in the incoming

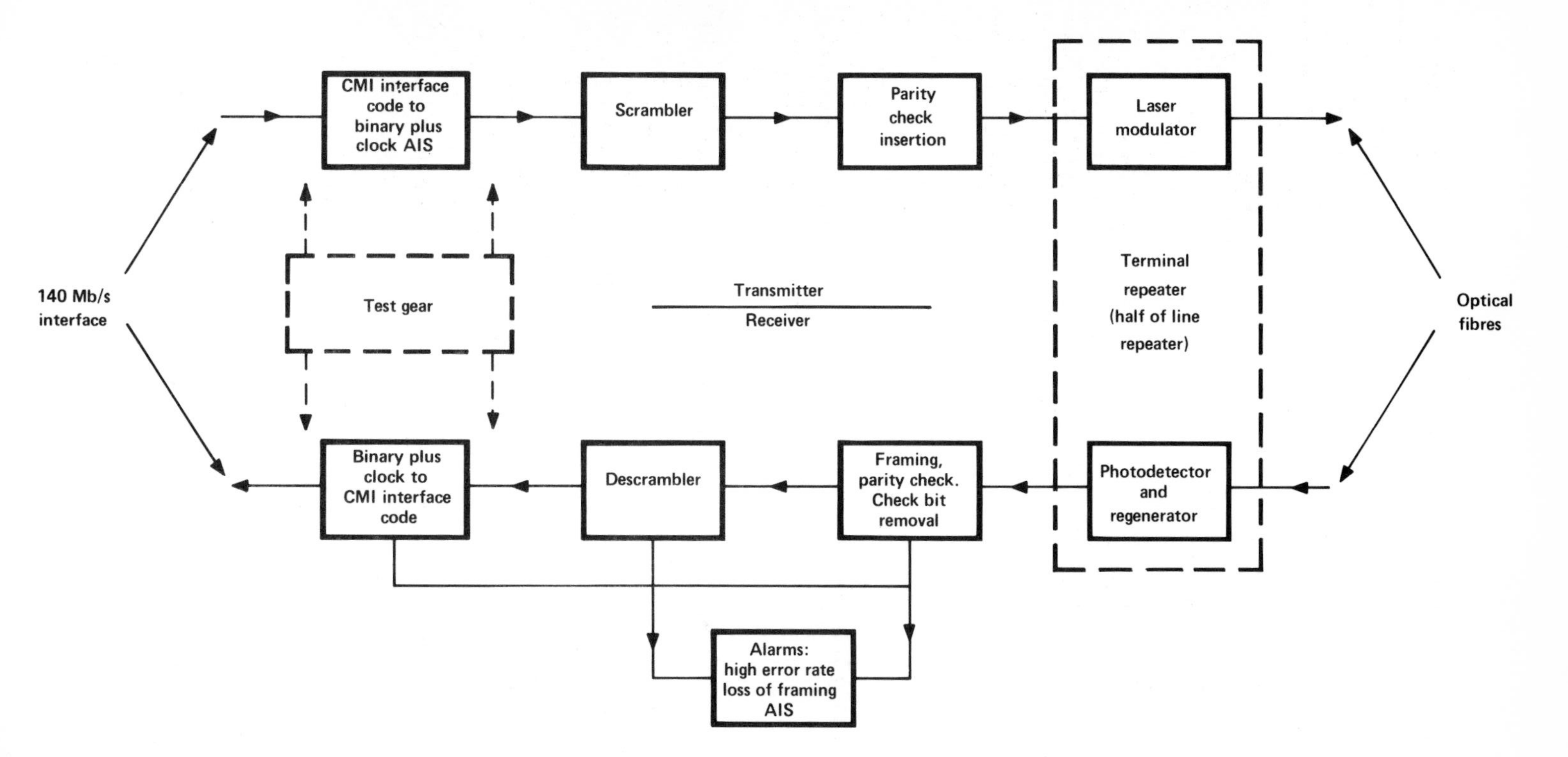

Figure 233. Block schematic diagram of the line terminal equipment showing major functional blocks

Figure 234. An engineer aligning an optical connector on a terminal repeater. The line terminal equipment is situated above the repeater

direction from the line. Figure 234 shows the terminal equipment. An engineer is adjusting the optical connectors on the terminal repeater. Above the repeater is the line terminal equipment and below is the 34/140 Mbit/s multiplex equipment built for this demonstration by Bell Telephone Manufacturing Company of Antwerp. To the left of the line terminal equipment is the 8/34 Mbit/s and 2/8 Mbit/s multiplex equipment produced by FACE

Milan. The 19-in. rack to the right of the line terminal equipment contains power supply units, engineering order wire circuits, power feed units, single channel multiplex equipment, etc. Figure 234 shows the terminal equipment including the line terminal.

REPEATER

The purpose of the repeater is to accept the attenuated optical signal, convert this into an equivalent electrical signal which can be amplified, reshaped and retimed and then reconverted to an optical signal which can be launched into a fibre. From Figure 231, the block schematic of the repeater it can be seen that the optical input is incident upon an avalanche photodiode biased in a constant current mode, as described in an earlier paragraph, providing both temperature compensation and a range of AGC to compensate for up to 10 dB of optical input. The APD is followed by a low noise transimpedance amplifier. After further gain a range of electrical AGC is provided to compensate for another 15 dB of optical input. Additional dynamic range is achieved by the provision for inserting neutral density filters in front of the detector. The following equalizer is of the adjustable single tap transversal type combined with a variable roll-off filter. Timing extraction is effected by LC tank circuits having Q values of approximately 100. After threshold detection and regeneration the signal should ideally be a replica of the original binary signal and as such forms the drive signal for the laser modulator. The optical output level from the laser is controlled by means of one of the feedback circuits described in a previous paragraph (both options being used in the field so that each could be evaluated).

As can be seen from Figure 231, there is provision for breaking the clock and signal path at the input to the modulator, thus enabling the same repeater to be used both as a terminal repeater, where optical input signal and electrical output is required and vice versa, and as a line repeater.

At each of the dependent repeaters there is a d.c. to d.c. converter which provides the required power lines for a two-way repeater from the 49 mA power feed provided over the coppers within the optical cable. The power feed coppers also serve as one of the pairs in a four-wire engineering order wire circuit, it is therefore necessary to provide transformers to separate the power and speech signal, as shown in the diagram of the line repeater.

INSTALLATION

In demonstrating the viability of a new technology it is of prime importance to be able to demonstrate its capability of being installed in a practical environment without significant degradation and without having to take undue precautions; accordingly, great emphasis was placed upon the importance of the installation phase of the 140 Mbit/s demonstration system.

Cable Installation

In conjunction with the British Post Office a route was chosen which is of greater difficulty than is typical from the viewpoint of cable installation. As can be seen from Figure 235 there are many bends, both in the ducts and in manholes. Over one-third of the ducts already had conventional telephone cables. These cables include 23 mm (150 pairs), 28 mm (250 pairs), 49 mm (400 pairs) and some lengths of duct contained up to four other cables before the optical cable was installed.

To enable pulling-in tensions to be recorded, an instrumented installation winch was designed and used throughout the installation programme. In addition to continuously recording the pulling tension a cutout was provided to limit the pulling tensions to a preset value—in this case 80 kg.

Before cables were installed all the ducts were cleaned using a brush and mandrel and lubricated with liquid paraffin. Several blockages were discovered, some of which were cleared with the mandrel and some of which required renewal of short lengths of duct.

An STC installation team installed all the cables using the instrumented winch with the exception of where the nature of the route dictated that it was preferable to install the cable in two directions from an intermediate manhole, then it was frequently more convenient to pull in the short end of the cable by hand. The winch is shown in Figure 236 being used during typical installation weather!

The tensions measured during installation are shown in Figure 237 from which it can be seen that in spite of the difficulty of the route only modest forces were required, even on the longest lengths. Only on one occasion was the cutout operated and this was due to the swivel catching on the edge of the duct.

Installation was completed without a single fibre break, and in fact the optical performance was only marginally changed from that measured on the cable drums. On average there was no change in dispersion and the attenuation decreased by 0.1 dB/km. The change of performance is shown in Figure 238 from which it is apparent that the spread of change in parameters is very small and much of which is attributable to the random errors in the measurements.

In the installation of a mature system in a ducted route it will be essential to carry out splicing in manhole and footway boxes and consequently the ability to do this was demonstrated by doing all splices in the aforesaid situations.[255]

Most of the footway boxes and manholes contained very considerable quantities of water and sludge. However, the only preparation deemed necessary was to pump the water away. Under these conditions it was found practicable, although frequently very uncomfortable, to carry out splicing; indeed, every splice was completed satisfactorily and the average of the measured splice losses was less than 0.5 dB.

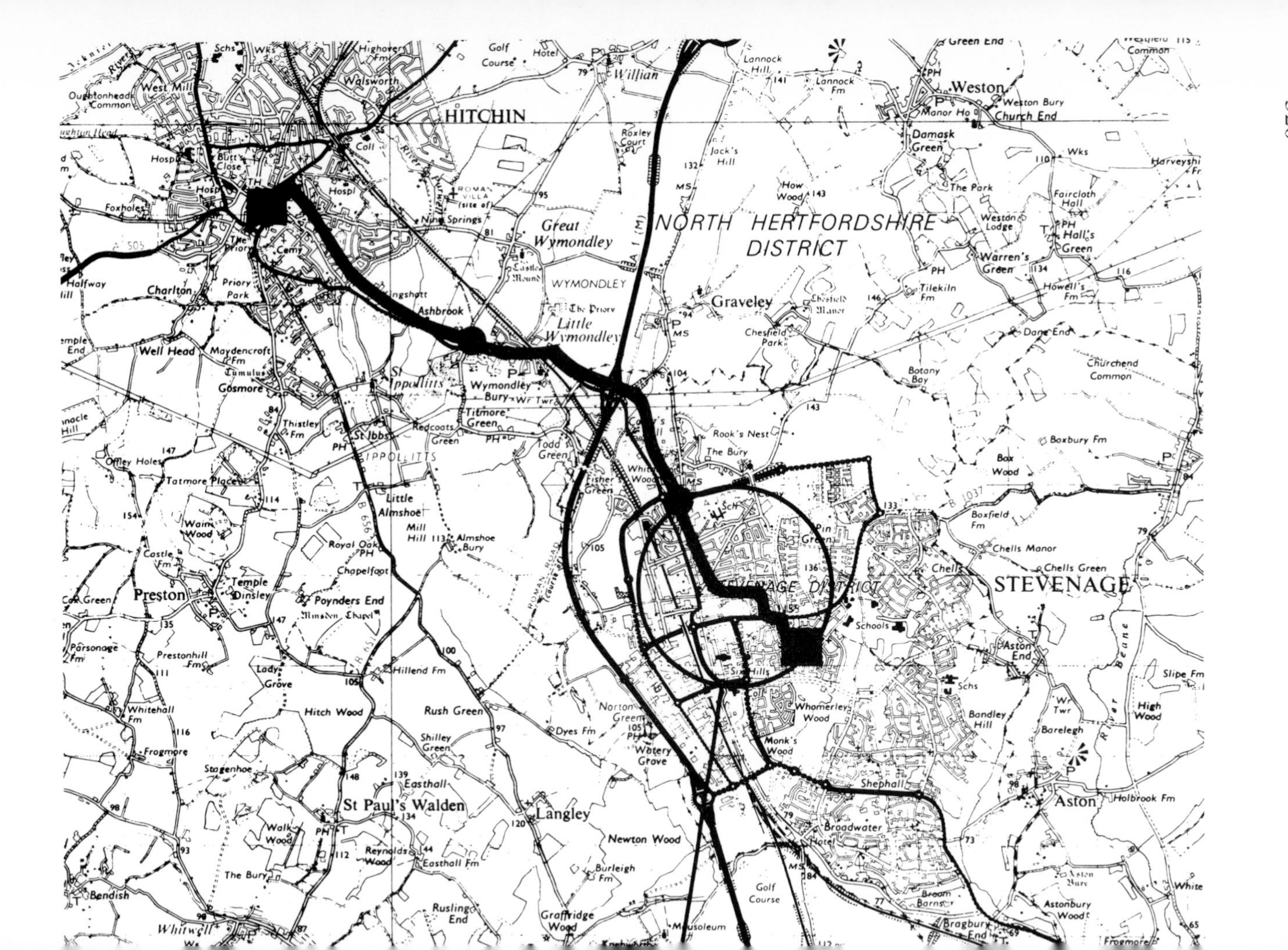
HITCHIN
Walsworth
West Mill
Oughtonhead Common
Golf Course
Hotel
William
Lannock Hill
Lannock Fm
Weston
Weston Bury
Church End
Manor Ho
Damask Green
Wks
The Park
Faircloth Hall
Weston Lodge
Hall's Green
Warren's Green
NORTH HERTFORDSHIRE
DISTRICT
Roxley Court
Jack's Hill
How Wood
Great Wymondley
WYMONDLEY
Little Wymondley
The Priory
Graveley
Chesfield Manor
Chesfield Park
Tilekiln Fm
Howell's Fm
Dane End
Churchend Common
Botany Bay
Boxbury Fm
Box Wood
Rook's Nest
The Bury
Nine Springs
ROMAN VILLA (site of)
Foxholes
Charlton
Priory Park
Halfway Hill
Ashbrook
Well Head
Maydencroft Fm
Tumulus
Gosmore
Wymondley Bury
Titmore Green
Redcoats Green
Todd Green
Fisher Green
Thistley Fm
St Ibbs
IPPOLLITTS
Offley Holes
Tatmore Place
Little Almshoe
Mill Hill
Almshoe Bury
Wain Wood
Castle Fm
Royal Oak PH
Chapelfoot
Temple Dinsley
Preston
Poynders End
Minsden Chapel
Parsonage Fm
Prestonhill Fm
Lady Grove
Hillend Fm
ROMAN ROAD
Whitehall Fm
Frogmore
Hitch Wood
Rush Green
Shilley Green
Dyes Fm
Norton Green
Watery Grove
Stagenhoe
Easthall
St Paul's Walden
Langley
Newton Wood
Walk Wood
Reynolds Wood
Easthall Fm
The Bury
Bendish
Whitwell
Burleigh Fm
Rusling End
Graffridge Wood
Mausoleum
Golf Course
Boxfield Fm
Chells Manor
Chells Green
Chells
STEVENAGE
STEVENAGE DISTRICT
Schools
Aston End
Schs
Six Hills
Whomerley Wood
Monk's Wood
Bandley Hill
Wr Twr
Barelegh
River Beane
Slipe Fm
High Wood
Shephall
Broadwater
Hotel
Aston
Holbrook Fm
Aston Bury
Astonbury Wood
Broom Barns
Bragbury End
Frogmore
White

Figure 235. Map of the Hitchin–Stevenage area with the cable route marked, below which is shown in more detail the section of the route between amplifier point 2 (AP2) and the Stevenage terminal building, illustrating the complexity of the duct route

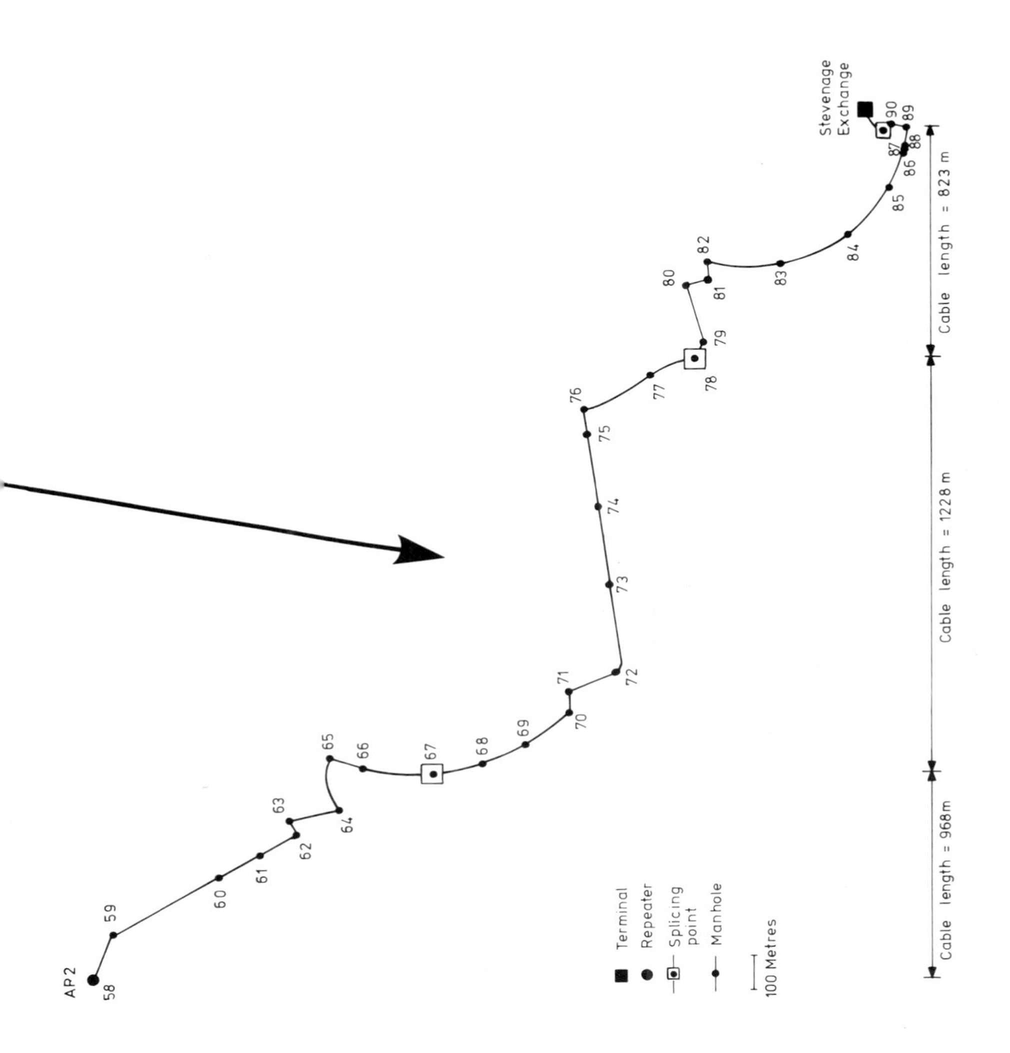

Figure 236. The specially designed cable pulling winch in use at amplifier point 1 (AP1) in weather typical of that encountered

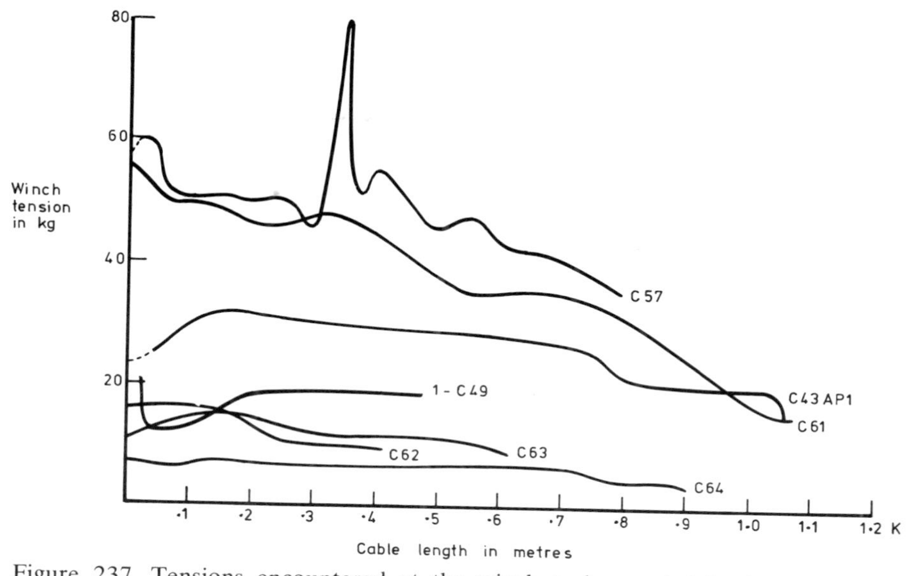

Figure 237. Tensions encountered at the winch and recorded by its specially designed instrumentation. The numbers adjacent to each curve are cable serial numbers

To ensure that the splicing technique described above was suitable for a field environment, a British Post Office splice sleeve type 31 was selected to envelope the splice. The optical cables were potted into the base of the sleeve using standard cable potting compounds. A metre or so of spare fibre was stored on a former within the sleeve to facilitate future measurements and the splices were fixed to supports potted into the base as shown in the photograph of the completed splice (Figure 65).

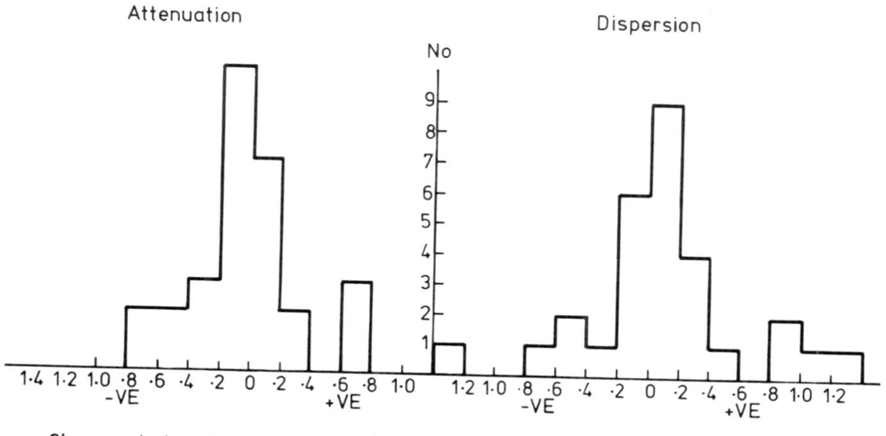

Figure 238. Histograms of the changes of attenuation and dispersion measured when individual lengths of cable were removed from drums and installed in the ducts

Equipment Installation

Installation of the equipment in the exchange buildings was carried out by an STC equipment installation team and in fact no special difficulties were encountered. Before installing the optical equipment in the field the repeaters were adjusted to give the required mean output power (−3 dBm) and the required 'zero' level (by adjustment of the modulation current). Field installation of the equipment and optical connectors is shown in progress in Figure 239.

The adjustable connectors are optimized by means of the orthogonal screws. To aid optimization the repeater receivers have a power monitor facility. The output from this is connected to a voltage-controlled audio oscillator which drives a loudspeaker, the tone from which is relayed back to the transmitter over the engineering order wire circuit. Maximum power transfer to the fibre corresponds to maximum frequency and in practice this can be assessed with relative ease by the engineer attempting to align the connector. The large core fibre (~200 μm) in the receiver connector top makes it unnecessary to adjust the receiver connector.

Figure 239. An engineer installing optical connectors on a two-way repeater fitted in a standard BPO equipment enclosure in footway box adjacent to the carriageway

After the received signal has been optimized it is necessary to adjust the equalizer for optimum 'eye' opening. Preadjustment of this control is not possible because with present fibres it is not possible accurately to predict the cable characteristics as it would be with coaxial cable.

Results

As the lengths of cable for the field system became available during the development phase they were spliced together in the laboratory to facilitate realistic testing. When tests were carried out on the first 3-km section of cabled fibre very large variations of received signal with a wide frequency range (noise) were observed when the cable at the transmitter end of the route was flexed. The magnitude of these variations was sufficient to cause complete collapse of the system, although the received power level should have resulted in excess of 20 dB system margin. This noise phenomenon, now described in Chapter IX as modal noise, was at this time not understood. However, it was established that the effects could be minimized by ensuring that the laser drive conditions did not permit the laser operating conditions to stabilize. Consequently, at this stage it became necessary to change the laser drive signal from nonreturn to zero (NRZ) to return to zero (RZ) thus ensuring that the laser operating conditions did not have time to stablize during long runs of 'ones'. In addition, the laser bias point was changed from the threshold point to marginally below threshold, to further ensure that the operating conditions were changing during each pulse but not sufficiently below threshold to cause significant increase in turn-on delay.

This remedial action removed all noise effects of flexing the fibre and so, what at first appeared to be a completely disastrous situation, became a perfectly workable situation with no loss of performance.

However, it must be emphasized that had all joints, particularly those close to the source, been perfect, then modal noise would have been greatly reduced.

The measured sensitivity (without dispersion) of a repeater is shown in Figure 240, from which it can be seen that a bit error rate of 10^{-9} is achieved for a received power level of almost -49 dBm. In the field, error rates consistently averaging 2×10^{-12} have been measured over periods of several weeks for the 9-km system with repeaters at 3-km intervals. Although this is higher than expected for the received power levels, it is a more than adequate performance for a PTT trunk system; some of the errors may be attributable to the large noise voltages measured on the exchange power lines.

Values of $2°$ r.m.s. jitter have been measured through a repeater when passing pseudo-random data with a sequence length of $2^{15} - 1$. This performance is also better than is required for a trunk system.

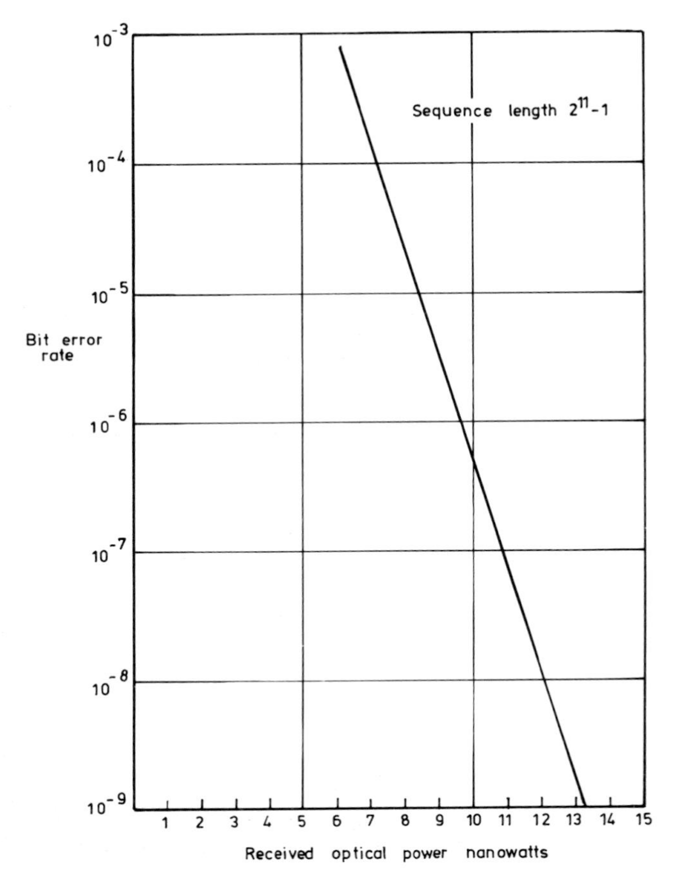

Figure 240. A curve of receiver sensitivity of a prototype repeater in terms of bit error rate versus power measured at the output end of the fibre

From the foregoing it is clear that the system performance is suitable for telephone trunk application but it is generally accepted that the most exacting test for a digital system is the transmission of highly structured signals, such as digitized TV. Tests on the transmission of broadcast quality using digitized TV over an 18-km route with repeaters at 3-km intervals have been carried out in co-operation with the British Broadcasting Corporation using their digital TV coding equipment.[257]

A block schematic of the sending equipment is shown in Figure 241, the equipment to the left of the dotted line being the British Broadcasting Corporation multiplex equipment for vision and sound and to the right the signal processing of the optical fibre system. In Figure 242 is shown the British Broadcasting Corporation multiplexing equipment (l.h.s), demultiplexing equipment (r.h.s.) with optical line terminal and telephone multiplexing equipment in the centre. During these tests the telephony multiplexing equipment was of course not in service.

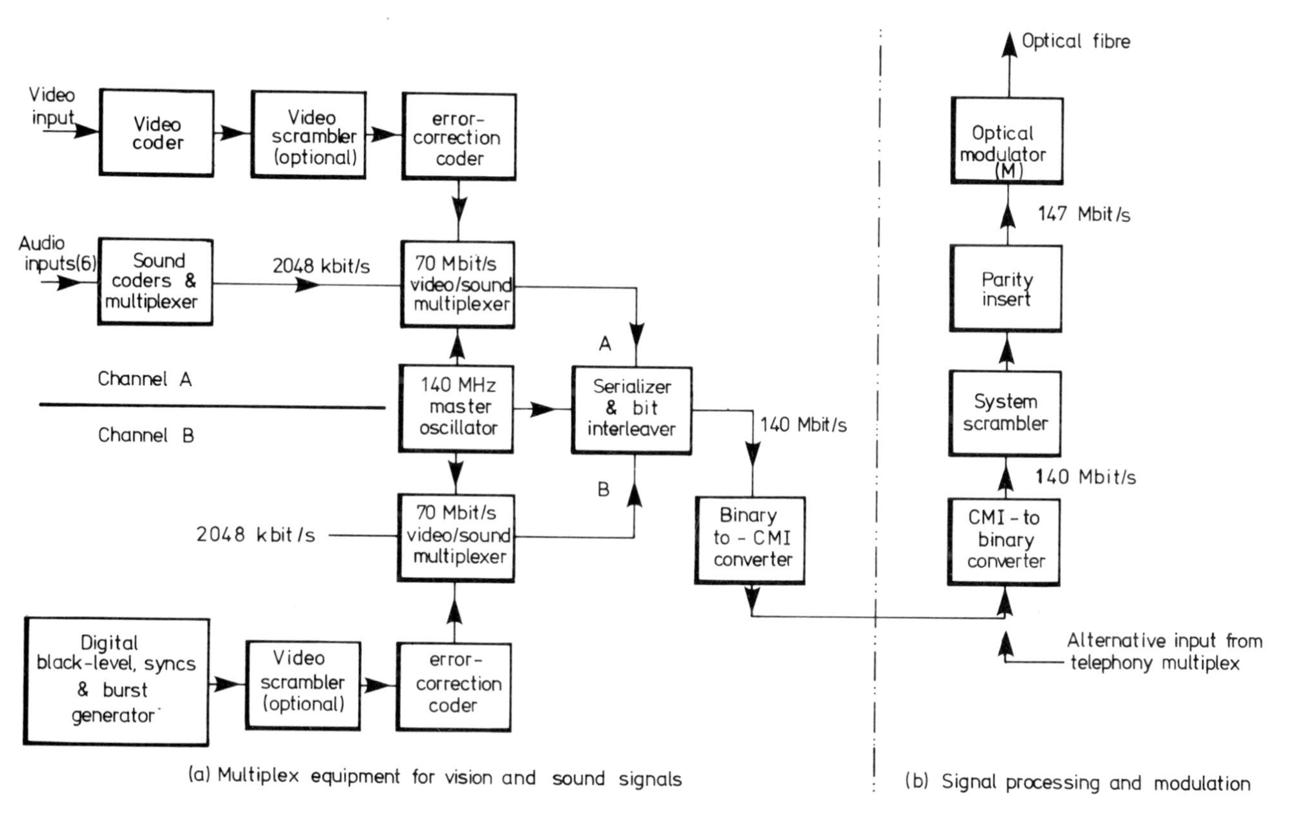

Figure 241. A functional block diagram of the equipment used for digitized television tests over an 18-km route

Figure 242. Equipment assembled in the Hitchin terminal building for the television transmission tests

The quality of transmission was determined both by qualitative assessment of the picture and by change in error rate. There were two methods of detecting errors; one by using the built-in system error detectors and the other by counting the errors corrected by the Wyner–Ash error corrector in the vision channel A. Errors were recorded for combinations of the following variables.

(a) Two different configurations of the system scrambler.

(b) Video information on channel A, e.g. 100 per cent chequerboard; sync, colour burst, and black level; horizontal grey scale. Channel B always transmitted digitally generated sync, black level, and colour burst.

(c) Coding schemes employed on channel A, e.g. d.p.c.m., p.c.m., and h.d.p.c.m.

(d) The four possibilities of scrambling or not scrambling the A and B video channels.

A slight increase in the system error rate was observed when television pictures or television test signals were being carried. This was well within the correction range of the video error correctors and error-free pictures were obtained; pattern dependence is not considered a problem. There appeared to be no correlation between error rate and video scrambler combinations.

Throughout the tests subjective assessments of picture were made. For all configurations and source material used there was no detectable degradation introduced by the transmission system.

These results give added confidence that the use of scramblers for a line code for such a system is viable but additional video scramblers may be desirable for the transmission of highly structured signals.

CONCLUSIONS

To progress from laboratory experiments to a system operating in a realistic environment is clearly an important step in any new technology and this field demonstration system has provided the means of making this progression. In so doing the necessary incentive was provided to establish new techniques and a number of selected approaches have been shown to be successful thus bringing nearer the time when PTT systems will be installed for normal service operation.

The reproducibility of both the optical and mechanical properties of optical fibre cables has been clearly demonstrated and the design of a cable with mechanical properties far exceeding those required for duct installation have also been shown.

It has been shown that using similar techniques to those used with conventional cables it has been possible to install optical cables without damage or, indeed, significant change in optical characteristics. In fact, a positive advantage in terms of the lengths of cable that can be installed in a single pull (1060 m) has been demonstrated.

Although the splicing technique used for field splicing is now being superseded by other methods (e.g. electric arc fusion) it was important to establish that low loss splices could be made in manholes without special precautions having to be taken and this was effectively done.

The development of circuits to enable both the APD and the laser to operate unattended for long periods in an uncontrolled environment whilst maintaining optimum performance is clearly a major achievement and a very necessary step in a move towards practicable systems.

The ability to transmit broadcast quality digitized TV over the system without degradation indicates the choice of scrambled binary to be a suitable choice of line codes and the inserted parity bit has been demonstrated as a simple and effective means of in-traffic error detection.

The possibility of achieving repeater spacings in the field with greater than 6 km while still preserving a system margin has been demonstrated and this is clearly controlled by the optical performance of the cable. Already cable performance exceeding that employed in this system is routinely produced making it possible to obtain even larger repeater spacings. The economic studies discussed in Chapter X show that optical systems with 6-km repeater spacing gives a clear advantage over coaxial systems which have 2-km repeater spacing at this bit rate. However, much greater spacing is possible with optical systems, leading to even greater cost advantages in the future.

In summary, a 140 Mbit/s optical fibre system has been built which is capable of being installed and used for the transmission of trunk telephony and broadcast quality television in a realistic environment. This has demonstrated that a high standard of performance can be maintained during service.

References

1. K. C. Kao and G. A. Hockham, Dielectric-fibre surface waveguides for optical frequencies, *Institution of Electrical Engineers Proceedings*, **113,** 7 (July 1966) 1151–1158.
2. A. H. Reeves, The future of telecommunications, *South African Institute of Electrical Engineers Transactions*, **61,** 9 (September 1970) 445–465.
3. D. Williams and K. C. Kao, Pulse communication along glass fibres, *Institute of Electrical and Electronics Engineers Proceedings*, **56,** 2 (February 1968) 197–198.
4. C. P. Sandbank, The challenge of fibre-optical communication systems, *Radio and Electronic Engineer*, **43,** 11 (November 1973) 665–674.
5. F. F. Roberts, Optical-fibre telecommunication transmission systems, *Post Office Electrical Engineers Journal*, **67,** 1 (April 1974) 32–36.
6. C. A. Burrus and R. W. Dawson, Small-area high-current-density GaAs electroluminescent diodes and a method of operation for improved degradation characteristics, *Applied Physics Letters*, **17,** 3 (August 1970) 97–99.
7. G. H. B. Thompson and P. A. Kirkby, Low threshold-current density in 5-layer-heterostructure (GaAl)As/GaAs localised-gain-region injection lasers, *Electronics Letters*, **9,** 13 (June 1973) 295–296.
8. H. Sasaki and R. M. de la Rue, Eoectro–optic 'Y' junction modulator switch, *Electronics Letters*, **12,** 18 (September 1976).
9. G. D. Pitt and A. M. E. Dooley, Oil content monitoring, Institute of Marine Engineers One-day Conference on Reduction of Pollution from Shipping, 1 December 1977.
10. J. Tyndall, On the colour of water, and on the scattering of light in water and in air, *Royal Institution of Great Britain Proceedings*, **6** (1870–1872) 188–199.
11. A. G. Bell, Selenium and the photophone, *Electrician*, **5** (1880) 214.
12. D. Hondros and P. Debye, Electromagnetic waves along long cylinders of dielectric, *Annalen der Physik*, **32,** 3 (June 1910) 465–476.
13. O. Schriever, Electromagnetic waves in dielectric wires, *Annalen der Physik*, **63,** 7 (December 1920) 645–673.
14. A. C. S. van Heel, A new method of transporting optical images without aberrations, *Nature*, **173** (January 1954) 39.
15. H. H. Hopkins and N. S. Kapany, A flexible fibrescope, using static scanning, *Nature*, **173,** (January 1954) 39–41.
16. T. Uchida, M. Furukawa, I. Kitano, K. Koizumi, and H. Matsumura, A light-focusing fiber guide, *Institute of Electrical and Electronics Engineers Journal of Quantum Electronics*, **QE5,** 6 (June 1969) 331.
17. A. W. Snyder, Asymptotic expressions for eigenfunctions and eigenvalues of a dielectric or optical waveguide, *Institute of Electrical and Electronics Engineers Transactions on Microwave Theory and Techniques*, **MTT-17,** 12 (December 1969) 1130–1138.
18. D. Gloge, Weakly guiding fibres, *Applied Optics*, **10,** 10 (October 1971) 2252–2258.
19. D. Gloge and E. A. J. Marcatili, Multimode theory of graded-core fibres, *Bell System Technical Journal*, **52,** 9 (November 1973) 1563–1578.
20. L. B. Felsen and S. Choudhary, A new look at evanescent waves, *Nouvelle Revue Optique*, **6,** 5 (Sept.–Oct. 1975) 297–301.

21. K. C. Kao and G. A. Hockham, Solution of an optical communication dielectric waveguide problem, Selected Papers from the U.R.S.I. Symposium on Electromagnetic Waves, Stresa, 24–29 June 1968, paper 7–14, pp. 381–384, in *Alta Frequenza*, **38,** Special Issue (May 1969).
22. S. Kawakami and S. Nishida, Characteristics of a doubly clad optical fiber with a low-index inner cladding, *Institute of Electrical and Electronics Engineers Journal of Quantum Electronics*, **QE-10,** 12 (December 1974) 879–887.
23. S. Kawakami and S. Nishida, Design theory of a doubly clad optical fibre, Digest of Technical Papers Presented at the Topical Meeting on Optical Fiber Transmission, Williamsburg, 7–9 January 1975, Washington, Optical Society of America, 1975, paper TuD5.
24. R. G. Smith, Optical power handling capacity of low loss optical fibres as determined by stimulated Raman and Brillouin scattering, *Applied Optics*, **11,** 11 (November 1972) 2489–2494.
25. R. H. Stolen, E. P. Ippen, and A. R. Tynes, Raman oscillation in glass optical waveguide, *Applied Physics Letters*, **20,** 2 (January 1972) 62–64.
26. E. P. Ippen and R. H. Stolen, Stimulated Brillouin scattering in optical fibres, *Applied Physics Letters*, **21,** 11 (December 1972) 539–541.
27. W. G. French, J. B. MacChesney, P. B. O'Connor, and G. W. Tasker, Optical waveguides with very low losses, *Bell System Technical Journal*, **53,** 5 (May–June 1974) 951–954.
28. P. Kaiser, Drawing-induced coloration in vitreous silica fibres, *Optical Society of America Journal*, **64,** 4 (April 1974) 475–481.
29. D. Marcuse, Mode conversion caused by surface imperfections of a dielectric slab waveguide, *Bell System Technical Journal*, **48,** 10 (December 1969) 3187–4215.
30. D. Marcuse, Radiation losses of dielectric waveguides in terms of the power spectrum of the wall distortion function, *Bell System Technical Journal*, **48,** 10 (December 1969) 3233–3242.
31. M. Ikeda, Propagation characteristics of multimode fibres with graded core index, *Institute of Electrical and Electronics Engineers Journal of Quantum Electronics*, **QE-10,** 3 (March 1974) 362–371.
32. T. Sumimoto *et al.*, National Convention of the Institute of Electronics and Communication Engineers of Japan, July 1974.
33. L. A. Jackson, J. E. Midwinter, M. H. Reeve, B. P. Nelson, and J. R. Stern, Propagation studies in sodium borosilicate glass fibres, Digest of Technical Papers Presented at the Topical Meeting on Optical Fiber Transmission, Williamsburg, 7–9 January 1975, Washington, Optical Society of America, 1975, paper TuCl.
34. E. A. J. Marcatili and S. E. Miller, Improved relations describing directional control in electromagnetic wave guidance, *Bell System Technical Journal*, **48,** 7 (September 1969) 2161–2188.
35. A. W. Snyder, Leaky-ray theory of optical waveguides of circular cross section, *Applied Physics* (Germany), **4** 4 (September 1974) 273–298.
36. M. W. Jones and K. C. Kao, Spectrophotometric studies of ultra low loss optical glasses: part 2: Double beam method, *Journal of Scientific Instruments* (*Journal of Physics E*), **2,** 4, (March 1969) series 2 331–335.
37. F. P. Kapron, D. B. Keck, and R. D. Maurer, Radiation losses in glass optical waveguides, *Applied Physics Letters*, **17,** 10 (November 1970) 423–425.
38. D. B. Keck, R. D. Maurer, and P. C. Schultz, On the ultimate lower limit of attenuation in glass optical waveguides, *Applied Physics Letters*, **22,** 7 (April 1973) 307–309.
39. W. G. French, J. B. MacChesney, P. B. O'Connor, and G. W. Tasker, Optical waveguides with very low losses, *Bell System Technical Journal*, **53,** 5 (May–June 1974) 951–954.

40. P. W. Black, J. Irven, K. Byron, I. S. Few, and R. Worthington, Measurements on waveguide properties of GeO_2-SiO_2-cored optical fibres, *Electronics Letters*, **10,** 12 (June 1974) 239–240.
41. D. N. Payne and W. A. Gambling, New silica-based low-loss optical fibre, *Electronics Letters*, **10,** 15 (July 1974) 289–290.
42. D. N. Payne and W. A. Gambling, Zero material dispersion in optical fibres, *Electronics Letters*, **11** 8 (April 1975) 176–178.
43. J. W. Fleming, Material dispersion in lightguide glasses, *Electronics Letters*, **14,** 11 (May 1978) 326–328.
44. G. J. Ogilvie, R. J. Esdaile, and G. P. Kidd, Transmission loss of tetrachloroethylene-filled liquid-core-fibre light guide, *Electronics Letters*, **8,** 22 (November 1972) 533–534.
45. J. Stone, Optical transmission loss in liquid-core hollow fibres, *Institute of Electrical and Electronics Engineers Journal of Quantum Electronics*, **QE-8,** 3 (March 1972) 386–388.
46. G. R. Newns, P. Pantelis, J. L. Wilson, R. W. J. Uffen, and R. Worthington, Absorption losses in glasses and glass fibre waveguides, *Opto-Electronics*, **5,** 4 (July 1973) 289–296.
47. T. Uchida, Preparation and properties of compound glass fibres, U.R.S.I. General Assembly Commission VI, Lima, 18 August 1975.
48. P. C. Schultz, Optical absorption of the transition elements in vitreous silica, *American Ceramic Society Journal*, **57,** 7 (July 1974) 309–313.
49. G. H. Sigel and B. D. Evans, Effects of ionizing radiation on transmission of optical fibers, *Applied Physics Letters*, **24,** 9 (May 1974) 410–412.
50. B. Scott and H. Rawson, Techniques for producing low loss glasses for optical fibre communications system, *Glass Technology* **14,** 5 (October 1973) 115–124.
51. C. E. E. Stewart, D. Tyldesley, B. Scott, H. Rawson, and G. R. Newns, High-purity glasses for optical-fibre communication, *Electronics Letters*, **9,** 21 (October 1973) 482–483.
52. J. G. Titchmarsh, Fibre geometry control with the double crucible technique, *Proceedings of the Second European Conference on Optical Fibre Communication*, Paris (1976), pp. 41–45.
53. C. E. E. Stewart and P. W. Black, Optical losses in soda-lime-silica-cladded fibres produced from composite rods; *Electronics Letters*, **10,** 5 (March 1974) pp. 53–54.
54. G. R. Newns, K. J. Beales, and W. J. Duncan, Low-loss glass for optical transmission, *Electronics Letters*, **10,** 10 (May 1974) 201–202.
55. K. J. Beales, C. R. Day, W. J. Duncan, and G. R. Newns, Low-loss compound glass optical fibre, *Electronics Letters*, **13,** 24 (November 1977) 755–756.
56. H. Imagawa and N. Ogino, Low loss silicate glass optical fibres with small and medium numerical apertures, *Proceedings of I.O.O.C. '77 Conference on Integrated Optics and Optical Fibre Communication*, Tokyo (1977), pp. 613–615.
57. S. Shibata and S. Takahashi, Effect of some manufacturing conditions on the optical loss of compound glass fibres, *Journal of Non-Crystalline Solids*, **23** (1977) 111–122.
58. H. M. J. Van Ass, R. G. Gossink, and P. J. W. Severin, Preparation of graded-index optical glass fibres in the alkali germanosilicate system, *Electronics Letters*, **12,** 15 (July 1976) 369–370.
59. K. Koizumi *et al.*, New light-focusing fibers made by a continuous process, *Applied Optics*, **13,** 2 (February 1974) 255–260.
60. K. Koizumi and Y. Ikeda, Low-loss light focusing fibers made by a continuous process, First European Conference on Optical Fibre Communication, London, 16–18 September 1975; *Institution of Electrical Engineers Conference Proceedings*, no. 132, pp. 24–26.

61. A Mühlich, K. Rau, and N. Treber, Preparation of fluorine doped silica preforms by plasma chemical technique, *Proceedings 3rd European conference on Optical Communication*, Munich (1977).
62. T. Izawa, S. Kobayashi, S. Sudo, and F. Hanawa, Continuous fabrication of high silica fibre preform, *Proceedings of I.O.O.C. '77 Conference on Integrated Optics and Optical Fibre Communication*, Tokyo (1977), pp. 375–378.
63. F. W. Dabby, D. A. Pinnow, L. G. Van Uitert, and I. Camlibel, A technique for making low-loss fused silica core-borosilicate clad fiber optical waveguides, *Materials Research Bulletin*, **10,** 5 (May 1975) 425–430.
64. W. G. French, A. D. Pearson, G. W. Tasker, and J. B. MacChesney, Low-loss fused silica optical waveguides with borosilicate cladding, *Applied Physics Letters*, **23,** 6 (September 1973) 338–339.
65. J. B. MacChesney *et al.*, Low-loss silica core-borosilicate clad fiber optical waveguide, *Applied Physics Letters*, **23,** 6 (September 1973) 340–341.
66. P. W. Black and J. Irven, Development of high-silica optical-fibre waveguides, *Proceedings of Electro-Optics International Conference*, Brighton (1974).
67. L. G. Van Uitert *et al.*, Borosilicate glasses for fibre optical waveguides, *Materials Research Bulletin*, **8,** 4 (April 1973) 469–476.
68. A. Mühlich, K. Rau, F. Simmat, and N. Treber; A new doped synthetic fused silica as bulk material for low-loss optical fibres, *First European Conference on Optical Fibre Communication*, London, 16–18 September 1975; *Institution of Electrical Engineers Conference Proceedings*, no. 132 (post-deadline paper).
69. R. D. Maurer and P. C. Schultz, Germania containing optical waveguide, US Patent 3884550.
70. D. A. Pinnow, F. W. Dabby, I. Camlibel, A. W. Warner, and L. G. Uitert, Preparation of high purity silica, *Materials Research Bulletin*, **10,** 12 (1975) 1263–1266.
71. D. Kuppers and J. Koenings, Preform fabrication by deposition of thousands of layers with the aid of plasma activated CVD, *Proceedings of the Second European Conference on Optical Fibre Communication*, Paris (1976) 49–54.
72. K. Fujiwara, N. Yoshioka, M. Hoshikawa, T. Miyashita, and H. Takata, Optical fibre fabrication by isothermal plasma activated deposition, *Proceedings of the Third European Conference on Optical Fibre Communications*, Munich, (1977), pp 15–17.
73. L. L. Blyler, Jr., A. C. Hart, Jr., R. E. Jaeger, P. Kaiser, and T. J. Miller, Low-loss, Polymer clad silica fibres, Produced by Laser Drawing Optical fibre *Topical Meeting on Optical Fiber Transmission, Williamsbury, Virginia*, paper TuA5-1.
74. K. Inada, T. Naruse, Y. Sugawara, and M. Kojima: 'Silfa' silicone clad optical fibre, *Fujikura Technical Review* (1977) 44–55.
75. K. Inada, T. Akimoto, and S. Tanaka, Pulse spread mechanism of long length silicone clad optical fibre, *Proceedings of the Second European Conference on Optical Fibre Communication*, Paris, (1976), pp. 157–161.
76. A. A. Griffith, The phenomena of rupture and flow in solids, *Philosophical Transactions of the Royal Society of London*, **A221** (1920) 163.
77. E. J. Friebele, M. E. Gingerich, and G. H. Sigel, Effect of ionizing radiation on the optical attenuation in doped silica and plastic fibre-optic waveguides, *Applied Physics Letters.*, **32,** 10 (1978) 619–621.
78. S. G. Foord, W. E. Simpson, and A. Cook, Some design principles for fibre optical cables, *Proceedings of the 23rd International Wire and Cable Symposium*, Atlantic City (1974), pp. 276–280.
79. D. Gloge, Optical-fiber packaging and its influence on fiber straightness and loss, *Bell System Technical Journal*, **54,** 2 (February 1975) 245–262.
80. W. B. Gardner, Microbending loss in optical fibers, *Bell System Technical Journal*, **54,** 2 (February 1975) 457–465.

81. P. W. Black, and A. Cook, Properties of optical fibre in cabling, *Proceedings of the First European Conference on Optical Fibre Communication*, London (1975); London, Institution of Electrical Engineers, *Conference Proceedings*, no. 132 (1975), pp. 67–69. Also other related papers at this conference.
82. U. H. P. Oestreich, Application of Weibull distribution to mechanical reliability of optical fibers for cables, *Proceedings of the First European Conference on Optical Fibre Communication*, London (1975); London, Institution of Electrical Engineers, *Conference Proceedings*, no. 132 (1975), pp. 73–75.
83. H. M. Liertz, Experimental determination of admissible mechanical loads of optical waveguides with respect to cabling process influences, *Proceedings of the First European Conference on Optical Fibre Communication*, London (1975); London, Institution of Electrical Engineers, *Conference Proceedings*, no. 132 (1975), pp. 76–78.
84. P. D. Wilbraham and D. A. Nelson, Coating optical fibres, UK Patent no. 1371740, 29 March 1973.
85. F. H. Buller, Pulling tension during cable installation in ducts or pipes, *General Electric Review*, **52,** (August 1949) 21–23.
86. R. A. Miller and M. Pomerantz, Tactical low loss optical fiber cable for Army applications, *Proceedings of the 23rd International Wire and Cable Symposium*, Atlantic City (1974), pp. 266–275.
87. R. J. Slaughter, A. H. Kent, and T. R. Callan, A duct installation of 2-fibre optical cable. *Proceedings of the First European Conference on Optical Fibre Communication*, London, (1975); London, Institution of Electrical Engineers, *Conference Proceedings*, no. 132 (1975) pp. 84–86.
88. M. I. Schwartz, Optical fiber cabling and splicing, *Topical Meeting on Optical Fiber Transmission, Williamsbury, Virginia*, 7–9 January 1975; Washington, DC, Optical Society of America, 1975, paper WA2.
89. D. L. Bisbee, Measurements of loss due to offsets and end separations of optical fibres, *Bell System Technical Journal*, **50,** 10 (1971) 3159.
90. F. L. Thiel *et al.*, Optical waveguide cable connection, *Applied Optics*, **15,** 11 (1976) 2785.
91. D. W. Gloge *et al.*, Optical fibre end preparation for low loss splices, *Bell System Technical Journal*, **52,** 9 (1973) 1579.
92. Y. Kohanzadeh, Hot splices of optical waveguide fibres, *Applied Optics*, **15,** 3 (1976) 793.
93. I. Hatakeyama and H. Tsuchiya, Fusion splices for optical fibres by discharge heating, *Applied Optics*, **17,** 12 (1978) 1959.
94. D. L. Bisbee, Splicing silica fibres with an electric arc, *Applied Optics*, **15,** 3 (1976) 796.
95. R. Jocteur *et al.*, Optical fibre splicing with plasma torch and oxyhydric microburner, *Proceedings of the Second European Conference on Optical Fibre Communications* (1976).
96. C. G. Someda, Simple, low loss joints between single mode optical fibres, *Bell System Technical Journal*, **52,** 4 (1973) 583.
97. E. L. Chinnock *et al.*, Preparation of optical fibre ends for low loss tape splices, *Bell System Technical Journal*, **54,** 3 (1975) 471.
98. D. Kunze *et al.*, Jointing techniques for optical cables, *Proceedings of the Second European Conference on Optical Fibre Communications*, Paris (1976).
99. H. Murata *et al.*, Connection of optical fibre cable, *Topical Meeting on Optical Fiber Transmission, Williamsbury, Virginia*, 7–9 January 1975; Washington, DC, Optical Society of America, 1975, paper WA5.
100. J. F. Dalgleish *et al.*, Splicing of optical fibres, *Proceedings of the First European Conference on Optical Fibre Communications*, London (1975).
101. C. M. Miller, Loose tube splices for optical fibres, *Bell System Technical Journal*, **54,** 7 (1975) 1215.

102. N. Kurochi *et al.*, A development study on design and fabrication of an optical fibre connector, *Proceedings of the Third European Conference on Optical Fibre Communication*, Munich (1977).
103. J. Guttmann *et al.*, Multipole optical–fibre connector, *IEE Electronic Letters*, **11,** 24 (1975) 582.
104. J. S. Cook and P. K. Runge, An exploratory fibreguide interconnection system, *Proceedings of the Second European Conference on Optical Fibre Communications*, Paris (1976).
105. P. Hensel *et al.*, Connecting optical fibres, *Electronics and Power*, **23,** 2 February (1977).
106. Y. Koyama *et al.*, Optical devices for optical fibre communication, *NEC Research and Development* No. 49 (1978) 51.
107. J. D. Archer, Single fibre optical connections, *New Electronics*, **11,** 2 (1978) 58.
108. M. Borner *et al.*, *Archiv Fur Elektronik Ubertragungstechnik*, **26,** 6 (1975) 288.
109. J. Guttmann *et al.*, A simple connector for glass fibre optical waveguides, *Archiv fur Elektronik Ubertragungstechnik*, **29,** 1 (1975) 50.
110. S. Zemon *et al.*, Eccentric coupler for optical fibres, a simplified version, *Applied Optics*, **14,** 4 (1975) 815.
111. V. Vucins, Adjustable single fibre connector with monitor output, *Proceedings of the Third European Conference on Optical Fibre Communication*, Munich (1977).
112. S. D. Personick, Time dispersion in dielectric waveguides, *Bell System Technical Journal*, **50,** 3 (March 1971) 843–859.
113. M. Ikeda *et al.*, Multimode optical fibres: Steady state mode exciter, *Applied Optics*, **15,** 9 (September 1976) 2116–2120.
114. M. Eve, *et al.*, Launching independent measurements of multimode fibres, *Proceedings of the Second European Conference on Optical Fibre Communications*, Paris, (1976) 143–146.
115. M. J. Adams, D. N. Payne, and F. M. E. Sladen, Mode transit times in near-parabolix-index optical fibres, *Electronics Letters*, **11,** 16 (August 1975) 389–391.
116. M. J. Adams, D. N. Payne, and F. M. E. Sladen, Length dependent effects due to leaky modes on multimode graded-index optical fibres, *Optics Communications*, **17,** 2 (May 1976) 204–209.
117. M. Eve, *et al.*, Transmission performances of three graded index fibre cables installed in operational ducts, *Proceedings of the Third European Conference on Optical Communication*, Munich (1977), pp. 53–55.
118. K. I. White and J. E. Midwinter, An improved technique for the measurement of low optical absorption losses in bulk glass, *Opto-electronics*, **5,** 4 (July 1973) 323–334.
119. D. N. Payne, F. M. E. Sladen, and M. J. Adams, Index profile determination in graded index fibres, *Proceedings of the First European Conference on Optical Fibre Communication*, London (1975), pp. 43–45.
120. K. Petermann, Uncertainties of the leaky mode correction for near square law optical fibres, *Electronics Letters*, **13,** 17 (August 1976) 513–514.
121. M. J. Adams, D. N. Payne, F. M. E. Sladen, and A. Hartog, Resolution limit of the near field scanning technique, *Proceedings of the Third European Conference on Optical Fibre Communication*, Munich (1977), pp. 25–27.
122. J. P. Hazan, Intensity profile distortion due to resolution limitation in fibre index profile determination by near field, *Electronics Letters*, **14,** 5 (March 1978) 158–160.
123. W. Eickhoff and E. Weidel, Measuring method for the refractive index profile of optical glass fibres, *Optical and Quantum Electronics*, **7,** 2 (March 1975) 109–113.
124. M. Ikeda, M. Tateda, and H. Yoshikiyo, Refractive index profile of a graded

index fibre: Measurement by a reflection method, *Applied Optics*, **14,** 4 (April 1975) 814–815.
125. J. Stone and H. E. Earl, Surface effects and reflection refractometry of optical fibres, *Optical and Quantum Electronics*, **8,** 5 (1976) 459–463.
126. W. J. Stewart, A new technique for measuring the refractive index profiles of graded optical fibres, *International Conference on Integrated Optics and Optical Fiber Communication*, Tokyo (1977), pp. 395–398.
127. P. D. Lazay, J. R. Simpson, W. G. French, and B. C. Wonsiewicz, Interference microscopy; automatic analysis of optical fibre refractive index profiles, *Proceedings of the First European Conference on Optical Fibre Communication*, London (1975), pp. 40–41.
128. J. S. Cook, Minimum impulse response in graded index fibres, *Bell System Technical Journal*, **56,** 5 (May 1977) 719–728.
129. J. A. Arnaud, Optimum profiles for dispersive multimode fibres, *Electronics Letters*, **12,** 25 (December 1976) 654–655.
130. R. Olshansky and D. B. Keck, Pulse broadening in graded-index optical fibres, *Applied Optics*, **15,** 2 (February 1976) 483–491.
131. M. Rousseau and L. Jeunhomme, Optimum index profile in multimode optical fibre with respect to mode coupling, *Optics Communications*, **23,** 2 (November 1977) 275–278.
132. E. Khular, A. Kumar, A. K. Ghatak and B. P. Pal, Effect of the refractive index dip on the propagation characteristics of step index and graded index fibre, *Optics Communications*, **23,** 2 (November 1977) 263–267.
133. D. Gloge, Dispersion in weakly guiding fibres, *Applied Optics*, **10,** 11 (November 1971) 2442–2445.
134. S. E. Miller, E. A. J. Marcatili, and T. Li, Research towards optical-fibre transmission systems, Part 1, *Proceedings of the IEEE*, **61,** 12 (December 1973) 1703–1726.
135. H. Yolota *et al.*, Long length single mode fiber with low attenuation in the dispersion free region, *Proceedings of the 2nd International Conference in Integrated Optics*, Amsterdam (September 1979), paper 5.4.
136. B. Luther-Davies, D. N. Payne, and W. A. Gambling, Evaluation of material dispersion in low loss phosphosilicate core optical fibres, *Optics Communications*, **13,** 1 (January 1975) 84–88.
137. C. R. Hammond, Silica based binary glass systems: Wavelength dispersive properties and composition in optical fibres, *Optical and Quantum Electronics*, **10** (March 1978) 163–170.
138. K. Daikoku and A. Sugimura, Direct measurement of wavelength dispersion in optical fibres—difference method, *Electronics Letters*, **14,** 5 (March 1978) 149–151.
139. L. G. Cohen, Shuttle pulse measurements of pulse spreading in an optical fibre, *Applied Optics*, **14,** 6 (June 1975) 1351–1356.
140. L. G. Cohen and S. D. Personick, Length dependence of pulse dispersion in a long multimode optical fibre, *Applied Optics*, **14,** 6 (June 1975) 1357–1360.
141. D. Gloge, E. L. Chinnock, and T. P. Lee, GaAs laser twin set-up to measure mode and material dispersion in optical fibres, *Applied Optics*, **13.** 2 (1974) 261.
142. C. Lin, L. G. Cohen, W. G. French, and V. A. Foertmeyer, Pulse delay measurements in the zero-material dispersion region for germanium and phosphorus doped silica fibres, *Electronics Letters*, **14,** 6 (March 1978) 170–172.
143. L. G. Cohen, Pulse transmission measurements for determining near optimal profile gradings in multimode borosilicate optical fibres, *Applied Optics*, **15,** 7 (July 1976) 1808–1814.
144. G. J. Cannell, Continuous measurement of optical fibre attenuation during manufacture, *Electronics Letters*, **13,** 5 (March 1977) 125–126.
145. K. C. Byron, G. J. Cannell, I. S. Few, and R. Worthington, Field measurement

of fibre optic cables: *Proceedings of the Third European Conference on Optical Communication*, Munich (1977) pp. 34–36.
146. I. S. Few, Instrument for testing telecommunications optical fibres, *Optical Engineering*, **15,** 3 (May–June 1976) 241–243.
147. R. L. Hartman and R. W. Dixon, Reliability of DH GaAs lasers at elevated temperatures *Applied Physics Letters*, **26,** 5 (March 1975) 239–242.
148. R. L. Hartman, N. E. Schumaker, and R. W. Dixon, Continuously operated (Al.Ga)As DH lasers with 70°C lifetimes as long as two years, *Applied Physics Letters*, **31,** 11 (Dec. 1977) 756–759.
149. A. R. Goodwin, P. A. Kirkby, M. Pion, and R. S. Baulcomb, Enhanced degradation rates in temperature sensitive $Ga_{1-x}Al_xAs$ lasers, *IEEE Journal of Quantum Electronics*, **QE13,** 8 (August 1977) 696–699.
150. L. A. D'Asaro, Advances in GaAs junction lasers with stripe geometry *Journal of Luminescence*, **7** (1973) 310–337.
151. A. R. Goodwin, J. R. Peters, M. Pion, G. H. B. Thompson, and J. E. A. Whiteaway, Threshold-temperature characteristics of double heterostructure $Ga_{1-x}Al_xAs$ lasers, *Journal of Applied Physics*, **46,** 7 (July 1975) 3126–3131.
152. B. C. De Loach, B. W. Hakki, R. L. Hartman, and L. A. D'Asaro, Degradation of C. W. GaAs DH lasers at 300 K, *IEEE Proceedings* **61,** 7 (1973) 1042–1044.
153. H. Yonezu, I. Sakuma, T. Kamejima, M. Ueno, K. Nishida, Y. Nannichi, and I. Hayashi, Degradation mechanism of (Al.Ga)As DH laser diodes, *Applied Physics Letters*, **24** 1 (1974) 18–19.
154. K. Kobayashi, R. Lang, H. Yonezu, Y. Matsumoto, T. Shinohara, I. Sakuma, T. Susuki, and I. Hayashi, Unstable horizontal transverse modes and their stabilization with a new stripe structure, *IEEE Journal of Quantum Electronics*, **QE13,** 8 (August 1977) 659–661.
155. M. D. Campos, C. J. Hwang, R. I. Bossi, and J. E. Ripper, Cavity competition in anomalous emission intensity in DH lasers, *IEEE Journal of Quantum Electronics*, **QE13,** 8 (August 1977) 687–691.
156. R. Lang, Horizontal mode deformation and anomalous lasing properties of stripe geometry injection lasers—theoretical model, *Japan Journal of Applied Physics* (Japan), **16,** 1 (January 1977) 205–206.
157. G. H. B. Thompson, D. F. Lovelace, and S. E. H. Turley, Kinks in the light/current characteristics and near-field shifts in (GaAl)As-heterostructure stripe lasers and their explanation by the effect of self focussing on a built-in optical waveguide, *Solid State and Electron Devices*, **2,** 1 (January 1978) 12–30.
158. R. W. Dixon, F. R. Nash, R. L. Hartman, and R. T. Hepplewhite, Inproved light-output linearity in stripe geometry double heterostructure (Al,Ga)As lasers, *Applied Physics Letters*, **29,** 6 (September 1976) 372–374.
159. T. Kobayashi, H. Kawaguchi, and Y. Furukawa, Lasing characteristics of very narrow planar stripe lasers, *Japan Journal of Applied Physics*, **16,** 4 (April 1977) 601–607.
160. W. Susaki, T. Tanaka, H. Kan, and M. Ishii, New structures of GaAlAs lateral injection lasers for low threshold and single mode operation, *IEEE Journal of Quantum Electronics* **QE13,** 8 (August 1977) 587–591.
161. P. A. Kirkby, P. R. Selway and L. D. Westbrook, Photoelastic waveguides and their effect on stripe-geometry $GaAs/Ga_{1-x}Al_xAs$ lasers, *Journal of Applied Physics*, **50,** 7 (1979) 4567–4579.
162. P. A. Kirkby, A. R. Goodwin, G. H. B. Thompson, and P. R. Selway, Observations of self-focusing in stripe geometry semiconductor lasers and the development of a comprehensive model of their operation, *IEEE Journal of Quantum Electronics*, **QE13,** 8 (August 1977) 705–719.
163. D. D. Cook and F. R. Nash, Gain-induced guiding and astigmatic output beam

of GaAs lasers, *Journal of Applied Physics*, **46,** 4 (1975) 1660–1672.
164. T. Ikegami and Y. Suematsu, Direct modulation of semiconductor junction laser, *Electronics Communications* (Japan), **51–B,** 2 (1968) 51–58.
165. M. J. Adams, Rate equations and transient phenomena in semiconductor lasers, *Opto-electronics* **5** (1973) 201–215.
166. Y. Suematsi, T. Hong, and K. Furuya, Reduction of resonance-like peak in direct modulation of injection lasers due to carrier diffusion and external circuit, *Proceedings of the Third European Conference on Optical Fibre Communications*, Munich (1977).
167. T. Ikegami, K. Kobayashi, and Y. Suematsu, Transient behaviour of semiconductor injection lasers, *Electronics and Communications* (Japan), **53-B,** 5 (1970) 82–89.
168. T. Ozeki and T. Ito, Pulse modulation of DH (GaAl)As lasers, *IEEE Journal of Quantum Electrontics*, **QE9,** 2 (1973) 388–391.
169. T. Ikegami, Spectrum broadening and tailing effect in directly modulated injection lasers, *Proceedings of the First European Conference on Optical Fibre Communication*, London (1975).
170. J. Buus and M. Danielson, Carrier diffusion and higher order transversal modes in spectral dynamics of the semiconductor laser, *IEEE Journal of Quantum Electronics*, **QE13,** 8 (1977) 669–674.
171. P. R. Selway, P. A. Kirkby, A. R. Goodwin, and G. H. B. Thompson, Dynamics of self-focusing in stripe-geometry semiconductor lasers, *Solid State and Electron Devices*, **2,** 1 (1978) 38–40.
172. M. Pion, A. R. Goodwin, P. A. Kirkby, and R. S. Baulcomb, Degradation characteristic of c.w. lasers with specific growth defects, *Proceedings of the 6th International Symposium on GaAs and Related Compounds*, Edinburgh (1976) (published by the Institute of Physics).
173. I. Ladany, M. Ettenberg, H. F. Lockwood, and H. Kressel, Al_2O_3 half-wave films for long-life c.w. lasers, *Applied Physics Letters*, **30,** 2 (1977) 87–88.
174. N. Chinone, H. Nakashima, and R. Ito, Long term degradation of GaAs–$Ga_{1-x}Al_xAs$ DH lasers due to facet erosion, *Journal of Applied Physics*, **48,** 3 (March 1977) 1160–1162.
175. S. Ritchie, R. F. Godfrey, B. Wakefield, and D. H. Newman, The temperature dependence of degradation mechanisms in long-lived (GaAl)As DH lasers, *Journal of Applied Physics*, **49,** 6 (1978) 3127–3132.
176. M. R. Matthews and A. G. Steventon, Spectral and transient response of low-threshold proton-isolated (GaAl)As lasers, *Electronics Letters*, **14,** 19 (1978) 649–651.
177. T. Tsukada, GaAs–$Ga_{1-x}Al_xAs$ buried heterostructure injection lasers, *Journal of Applied Physics*, **45,** 11 (1974) 4899–4906.
178. P. A. Kirkby and G. H. B. Thompson, Channeled substrate buried heterostructure GaAs–(GaAl)As injection lasers, *Journal of Applied Physics*, **47,** 10 (1976) 4578–4589.
179. K. Aiki, M. Nakamura, T. Kuroda, J. Umeda, R. Ito, N. Chinone, and M. Maeda, Transverse mode stabilized (AlGa)As injection lasers with channelled substrate planar structure.
180. G. H. B. Thompson, D. F. Lovelace and S. E. H. Turley, Deep Zn-diffused (GaAl)As heterostructure stripe laser with twin transverse junctions for low threshold current and kink-free light characteristics. *IEEE Journal of Quantum Electronics*, **QE15,** 8 (1979) 772–775.
181. J. J. Hsieh and C. C. Shen, Room temperature c.w. operation of buried stripe double heterostructure GaInAsP/InP diode lasers, *Applied Physics Letters*, **30,** 8 (1977) 429–431.
182. H. Melchior and W. T. Lynch, Signal and noise response of high speed

germanium avalanche photodiodes, *IEEE Transactions, Electron Devices,* **ED-13,** 12 (1966) 829.
183. S. M. Sze, *Physics of Semiconductor Devices* (Wiley–Interscience, New York, 1969).
184. R. J. McIntyre, The distribution of gains in uniformly multiplying avalanche photodiodes, theory, *IEEE Transactions, Electron Devices,* **ED-19** (June 1972) 702–713.
185. S. D. Personick, Statistics of a general class of avalanche detectors with applications to optical communication, *BSTJ,* **50,** 10 (1971) 3075–3095.
186. J. A. Raines, Noise performance of silicon avalanche diodes, *SERL Technical Journal,* **21,** 1, (February 1971) 4.1–4.11.
187. F. N. H. Robinson, *Noise in Electrical Circuits* (Oxford University Press, London, 1962).
188. R. G. Plumb and J. E. Carroll, Thin silicon ion-implanted p–i–n photodiodes, *IEEJ Solid State and Electron Devices,* **1,** 3 (April 1977) 89–91.
189. M. V. Schneider, Schottky barrier photodiodes with anti-reflection coating, BSTJ, **45,** 9 (1966) 1611.
190. L. K. Anderson *et al.*, Microwave photodiodes exhibiting microplasma-free carrier multiplication, *Applied Physics Letters,* **6,** 4 (1965) 62.
191. H. W. Ruegg, An optimised avalanche photodiode, *IEEE Transactions, Electron Devices,* **ED-14,** 5 (1967) 239.
192. C. J. Nuese, III–V Alloys for opto-electronic applications, *Journal of Electronic Materials,* **6,** 3 (1977) 253–293.
193. P. K. Ajmera and J. R. Hauser, Ion implanted photodiode detectors in epitaxial $(Ga_xIn_{1-x})As$, *Applied Optics,* **14,** 12 (December 1975) 2905–2910.
194. R. C. Eden, Heterojunction III–V Alloy photodetectors for high sensitivity 1.06 μm optical receivers, *Proceedings of the IEEE,* **63,** 1 (January 1975) 32–37.
195. J. G. Edwards and R. Jefferies, Response time of F4000 biplanar photocell in various holders, *Journal of Scientific Instruments* (*Journal of Physics E*) series 2, **2,** 12 (1969) 1126–1129.
196. O. L. Gaddy and D. F. Holshouser, High gain dynamic microwave photomultiplier, *Proceedings of the IRE,* **50,** 2 (1962) 207–208.
197. M. DiDomenico Jr. and O. Svelto, Solid state photodetection comparison between photodiodes and photoconductors, *Proceedings of the IEEE,* **52,** 2 (1964) 136.
198. Roundy and Byer, sub-nanosecond pyroelectric detector, *Applied Physics Letters,* **21,** 10 (1968) 512–515.
199. M. Ross, Laser Receivers—Devices, Techniques, Systems (Wiley, New York, 1966).
200. J. E. Goell, An optical repeater with high-impedance input amplifier, *Bell System Technical Journal,* **53,** 4 (April 1974) 629–643.
201. W. E. Heinlein and H. R. Trimmel, Repeater spacings of 8 Mbit/s and 34 Mbit/s transmission system using multimode optical waveguides and LEDs, *Proceedings of the First European Conference on Optical Fibre Communication,* London (1975); London, Institution of Electrical Engineers, *Conference Proceedings,* no. 132 (1975), pp. 177–178.
202. P. K. Runge, A 50 Mb/s repeater for a fiber optic PCM experiment, *Proceedings of the International Conference on Communications,* Minneapolis (1974), pp. 17B 1–3.
203. T. Ogawa, T. Yamashita, Y. Mochida, and K. Yamaguchi, Low noise 100 Mbit/s optical receiver, *Proceedings of the Second European Conference on Optical Fibre Communications,* Paris (1976), pp. 357–363.
204. E. E. Basch and R. A. Beaudette, The GTE fibre optic system, *NTC Conference Record,* part 1, Los Angeles (1977) p. 14.

205. J. E. Midwinter, Optical fibre transmission systems, *Post Office Electrical Engineers Journal*, **70,** 3 (October 1977).
206. D. D. Sell and T. L. Maione, Experimental fibre optic transmission system for interoffice trunks (Bell Laboratories Atlanta experiment), *IEEE Transactions on Communications*, **COM 25,** 5 (May 1977) 517–523.
207. T. Ozeki and E. H. Hara, Measurement of nonlinear distortion in photodiodes, *Electronics Letters*, **12,** 3 (February 1976) 80–81.
208. W. Horak, Analog TV transmission over multimode optical waveguides, *Siemens Research and Development Reports*, **5,** 4 (1976).
209. C. C. Timmerman, A fibre optical system using pulse frequency modulation, *NTZ*, **30,** 6 (1977).
210. D. Chan and T. M. Yuen, Systems analysis and design of a fiber optic VSB-FDM system for video trunking, *IEEE Transactions on Communications*, **COM 25,** 7 (July 1977) 680–686.
211. H. Takanashi, T. Yamaoka, M. Takusagawa, and T. Misugi, Sources and detectors, *NTC Conference Record* (1977) p. 061 1/1–3.
212. K. Asatani and T. Kimura, Nonlinear phase distortion and its compensation in LED direct modulation, *Electronics Letters*, **13,** 6 (March 1977) 162–163.
213. Y. Ueno and M. Kajitani, Colour TV transmission using light emitting diode, *NEC Research and Development*, **35** (October 1974) 15–20.
214. J. Straus and O. Szentes, Linearisation of optical transmitters by a quasifeed-forward compensation technique, *Electronics Letters*, **13,** 6 (March 1977) 158–159.
215. M. M. Ramsay, A. W. Horsley, and R. E. Epworth, Subsystems for optical fibre communication field demonstrations, *Proceedings of the IEE*, **123** 6 (June 1976) 633–641.
216. R. E. Epworth, Subsystems for high speed optical links, *Proceedings of Second European Conference on Optical Communications*, Paris (1976).
217. S. D. Personick, Receiver design for optical fiber systems, *Proceedings of the IEEE*, **65,** 12 (December 1977) 1670–1678.
218. M. Treheux, R. Bouillie, and C. Boisrobert, Telecommunication research and development programme: evolution of optical fibre communication in France, *Proceedings of the IEE*, **123,** 6 (June 1976).
219. P. R. Selway, Semiconductor lasers for optical communications, *Proceedings of the IEE*, **123,** 6 (June 1976) 609–618.
P. R. Selway and A. R. Goodwin, Effect of d.c. bias level on the spectrum of GaAs lasers operated with short pulses, *Electronics Letters*, **12,** 1 (1976) 25–26.
220. B. Crosignani and P. DiPorto, Propagation of coherence and very high resolution measurements in optical fibers, SPIE, **77,** Fibers and Intergrated Optics (1976) 49–56.
221. B. Crosignani, B. Daino, and P. DiPorto, Measurement of very short optical delays in multimode fibers, *Applied Physics Letters*, **27,** 4 (1975) 237–239.
222. B. Crosignani, B. Daino, and P. DiPorto, Coherence properties of electromagnetic fields propagating in multimode waveguides and application to the measurement of dispersion, *Proceedings of the Second European Conference on Optical Fibre Communication* (1976).
223. D. E. N. Davies and S. A. Kingsley, A novel optical fibre telemetry highway, *Proceedings of the First European Conference on Optical Fibre Communication* (1975).
224. B. Culshaw, D. E. N. Davies, and S. A. Kingsley, Acoustic sensitivity of optical-fibre waveguides, *Electronics Letters*, **13,** 25 (1977) 760–761.
225. D. A. Kahn, Colour multiplexing techniques and applications in optical waveguide links, AGARD Conference Proceedings no. 219, *Optical Fibres, Integrated Optics and their Military Applications*.

226. S. Sugimoto *et al.*, Wavelength division two-way fibre-optic transmission experiments using micro-optic duplexers, *Electronics Letters,* **14,** 1 (January 1978) 15–17.
227. M. C. Hudson and F. L. Thiel, The star coupler: A unique interconnection component for multimode optical waveguide communications systems, *Applied Optics,* **13,** 11 (November 1974) 2540–2545.
228. D. R. Porter and I. R. Reese, A hybrid configured fibre optic data bus system, *Proceeding of the Second European Conference on Optical Fibre Communication,* Paris (1976), pp. 421–427.
229. J. G. Farrington and M. Chown, An optical fibre multi-terminal data system for aircraft, *Fiber and Integrated Optics,* **2,** 2 (1979).
230. S. D. Hersee and R. C. Goodfellow, The reliability of high radiance LED fibre optic sources, *Proceedings of the Second European Conference on Optical Fibre Communications,* Paris (1976), pp. 213–216.
231. R. J. McIntyre, Multiplication noise in uniform avalanche diodes, *Institute of Electrical and Electronics Engineers Transactions on Electron Devices,* **ED-13,** 1 (January 1966) 164–168.
232. S. D. Personick, Receiver design for digital fiber optic communication systems—I, *Bell System Technical Journal,* **52,** 6 (July–August 1973) 843–874.
233. S. D. Personick, Receiver design for digital fiber optic communication systems—II, *Bell System Technical Journal,* **52,** 6 (July–August 1973) 875–886.
234. S. D. Personick, Statistics of a general class of avalanche detectors with applications to optical communication, *Bell System Technical Journal,* **50,** 10 (December 1971) 3075–3095.
235. M. Chown, J. G. Farrington, and R. G. Plumb, Sensitivity of optical receivers, *Proceedings of the Third European Conference on Optical Communications,* Munich (1977).
236. A. Jessop, Error detection in digital transmission systems, British Patent Application 22849/76, 1976.
237. M. Chown and K. C. Kao, Some broadband fibre system design considerations, *Proceedings of the IEEE International Conference on Communications,* Philadelphia Pa. (1972).
238. M. Horiguchi and H. Osanai, Spectral losses of low OH-content optical fibres, *Electronics Letters,* **12,** 12 (1976) 310.
239. System description papers at the International Conference on Millimetric Waveguide Systems, *IEE Conference Publication,* no. 146 (1976).
240. M. E. Collier, Economic planning of transmission systems, Electronic Communication, **148,** 2 (1973).
241. T. L. Maione and D. D. Sell, Experimental fibre-optic transmission systems for inter-office trunks, *IEEE Transaction on Communication* **COM25,** 5 (1977).
242. S. Senmoto and K. Okura, Optical fibre cable transmission systems applied to telecommunication networks in Japan, *Second European Conference on Optical Fibre Communication,* Paris, 1976 (post-deadline paper).
243. F. Bigi, C. Colavito, and B. Catania, Optical fibres in local digital networks, *Proceedings of the Third European Conference on Optical Communication,* Munich (1977).
244. I. A. Ravenscroft, Feasibility trial of optical fibre transmission system *POEEJ,* **70,** 2 (July 1977).
245. Kawahata, Hi-OVIS (Higashi Ikoma Optical Visual Information System) Development Project, *Proceedings of the International Conference on Optical Fibre Communications,* Tokyo (1977).
246. J. G. Farrington and M. Chown, An optical fibre, multiterminal data system for aircraft, *AGARD Conference on Optical Fibres, Integrated Optics and their Military Applications,* London, May 1977.

247. R. A. Greenwell and G. M. Holma, A-7 ALOFT economic analysis and EMI-EMP test results, *AGARD Conference on Optical Fibres, Integrated Optics and their Military Applications*, London, May 1977.
248. Jefferies *et al.*, An experimental application of optical fibre cable on the electricity grid system, *IEE Colloquium Digest*, no. 32 (1977).
249. Cutler, A $\frac{1}{2}$ kilometre television link using optical fibre, IEE Lecture synopsis, 29 November 1976.
250. T. Tamir, editor, *Integrated Optics*, Vol. 7 of Topics in Applied Physics Springer-Verlag (1975).
251. S. Sugimoto *et al.*, High speed signal transmission experiments by optical wavelength division multiplexing, *Proceedings of the International Conference on Optical Fibre Communication*, Tokyo (1977).
252. E. H. Hara, Conceptual design of a switched television-distribution system using optical fibre waveguides, IEEE Trans. on Cable Television, *CATV*-2, 3 (July 1977).
253. K. Nakagawa, *et al.*, 32 Mb/s optical fibre transmission experiment with 53 km long repeater spacing, *Proceedings of the Fourth European Conference on Optical Communication*, Genoa, (1978).
254. P. W. Black, A. Cook, A. R. Gilbert, M. M. Ramsay, and J. R. Stern, The manufacturing testing and installation of rugged fibre-optic cables, *Proceedings of the Third European Conference on Optical Communication*, Munich (1977).
255. D. G. Dalgoutte, Collapsed sleeve splices for field jointing of optical fibre cable, *Proceedings of the Third European Conference on Optical Communication*, Munich (1977).
256. C. Game and A. Jessop Random coding for digital optical systems, *Proceedings of the First European Conference on Optical Fibre Communication*, London (1975).
257. J. D. Weston and N. H. C. Gilchrist, Joint STC/BBC field trials on an experimental 140 Mbit/s optical fibre digital transmission system, *International Broadcasting Convention*, London 1978.

Index